ARCHITECTURAL
DRAFTING
FOR INTERIOR DESIGNERS

ARCHITECTURAL
DRAFTING
FOR INTERIOR DESIGNERS

LYDIA SLOAN CLINE

THOMSON

DELMAR LEARNING

Australia   Brazil   Canada   Mexico   Singapore   Spain   United Kingdom   United States

THOMSON
DELMAR LEARNING

Architectural Drafting for Interior Designers
Lydia Sloan Cline

Vice President, Technology and Trades ABU:
David Garza

Director of Learning Solutions:
Sandy Clark

Managing Editor:
Larry Main

Acquisitions Editor:
James Devoe

Product Manager:
John Fisher

Marketing Director:
Deborah Yarnell

Marketing Manager:
Kevin Rivenburg

Marketing Coordinator:
Mark Pierro

Director of Production:
Patty Stephan

Production Manager:
Stacy Masucci

Content Project Manager:
Michael Tubbert

Technology Project Manager:
Kevin Smith

Editorial Assistant:
Tom Best

Art Director:
Bethany Casey

Cover Designer:
Bethany Casey

Cover Image:
Veer, Inc. / Digital Vision Photography Collection

Library of Congress Cataloging-in-Publication Data:
Cline, Lydia Sloan.
  Architectural drafting for interior designers / Lydia Sloan Cline.—1st ed.
      p. cm.
  Includes bibliographical references and index.
  ISBN-13: 978-1-4180-3297-5 (alk. paper)
  ISBN-10: 1-4180-3297-2 (alk. paper)
  1. Architectural drawing—Technique. 2. Interior decoration—Designs and plans. I. Title.
NA2708.C58 2007
720'.284—dc22   2007044591

Card Number:
1-4180-3297-2

NOTICE TO THE READER

Publisher does not warrant or guarantee any of the products described herein or perform any independent analysis in connection with any of the product information contained herein. Publisher does not assume, and expressly disclaims, any obligation to obtain and include information other than that provided to it by the manufacturer.

The reader is expressly warned to consider and adopt all safety precautions that might be indicated by the activities herein and to avoid all potential hazards. By following the instructions contained herein, the reader willingly assumes all risks in connection with such instructions.

The publisher makes no representation or warranties of any kind, including but not limited to, the warranties of fitness for particular purpose or merchantability, nor are any such representations implied with respect to the material set forth herein, and the publisher takes no responsibility with respect to such material. The publisher shall not be liable for any special, consequential, or exemplary damages resulting, in whole or part, from the readers' use of, or reliance upon, this material.

To all that study, mentor, and commission the creation of beauty in our built environment.

CONTENTS

PREFACE

Interior design is a field that has seen explosive popularity in the past 20 years. With TV shows promoting the profession and families having more disposable income, interior designers are being hired in ever greater numbers worldwide to make our homes and offices satisfying to the eye and soul.

It is a profession that requires a broad range of knowledge, critical to which is the ability to communicate ideas. There are two ways to do this: verbally and visually. Visual, or *graphic,* communication is largely done via manual or computer drafting. However, most drafting texts are written for architecture and engineering students. While there is overlap between what they need to know and what interior designers need to know, their content largely consists of building exteriors and mechanical pieces. *Architectural Drafting for Interior Designers* is written specifically for the beginning interior design student, using content and examples relevant to the field. It is written to comply with the standards of the National Council for Interior Design Qualification (NCIDQ) and the Council for Interior Design Accreditation (CIDA) the governing authorities in interior design education and practice.

Computer software is used for the bulk of drawing production in today's offices. However, in this book the execution of drawings is shown on a board, because this book is about the critical thought processes employed to create a drawing. These thought processes are a skill set that is independent of, and precedes, computer drafting. It is also often easier for the student to grasp concepts when they are not simultaneously manipulating an operating system and software. Further, manual drafting skills are still required for the NCIDQ test. Once the thought processes needed to create a drawing have been mastered, learning to use computer-aided drafting software is a logical and important second step.

No prior background or experience in drafting is assumed. The reader starts by learning how to use a mechanical pencil and progresses through sketching, orthographic projection theory, floor plans, dimensioning, elevations, sections, reflected ceiling plans, and building components until he or she is ready to create 3-D drawings. The relationship between 2-D and 3-D views is frequently discussed and reinforced.

Each chapter begins with objectives and keywords and ends with suggested activities. Most include further resources and exercises that reinforce chapter concepts. Sidebars with information related to the chapter topic are scattered throughout the text; step-by-step instructions are given for difficult concepts; and the text is heavily illustrated with professional and student work. The book is arranged so that material builds from one chapter to the next. Thus it is best to read the book from front to back.

We hope the student will discover that drafting is not only a necessary skill to the design field but can be an enjoyable art and part of what makes the practice of design fun.

Supplements

Solutions Manual

A solutions manual is available with answers to the end-of-chapter review questions and solutions to the worksheet problems. (ISBN 1-4180-3297-2)

This is an educational resource that creates a truly electronic classroom. It is a CD-ROM containing instructional resources that will enrich your classroom and make your preparation time shorter. The elements of the e.resource link directly to the text and tie together to provide a unified instructional system. With the e.resource you can spend your time teaching, not preparing to teach. (ISBN 1-4180-3297-2)

Features include:

- PowerPoint® Presentations: Slides for each chapter of the text provide the basis for a lecture outline that helps you to present concepts and material. Key points and concepts can be graphically highlighted for student retention.

- Exam View Computerized Test Bank: Over two hundred questions of varying levels of difficulty are provided in true/false and multiple choice formats so you can assess student comprehension.

- Image Library: Many figures from the text are available in a searchable database to be used for modifying the PowerPoint presentation or for use as handouts or transparencies.

Acknowledgments

I would like to extend a special thank you to Professor R.E. Budd Langley of Johnson County Community College, who performed a technical edit on the entire manuscript. I would also like to thank the many professionals who reviewed the manuscript in detail:

Sue Ballard de Ruiz, Tennessee State University, Nashville, TN

Associate Professor M. Jean Edwards, University of Louisiana, Lafayette, LA

Assistant Professor Stephen Huff, High Point University, High Point, NC

David LaComb, AIA, Sage Colleges, Albany, NY

Professor Marie Planchard, Massachusetts Bay Community College, Wellesley, MA

Professor Emeritus Jack Harris, Johnson County Community College, Overland Park, KS.

Thanks also to Associate Professor Margaret Davis of Johnson County Community College for her input and to Sonia Levin and Darcelle Bunn for contributing their drafting skills to the book.

CHAPTER 1

What is Drafting?

OBJECTIVE

- This chapter discusses what drafting is and why it is used by interior designers.

KEY TERMS

drafting
presentation drawings
production drawings

shop drawings
specifications

Drafting, also known as graphic communication, is the art of putting ideas to paper in picture form to create instructions. Drafters use universally recognized symbols and protocols to enable all involved in a design's implementation to interpret the drawings in a consistent manner, no matter what country the drawings are made or read in. Pictures convey ideas in a way that written descriptions cannot, hence drafting is an important skill for anyone who makes a living showing others how to put things together. All manufactured items—whether they are houses, clothing, cars, toys, barbeque grills, or furniture—started with drafted pictures. We see these pictures everywhere. For example, the pictures in product owner manuals are

Figure 1-1
Floor plan and samples mounted on boards for presentation purposes. *Courtesy of Rhiannon Craven, Craven Design.*

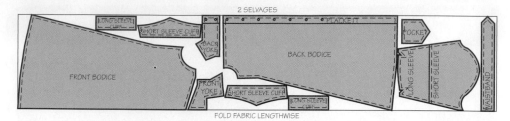

Figure 1-2
Clothing pattern. It is a drafted set of instructions.

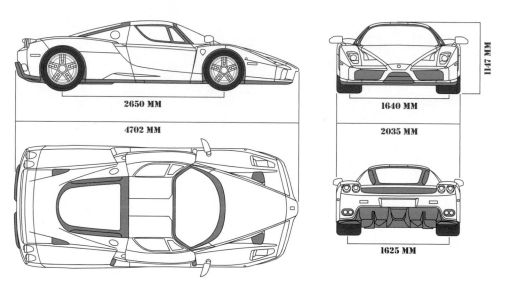

Figure 1-3
2-D drawings (top and front views) of a Ferrari. *Courtesy of http://www.blueprints.ononvanbraam.com.*

drafted instructions showing how the item goes together. So are clothing patterns. Most of us have drafted simple pictures ourselves, such as a sketched map giving directions to our home, or a bar chart to show a math concept. We imagine our surroundings before we create them. Drafting enables us to communicate those imaginings to others.

While artistic skill will contribute to making drafted drawings nicer, if you are not artistic, you can still draft. Anyone can learn how to use instruments to create hard-line drawings; non-artistic people can use straightedges just as skillfully as artistic people do.

History of Drafting

Drafting originated when the first designer needed to give instructions. The ancient Egyptians, Greeks, and Romans sketched elaborate structures onto papyrus (paper made from reed pulp) using reeds dipped in ink. Drawings were also sometimes carved into flat stone panels that were later installed into the base of a building. Since ancient structures could take centuries to build, this served the purpose of showing future workers the designer's intent. Examples of these "blueprints" are found in Egyptian, Greek, and Roman buildings, as well as medieval castles. (Construction plans are etched into the walls and floors of the cathedrals at York, Chartres, and Rheims.)

For thousands of years there was little change in the instruments that craft guild members used to draw their pictures; in fact, the T square and compass were carefully guarded secrets.

Figure 1-4
Drafting classroom circa 1906. *Courtesy of Colorado College, Colorado Springs, CO.*

The tools of modern board drafting have been largely unchanged for the past 200 years. While some tools have been modernized, e.g., technical pens have replaced inkwell-dipped quill pens, others, such as triangles, scales, and dividers, have remained the same. In the early 1980s, computer-aided drafting (CAD) software began replacing board drafting because of CAD's speed and ease of use. AutoCAD by Autodesk is the drafting industry standard for CAD software. There are also many specialized programs such as 20/20, Chief Architect, Revit, SketchUp, and programs for millwork design.

In 1979, a German archaeologist noticed dozens of thin lines etched into some lower walls in the ruined temple of Apollo at Didyma. He found straight lines, circles, quarter circles, and more complicated shapes. While some had been washed away by rain and others covered by mineral deposits, most were still visible even after thousands of years of exposure. They turned out to be construction drawings for the temple, and were done so precisely that they were clearly made by experienced draftsmen. It was even evident where some of the designs had been changed, and where small, deliberate mistakes had been made, possibly in keeping with some cultures' beliefs that humans should not aspire to perfection. Most of the drawings were full size.

How Drafting is Used in Interior Design

An interior designer may be asked, among other things, to lay out a floor plan, show a furniture arrangement, custom-design kitchen cabinetry, or describe a window treatment. Drafting is used to create drawings of ideas. Such drawings will show specifics, such as sizes, locations, and features. These drawings are shown to the client, which helps sell the ideas, and to whoever will manufacture the ideas. Simply put, drafting helps an interior designer imagine, visualize, create, and present a space.

Types of Drawings

There are two umbrella categories of drawings that interior designers create: presentation drawings and production drawings. **Presentation drawings** (also called general purpose drawings) are show drawings, typically used to sell ideas to clients or as exhibits. These drawings contain information of interest to the client, given in a manner that a layperson can understand. They most commonly consist of floor plans, interior elevations, and perspective drawings. Room and building layouts, fixed features, furniture arrangements, and other details that the designer feels appropriate are included. Dimensions are simplified, and may be presented as a note giving the length and width of a room. They can either be computer generated or produced on hard-copy media such as vellum, plastic film, or board. Such drawings are often paired with fabric swatches and material samples to completely describe the designer's idea. **Production drawings**, also called contract documents or construction or working drawings, are aimed at the tradespeople who will implement the ideas. These are technical and highly detailed and often contain information different than what is given in presentation drawings. They consist of floor plans plus other descriptive drawings such as details, sections, and elevations. Dimensioning is more complex and symbols representing materials and reference points are present. They are overall more technical, as they are the instructions for building.

A third type of drawing, typically called a **shop drawing** but also referred to as a fabrication or millwork drawing, is even more detailed than a production drawing. A shop drawing is typically done by the person who will be doing the actual work (often a subcontractor) and its purpose is to show the designer precisely how the manufacturing of the ideas communicated in the production drawings will be done. The production drawing might call for two pieces of steel to be welded together and the shop drawings will show the type of weld and its symbol.

Interior designers also provide written instructions called **specifications** for their projects. These complement the drawings in that they carefully describe the quality of the materials that the drawings show. For instance, a floor plan would show where carpet is to be laid. The specifications will describe the carpet material, thickness, backing, fire rating, and other particulars. These are details that would clutter up the drawings so they are presented in a separate document.

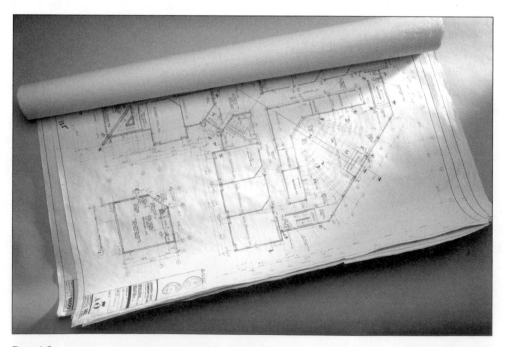

Figure 1-5
Construction drawings. *Courtesy of Edwin Korff, AIA, Prairie Village, KS.*

2-D vs. 3-D

Whether presentation or production, all drawings are either two-dimensional or three-dimensional. That is, some drawings show all three of an object's physical dimensions—length, width and height—and some show two of those dimensions. The floor plan is a two-dimensional drawing, showing length and width but not height. An interior perspective is a three-dimensional drawing. Three-dimensional (3-D) drawings are useful for presentation, since they are closest to how we actually see the space; thus a client can easily understand them. Two-dimensional (2-D) drawings are typically more difficult for laypeople to read. While there is some overlap, 2-D drawings are most commonly used for production and 3-D for presentation purposes.

Why Draft on Boards?

Why, in this age of computer-aided drafting, is manual drafting still needed? Well, for the actual production of drawings, it isn't. But drafting is all about problem solving, and it is easiest for most beginning students to problem solve without simultaneously learning a new software package. While board drafting skills are largely out of style, board thinking and problem-solving skills will never be.

The relationship between board and computer drafting is analogous to that between longhand essay writing and word processing. The latter is efficient for cutting, pasting, moving large blocks of text around, and spell checking. But word-processing software will not write your essay for you. Until you input data, all you have is a blinking cursor. Similarly, a computer drafting program will make drawing and revising easy. But it won't draw anything for you. It is merely a tool to make drawing easier. You must know how to construct a floor plan, how to use it to create an elevation, and where to place everything inside. A computer program will not figure that out; it simply makes your job easier once you have figured it out.

Figure 1-6
Presentation Drawing. *Courtesy Aaron White and Rich Dunham, Marshall Erdman, Madison, WI.*

Learning board drafting first will give you a solid foundation for computer drafting. The difference between drafters who learned board drafting prior to computer drafting and drafters who only learned computer drafting is often apparent. The former will know how to use a scale, how to orthographically project lines from a floor plan to make a section, and how to sketch a space. The latter often won't.

Importance of Legible Drawings

Graphic instructions are carefully drafted to industry standards rather than sketched to personal standards because many people are often involved in one project. Hence, clear, legible drawings with carefully lettered notes will reduce the chances of misinterpretation. Drafted documents are also legally binding, so having clear instructions can help in case of conflict.

WHAT DOES "STANDARD" MEAN?

The dictionary defines "standard" as something established by authority, custom, or general consent as a model or example. It's a recognized conformity to criteria that will be used consistently in an item's production. Such conformity is necessary for a common understanding of what to expect from products. When mass-produced items are standardized we can be reasonably confident that, even when produced by different manufacturers, they will be of similar quality and will connect as needed to other products.

People have standardized for thousands of years; weights, measurements, money, distance, and time are examples. But the mass production of goods that began in the Industrial Revolution made the need to standardize consumer goods imperative. For example, if light bulbs and sockets or notebook paper and binders were not standardized, we couldn't buy them with any certainty the bulbs and paper would fit. Standards on architectural drawings include methods of dimensioning, labeling, symbols, and drawing protocols.

The American National Standards Institute (ANSI) is an organization founded in 1918 that administers conformities for most U.S. industries. When you see the phrase "meets ANSI standard," it means the product's size or other features are recognized by ANSI. There are other standards groups, such as the International Standards Organization (ISO), which establishes conventions for measurements, the National Kitchen and Bath Association (NKBA), which establishes conventions for kitchen and bath design, and the American Institute of Architects (AIA), which establishes standards for architectural drawings.

Summary

Drafting uses both analytic thinking and drawing to explain ideas. These graphic instructions are tailored to different audiences to be most appropriately presented. Standardized ways of presenting information have evolved to make drawings readable no matter where they are created or read.

Suggested Activities

1. Obtain some production drawings and browse through them. Notice how carefully they are done and the different symbols and types of drawings. Obtain the accompanying specifications to see how the written instructions complement the graphic ones.

2. Obtain some presentation drawings and note how the graphics on them differ from similar graphics on production drawings. Note what else is included on the drawings.

Questions

1. Why is drafting used in industry?
2. How was drafting used in past centuries?
3. What is the difference between a presentation drawing and a production drawing?
4. What is an example of a two-dimensional drawing?
5. Why are manual drafting tools and boards still useful?

CHAPTER 2

Tools and How to Use Them

OBJECTIVE

- This chapter discusses the tools needed for manual drafting and how to select and use them.

KEY TERMS

Ames lettering guide	drafting board	irregular curve	parallel bar
architect's scale	drafting brush	layout tape	proportional scale
blueprints	drafting pencil	lead	protractor
compass	drafting tape	mechanical eraser	triangle
construction lines	eraser shield	mechanical pencil	T square
digital copies	film	metric scale	vellum
dividers	flexible curve	object lines	

All jobs require the right tools, and drafting is no exception. As with crafts and construction projects, drafting projects are best done with the proper equipment. It is true that the advent of computer-aided drafting has made drafting as fine art largely unimportant. In fact, with the increasing obsolescence of hand drafting for creating production drawings, some old architectural drawings have even become collector's items. However, for the production of whatever hand-drafted documents are still required, today's board drafter can utilize a more streamlined set of tools than his or her predecessors used (see Figure 2-1).

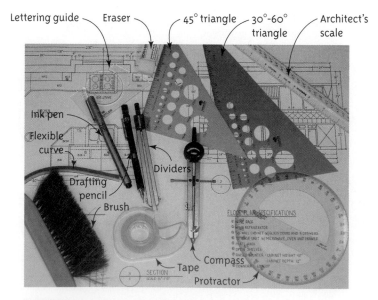

Figure 2-1
An assortment of tools is needed to draft.

The Basics

Drafting Board

A **drafting board** is a large, smooth surface atop a base. There are many different models and sizes. A minimum size of 2' × 3' with tilt and height adjustment and a replaceable vinyl cover is desirable. There are also small boards that consist of a drawing surface only, with no base; these can be placed on a desk or kitchen table. Although they are convenient and portable, they do not offer one the ability to draw on large sheets of paper (see Figure 2-2).

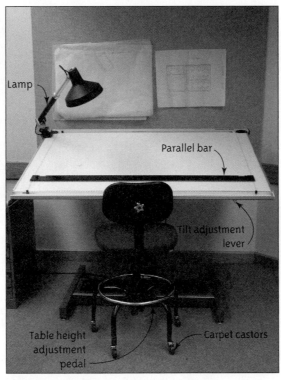

Figure 2-2
This board with hydraulic height adjustment, lever for tilt adjustment, parallel bar, lamp, and cushioned chair are conducive to good drafting.

Do not use a kitchen table or other makeshift surface as a drafting table. Even if it is square and accommodates a T square, this arrangement isn't conducive to good work. A pencil needs the cushion of a quality vinyl cover that hard surfaces don't provide. A sharp eye can discern work that was done on a board and work that wasn't.

Chair or Stool

Because comfort is important for long drafting sessions, a cushioned chair, with resilient casters and height adjustment, is desirable.

Lighting and Magnification

Choices include fluorescent, incandescent, or halogen lamps, or a combination of these. Some lights have a built-in magnifying glass. A light that clamps to the board and is adjustable in length is desirable.

Parallel Bar

A **parallel bar** is a long, straight tool used for drawing horizontal lines. Hardware is installed directly onto the board and the bar slides up and down on wire cables threaded through it. Some bars are lockable, meaning they can be held in place on a tilted board. Parallel bars come in different sizes and can be installed on both regular and tabletop boards. One that runs the full length of the board is desirable.

The bar's purpose is to help you draw horizontal lines, so use it for this. Don't draw horizontal lines by placing a **triangle** or scale on the board and "eyeballing" it straight. This is poor drafting practice. Whether you're drawing a line for the first time or tracing it, use the bar to ensure consistently horizontal lines (see Figures 2-3 and 2-4).

Figure 2-3
Properly drawn horizontal line. Note how the pencil glides directly on the parallel bar and is held perpendicular to the board.

Figure 2-4
Improperly drawn horizontal line. Straight-edge tool is ignored and a triangle is eyeballed horizontal. Pencil is inclined to the board.

T Square

The **T square** is an alternative to the parallel bar for drawing horizontal lines. It consists of a head and blade and is not installed on the board. T squares are available in different lengths and materials (plastic, metal, or a wood/plastic combination). While it is cheap and portable, the T square is not the most desirable tool to draw lines with because during use, the head must be firmly held against the board at all times to keep the bar straight.

Triangle

The *triangle* is used for drawing vertical or angled lines. There are three types:

- *45°* Draws lines that angle 45° and 90° to the horizontal.
- *30°–60°* Draws lines that angle 30°, 60°, and 90° to the horizontal.
- *Adjustable* Can be set to draw all angles between 0°–90°.

Triangles come in different sizes suitable for different drafting chores (see Figure 2-5). For instance, 36" ones are useful for perspective drawing and 6" ones work well for lettering. Medium-height triangles (8–12 inches) work for most general drafting purposes. The proper way to use the triangle is to place it on the parallel bar and slide it back and forth (Figure 2-6). Don't hold a triangle (or any other straightedge) on your paper and "eyeball" it vertical (Figure 2-7). This will not result in consistently vertical lines (see Figures 2-5, 2-6, 2-7).

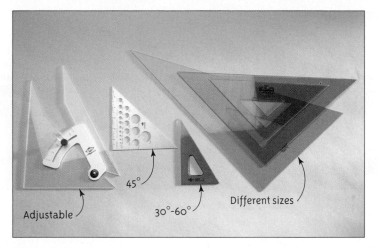

Figure 2-5
Assorted triangles and T square. The adjustable triangle can be set to any right angle. The rest draw lines 30°, 45°, 60°, and 90° to the horizontal.

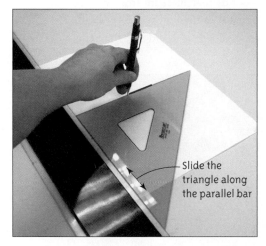

Figure 2-6
Properly drawn vertical line. The triangle rests on the parallel bar and slides left and right. The pencil is held perpendicular to the board.

Figure 2-7
Improperly drawn vertical line. The parallel bar is ignored and the triangle is eyeballed vertical. The pencil is not perpendicular to the board.

To draw 30°- or 45°-angled lines an equal distance apart, draw a line on the triangle the desired distance from the edge. Draw the line on the paper, then slide the triangle over until the line drawn on it aligns with the line on the paper. Draw another line (see Figure 2-8).

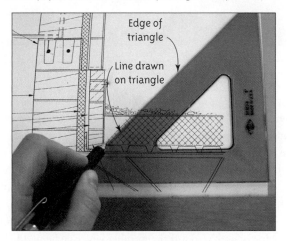

Figure 2-8
Using one triangle to draw angled lines a constant distance apart.

To draw parallel angled lines, follow the steps in figures 2-9, 2-10, 2-11.

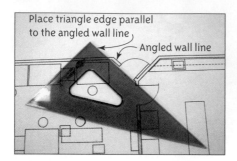
Figure 2-9
Place a triangle parallel to the line.

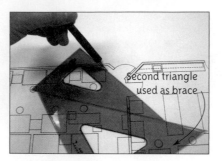
Figure 2-10
Brace the triangle by placing a second one, or any other straightedge, under it.

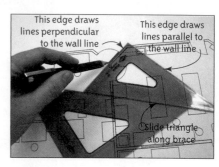
Figure 2-11
Slide the triangle along the brace to draw lines parallel and perpendicular to the line desired.

Protractor

A **protractor** is used for drawing angles. Some are half-rounds that measure from 0°–180° (Figure 2-12). Others measure 360° (a full circle). A protractor normally has scales that run from left to right as well as right to left so you can draw an angle on either side.

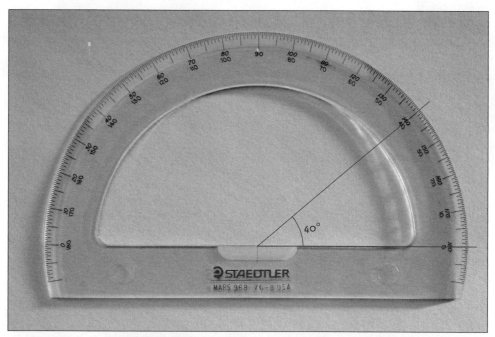
Figure 2-12
Half-round protractor showing an acute angle of 40°.

To draw an angle, align the protractor horizontally with the parallel bar. Find the crosshair or hole in the center. Draw a line from it to a number on the curved part. That number is the angle. To measure an existing angle, align the protractor's flat side with one line, placing the hole/crosshair at the vertex (the point where the two lines intersect). Note where the other line intersects the protractor's curve; this number is the angle. To decide whether to read the upper or lower row of numbers, use the number that is greater than 90° for an obtuse angle and the number that is less than 90° for an acute angle (see Figure 2-13).

To draw angles larger than 180° with a half-round protractor, draw a horizontal line, flip the protractor upside down, and mark off the number which, when added to 180, is the angle wanted. Then draw a line from the center hole through it.

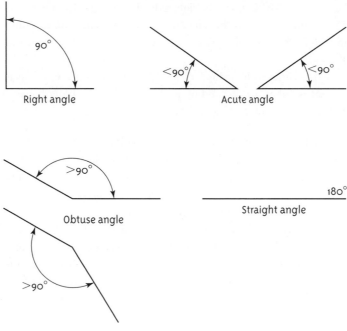

Figure 2-13
A right angle is 90°. An acute angle is less than 90°. An obtuse angle is between 90° and 180°. A horizontal line is 180°.

Scales

A scale is a tool that enables a drafter to draw large objects small enough to fit on a small sheet of paper while keeping them proportionately accurate. Objects can also be enlarged (drawn larger than true size) while remaining proportionately accurate. The phrase "drawn to scale" means the reader can accurately measure dimensions and distances on the drawing, as opposed to "not to scale," which means the drawing is not proportionately accurate, so dimensions and distances cannot be measured. All pictures drafted for instructional and presentation purposes are drawn to scale. The only drawings not drawn to scale are rough sketches, and even then, eyeballing for proportional accuracy is desirable.

Scales are usually triangular and 12 inches long. Some are flat and 6 inches long for greater portability. It is incorrect to call a scale a "ruler," although most scales are marked with an imperial (full-scale) rule on one side.

Architect's scale An **architect's scale** measures in units of feet and inches. Triangular models have six edges, with a total of 11 scales. All edges except one have two scales printed on them. One scale reads left to right, the other right to left. The left-to-right scale is half the size of the right-to-left. Select a scale based on the size and detail of what is being drawn. Large items, such as floor plans, are drawn with a small scale, e.g., 1/4" = 1'-0", to fit on the paper. Small, intricately detailed items, such as windows, are drawn with a large scale, e.g., 3/4" = 1'-0", to adequately show their features (see Figure 2-14).

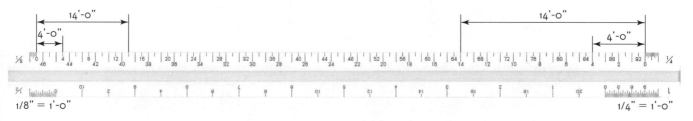

Figure 2-14
Identical measurements shown on the 1/8' = 1'-0" and 1/4" = 1'-0" scales.

Look at the 1/4" = 1'-0" scale on the right side of Figure 2-14.

Starting from 0 and reading to the left, each mark represents one real-life, or full (true)-size, foot. A line drawn from 0 to 4 represents a line 4' long. A line drawn from 0 to 14 represents a line 14' long.

Don't confuse these lines with the lines on the 1/8" = 1'-0" scale. One way is to keep them apart is to note that numbers corresponding to a scale will ascend (get bigger) from its zero. Always start from zero when measuring or drawing a line.

Subdivided Scale The architect's scale is open-divided, meaning whole feet are marked off, but not inches. To measure inches on the 1/4" = 1'-0" scale, look at the marks to the *right* of the zero as shown in Figure 2-15. Those marks show a fully-divided foot. Every scale's zero has a fully divided foot to its left or right. One-quarter of the total distance is 3 inches; 6 inches is in the middle, and three-quarters of the way is 9 inches.

On larger scales we can easily see the 3", 6" and 9" marks, as well as 1" increment marks between them. When confused, remember that the distance between zero and the last mark is always 12 inches, so the 6" mark will always be in the middle. The 3 inch mark is exactly between the 0" and 6" marks, and the 9" mark is exactly between the 6" and 12" marks. Note how much larger the subdivided 3/4" scale is than the 1/4" scale (Figures 2-15 and 2-16).

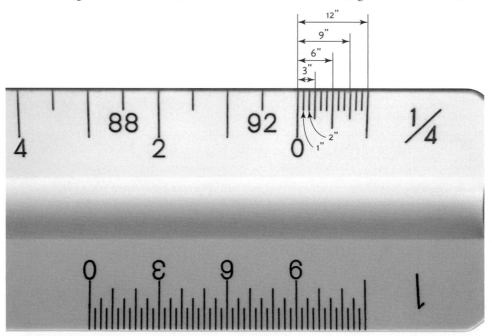

Figure 2-15
The subdivided 1/4" = 1'-0" scale.

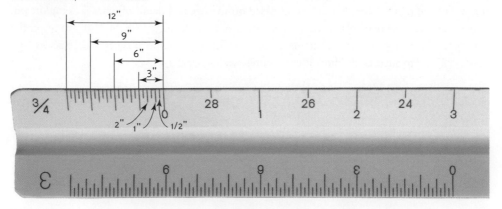

Figure 2-16
The subdivided 3/4" = 1'-0" scale.

The subdivided 3/4" = 1'-0" scale has 1/2" increments as well as 1" increments. Find the 6" mark, then the 3" mark. Notice that there are even half-inch increments. These are suitable for drawing objects that must be dimensioned to a fraction of an inch.

Look at the 1½" = 1'-0" scale in Figure 2-17. It is easy to confuse the lines delineating a large scale like this with the lines delineating the scale opposite it. One trick for keeping the scales' lines apart is to put your thumb and forefinger on the subdivided inch scale, as shown in Figure 2-17, and holding that position, move them to the open-divided scale as shown on Figure 2-18. The distance between your thumb and forefinger represents one foot.

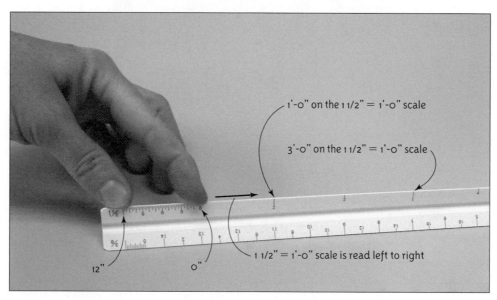

Figure 2-17
To verify which numbers go with the 1½" = 1'-0" scale, put your forefinger on the 0 and your thumb on the 12 of the subdivided scale. The distance between these two fingers is twelve inches.

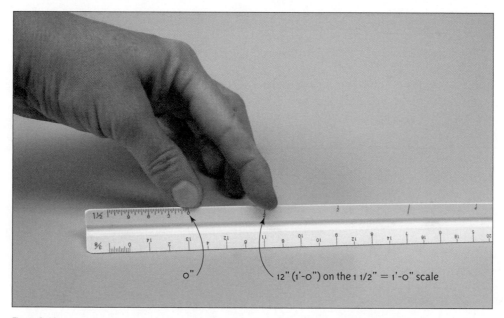

Figure 2-18
Holding that position, move your fingers so your thumb lands on the zero. Your forefinger will land on the one foot increment on the open scale.

Although single inches are delineated on the larger scales, the tiny 1/8" and 3/32" scales lack room for this. Therefore, each mark on their subdivided scales represents 2 inches.

A relatively small scale that is suitable for drawing large items such as plot plans or large commercial plans is 1/8" = 1'-0" (Figure 2-19). However, if great detail and fractional inch measurements are needed, this is not a good choice.

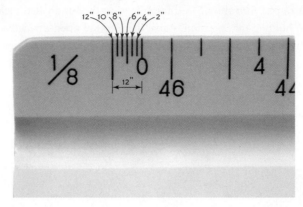

Figure 2-19
The subdivided 1/8" = 1'-0" scale.

Look at the side marked 16 (Figure 2-20). This is the imperial, or full, scale. It enables drafting at life, or true, size. It is labeled 16 because each inch is subdivided into 16 segments. Each segment is 1/16th of an inch, thus two segments are 1/8th of an inch. Don't confuse this scale with the 1" = 1'-0" scale, where each inch represents a foot (see Figure 2-21 and 2-22).

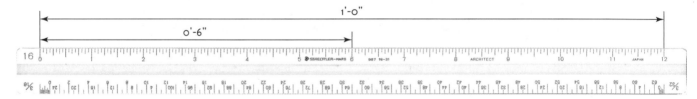

Figure 2-20
Imperial rule, or full (true) scale. This is used for drawing objects life size.

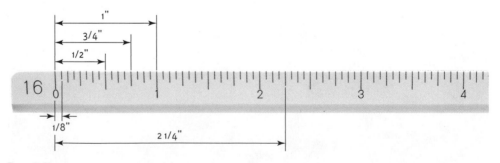

Figure 2-21
Inch increments on the imperial rule.

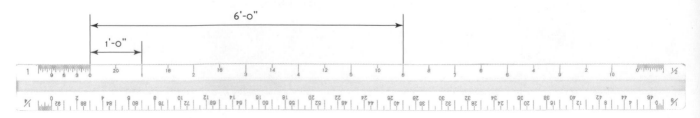

Figure 2-22
The 1' = 1'-0" scale.

Proper Architectural Notation Look at the dimension notes on Figures 2-20, 2-21, and 2-22. They are written in proper architectural format. For instance, 5 feet is written 5'-0". It is incorrect to write 5' or 5 ½ feet or 5.5'. The first way is wrong because a hyphen and inch symbol should be included for clarity. The second way is wrong because it could be mistaken as 5 feet, half an inch (which would be written 5'-0½"). The third way is wrong because decimals are not used on the architect's scale. Decimals are based on units of 10 and the architect's scale is based on units of 12. The following are examples of dimension notes written in proper architectural format:

0'-3"

0'-4 ½"

6'-7 ¾"

9'-0"

100'-7"

When writing the scale itself, it is incorrect to write just the fraction. The scale notation should be written. For example

Scale: 1/4" = 1'-0", 1/8" = 1'-0", 1' = 1'-0"

See Table 2-01 for some reductions achieved by the use of different architect's scales.

1" = 1'-0"	makes a drawing	1/12th true scale
1½" = 1'-0"	makes a drawing	1/8th true scale
3" = 1'-0"	makes a drawing	1/4th true scale
3/4" = 1'-0"	makes a drawing	1/16th true scale
1/2" = 1'-0"	makes a drawing	1/24th true scale
1/4" = 1'-0"	makes a drawing	1/48th true scale
1/8" = 1'-0"	makes a drawing	1/96th true scale

Table 2-01

Put a clip on the side of the scale you use most frequently to avoid having to constantly hunt for it.

SCALE CONVERSIONS

To convert a number from architectural to decimal units, divide the inch portion by 12. For instance, 7'-3" on the architect's scale is 7.25' in decimal units (3 divided by 12 is 0.25). 8'-9" is 8.75' (9 divided by 12 is 0.75). To convert from decimal to architectural units, set up an equation and cross-multiply to solve (Table 2-02).

Example 1: What is 0.5' in architectural units?	Example 2: What is 0.9' in architectural units?
5/10 = x/12	9/10 = x/12
10x = 60	10x = 108
x = 60/10	x = 108/10
x = 6"	x = 10.8
	x = 10 2/3" (divide 8 by 12 to get 2/3)
Example 3: What is .8' in architectural units?	Example 4: What is .2' in architectural units?
8/10 = x/12	2/10 = x/12
10x = 96	10x = 24
x = 96/10	x = 24/10
x = 9.6	x = 2.4
x = 9½" (divide 6 by 12 to get 1/2)	x = 2⅓" (divide 4 by 12 to get 1/3)

Table 2-02
Examples of converting from imperial to architectural units.

Metric Scale A **metric scale** is used for the same purpose as an architect's scale; the difference is that the metric scale measures System International (SI) or metric units. Meters and millimeters are the most common units for architectural dimensioning. Look at the 1:100 scale in Figure 2-23.

Its parts can represent any metric unit, but we will use meters for our discussion look at figure 2-23. The distance between 0 and the designated increment of 1 represents 1 meter. If the drawing scale is 1:100 m, there are 10 spaces between 0 and 1, so each of those spaces represents 0.1 (1/10) of a meter. A line from 0 to 5 represents 5 meters (5 m). Look at the marks between 5 and 6. The mark exactly between them represents 5.5 meters. The line next to it represents 5.6 m (5 $\frac{6}{10}$ meters). A line exactly between the 5.5 and 5.6 marks is 5.55 m (5 $\frac{55}{100}$ meters) long.

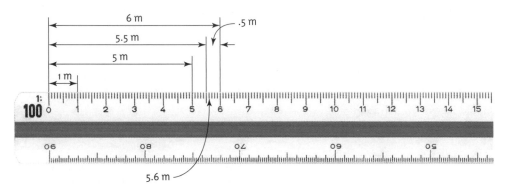

Figure 2-23
Increments on the 1:100 metric scale.

All the metric scales are read similarly. On the 1:200 scale, assuming no adjustment to the decimal location has been made, a line from 0 to 5 represents 5 meters and the four long marks between 0 and 5 are individual meter marks. The short marks between those meter marks are twentieths of a meter.

Decimal places can be added to or subtracted from each ratio to enlarge or reduce it. Look at the 1:100 scale again. Removing a zero (moving the decimal one place to the left) makes it 1:10 and turns the distance between 0 and 1 into 0.1 m (1/10th meter). The distance between 0 and 10 then becomes 1 meter, and each of the 10 marks in between them become 0.01 m (1/100 of a meter). Adding a zero (moving the decimal one place to the right) makes the scale 1:1000, which turns the distance between 0 and 1 into 10, so each mark in between represents 1 meter. The 1:1 ratio is the metric true scale. The distance between 0 and 1 not only represents 1/100th of a meter; it *is* a full scale 0.01m (1/100th meter). This is one centimeter, the metric "inch."

Proper Architectural Metric Notation Units are written lower case with no *s* or period after them. Leave a space between the number and the abbreviation. Place a zero in front of any number less than one unit. The following are examples

10 dm

4 km

89 m

7.65 m

1,000 mm

0.8 m

One meter is 39 1/3" long, slightly larger than a yard, which is 36". For proficiency with the metric system, *think* in metric. For visualization purposes, pace off a room's size in meters; an adult's long stride is roughly 1 meter. Helpful props are a metric tape measure (or a dressmaker's tape), some 20 cm and 30 cm rulers subdivided into millimeters, and centimeter grid paper.

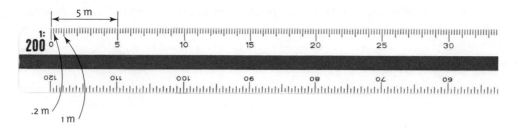

Figure 2-24
Increments on the 1:200 metric scale.

See Table 2-03 for different drawing types and the scales commonly used for them.

DRAWING	ARCHITECT'S SCALE	METRIC SCALE
Plot, commercial plans	1/8" = 1'-0"	1:100 m or 1:150 m
Residential floor plans; commercial enlargement plans	1/4" = 1'-0"	1:40 m, 1:5 m, 1:7 5m
Reflected ceiling plans	1/4" = 1'-0"	1:50 m
Electrical plans	1/4" = 1'-0"	1:50 m
Sections, details	3/4" = 1'-0" to 3" = 1'-0"	1:5 m to 1:10 m
Cabinetry	1/2" = 1'-0" to 1½" = 1'-0"	1:5 m, 1:10 m, 1:20 m
Interior Elevations	1/4" = 1'-0", 1/2"' = 1'-0"	1:20 m

Table 2-03

Proportional Scale A **proportional scale** enables you to enlarge or reduce a picture by a specific amount. For example, if you print a large picture or photo and want it to fit inside a small mat; use a proportional scale to figure out how much the picture should be decreased. This tool consists of a small wheel on top of a large wheel. On the perimeter of both wheels are measurements. To reduce the width of a picture from 14 to 10 inches, spin the small wheel until its 14" mark lines up with the outer wheel's 10" mark. Then look in the center window for the percentage by which it needs to be reduced (see Figure 2-25).

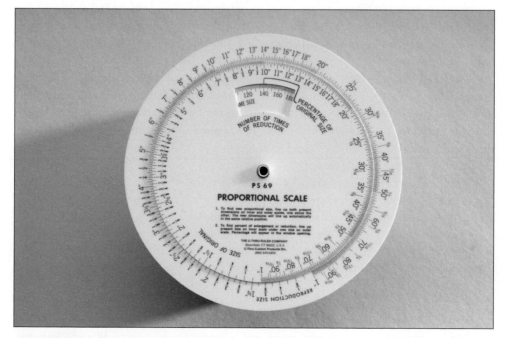

Figure 2-25
Proportional scale. To reduce a 14" picture to 10", reduce it 140%.

Unlike the scales discussed earlier, the proportional scale doesn't measure sizes. It simply does the math needed to enlarge or reduce a picture.

Working with Pencils and Leads

Pencils A **mechanical pencil** is a multipiece, refillable tool. It is available in 0.3 mm, 0.5 mm, 0.7 mm, 0.9 mm, and 2 mm sizes, which refer to the sleeve (silver tip) opening diameter (see Figure 2-26). There are also wood **drafting pencils** sold in a 2.0 mm size. Wood pencils are more cumbersome to use because they require frequent sharpening with a separate instrument called a lead pointer or a sandpaper block. The larger pencils' advantage is that different line thicknesses can be had by strategically sharpening and beveling the pencil point for different line thicknesses. This technique may be used for specialty drafting where a more aesthetically pleasing line quality is desired.

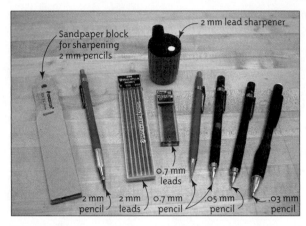

Figure 2-26
Different sizes and styles of drafting pencils and leads.

A mechanical pencil sold for drafting purposes is different from the many varieties of mechanical pencils sold for general writing purposes. Its sleeve is cylindrical and approximately 1/8" long, enabling it to glide smoothly along the edge of a parallel bar. Mechanical pencils are available in several price ranges. Inexpensive ones break easily and don't hold the lead as well as sturdier ones.

To fill a mechanical pencil with lead, remove the top cap and eraser, exposing a hollow area called the lead reservoir. This area will hold an entire package of leads. Empty a package into it, and then replace the eraser and cap. Click the cap until a lead protrudes through the opposite end. To change leads, empty the reservoir, then remove any lead already threaded through the tip by clicking the cap so enough lead protrudes that you can grasp and remove it. (Keep the cap depressed while removing the lead.)

When leads get short they will push back into the pencil when you write. When this happens, remove the short piece and click a new piece through. If a piece of lead becomes jammed inside, remove it by unscrewing the sleeve from the pencil and pulling the lead out.

To draw with a mechanical pencil, hold the pencil perpendicular to the board and rotate it as you pull it across the paper. This will give a consistent line width. Angle it slightly to create small variations in line width. When using a 2 mm pencil, wipe graphite dust off on a tissue after sharpening.

Leads The word **"lead"** is a misnomer, because in the drafting context it refers to sticks of powdered graphite (a form of carbon) mixed with a binder. Leads come in different diameters and grades or weights, meaning levels of hardness and softness. There are many different grades, but the ones most relevant to drafting are B, F, HB, H, 2H, 3H, and 4H. Soft leads such as B and F draw dark, easily smudged lines. Harder leads such as H, 2H, 3H, and 4H draw light, harder-to-smudge lines. HB is a medium-weight lead, but draws a line as dark as one created with a soft lead.

Hard leads are used for **construction lines.** Medium and soft leads are used for **object lines.** A construction line is a light line used for layout work; it leads to the creation of an object line, which is a line that defines the idea being communicated. For instance, the walls on floor plans are object lines. Projection lines on a floor plan that define the edges of elevations are construction lines. Object lines are intended to reproduce on copies; construction lines are not, or are intended to reproduce very lightly. Letters for notes and titles are object lines; the guidelines drawn between them are construction lines.

Because some construction lines do reproduce on copies, it is important to have sufficient contrast between the lead grades you use on a drawing. Most drafting students will find that working with 4H and HB leads will create sufficient contrast. To facilitate pencil drafting, it is convenient to keep two mechanical pencils on hand—one filled with hard lead and one filled with soft lead.

The amount of pressure placed on a pencil has an effect on the darkness of the line. Drafters who press down heavily might want to bump up their lead weights; for instance, if you've been using an H as your soft lead, try a 2H. Blue colored leads are also available. These don't reproduce (or reproduce very lightly) on copies; thus they can be used for construction lines. As mentioned, all drafting leads are made in sizes that correspond to the sizes of mechanical pencils. A 0.5 mm pencil needs a 0.5 mm lead; a 0.9 mm pencil needs a 0.9 mm lead; and so on. Filling pencils with the wrong size lead will lead to jamming or allow lead to slide out the tip. You can also purchase plastic leads, which combine the ease of pencil drafting with the dark look of ink.

Eraser Erasers are formulated for use on either **film** or **vellum** and may come in stick or block form. A white plastic eraser is formulated specifically for drafting on vellum (as opposed to bond paper) and works better on that medium than does the ubiquitous Pink Pearl®, which is a general-purpose eraser. For large amounts of erasing, electric or block erasers are useful. For fine erasing, try a **mechanical eraser**, also called a click eraser. This is a hollow, refillable stick in which long, thin erasers are placed. It is typically used with an **eraser shield**. An electric eraser is a more powerful tool for this purpose. Electric erasers are available in plug-in and battery-operated cordless models. The long, thin erasers that fit in click holders generally fit electric erasers, too.

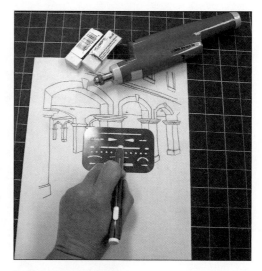

Figure 2-27
Proper use of a mechanical eraser and shield. Also pictured are electric and bar erasers.

Eraser shield An eraser shield is a metal template that has openings of different sizes and shapes. It allows the drafter to do fine erasing without removing surrounding lines (see Figure 2-27).

Drafting Brush Using a **drafting brush** to remove eraser crumbs from the paper or vellum is preferable to sweeping it by hand, which rubs oil and dirt onto the drawing and smears lines. Drafting brushes are available in different handle lengths. Their bristles are typically made of horsehair.

Tape

Drafting Tape **Drafting tape** resembles masking tape but is less sticky; thus it is less likely to damage paper when removed. Drafting tape is available in rolls and dots. Dots avoid the ragged edges that result from tearing roll tape; these edges tend to curl up under the parallel bar. When using roll tape, a dispenser is recommended.

Plastic Tape This tape is pressure-sensitive. It is available in different widths and printed patterns and in glossy and matte finishes. Tape with printed patterns can be used to represent specific line types such as poché symbols. Black tape is often used as **layout tape** to represent walls on presentation drawings or as a border. To do this, align plastic tape along a straight pencil line. Where the tape overlaps at the corners, miter with an artist's knife for a finished appearance. The tape should be burnished, or rubbed down hard, on the board to avoid peeling off. A burnishing tool is available for this purpose. There are also handheld, roll-on dispensers that make tape application easier (see Figure 2-28).

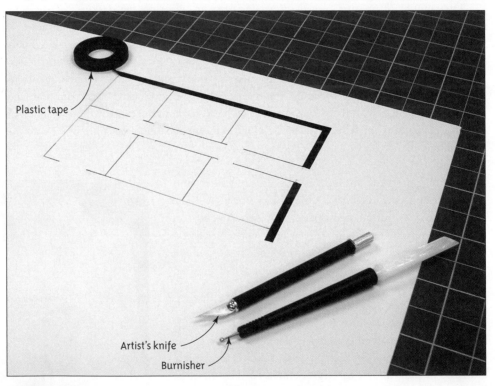

Figure 2-28
Plastic tape can be used for walls on presentation drawings. Apply with an art knife and burnishing tool.

Ames Lettering Guide

A lettering guide is used for drawing guidelines. Guidelines are construction lines that regulate letter height. The **Ames lettering guide**, which is described here, draws multiple rows of guidelines, each separated by a small space, This is useful when lettering rows of notes. It consists of a stationary frame and a rotating wheel. The numbers at (D) are imperial (U.S. customary) numbers, and model numbers represent numerators over the denominator 32. The #2 guide is 2/32" (1/16"), the #3 guide is 3/32, the #4 guide is 4/32 (1/8"), and so forth. Metric units are opposite the imperial numbers (see Figure 2-29).

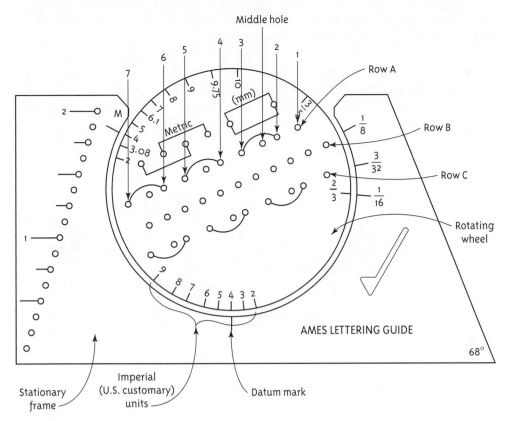

Figure 2-29

Ames Lettering Guide. The #4 is aligned with the datum mark, which will create guidelines 1/8" tall separated by spaces approximately 1/16" tall.

How to Use the Guide

1. Choose your guideline height and then set it by rotating the wheel so the mark under the number lines up with the datum mark on the frame. For example, to draw letters 1/8" tall, choose the 4 hole, because 4/32 is 1/8". Rotate the wheel until the mark under the 8 lines up with the datum mark.

2. Place the lettering guide on the parallel bar. Put the pencil in hole 2 and slide the lettering guide across the paper to draw a line. Because guidelines are construction lines, use a hard lead.

3. Place the pencil in hole 3 and slide the guide back. You have just drawn a second line 1/4" from the first. Drawing a line through the middle hole creates an intermediate line that regulates crossbars (such as are found on the letters B, P and R), and lower-case letter height.

4. Place the pencil in hole 4, and then slide the lettering guide along the bar. You've just drawn a spacer line 1/8" away from the second line.

5. Place the pencil in hole 5, and then slide the lettering guide along the bar again. You now have two rows of guidelines and a space between them.

6. Place the pencil in hole 6, and then slide.

7. Place the pencil in hole 7, and then slide. You now have three sets of guidelines, each 1/4" tall, separated by 1/8"spaces. Draw letters in the 1/4" spaces.

8. If more sets of lines spaced 1/8" from the last set are needed, place the pencil in hole 1 and then move the parallel bar and guide together so that hole 1 lines up with the last line drawn. Remove the pencil, place it in hole 2, and repeat steps 2–5.

The difference between rows A and C is in the height of the lower-case letters that their middle hole produces. Row A produces lower-case letters whose bodies are three-fifths the height of upper-case letters; Row C produces lower-case letters whose bodies are two-thirds the height of upper-case letters. Row B produces evenly spaced lines with a spacer row between them that is the letter height. Line the tool up with the fractions on the right to produce lines spaced a particular fraction apart. You can also draw lines through the slanted column of holes at the left for lines 1/8" apart with no spacer row between them (See Figures 2-29 and 2-30).

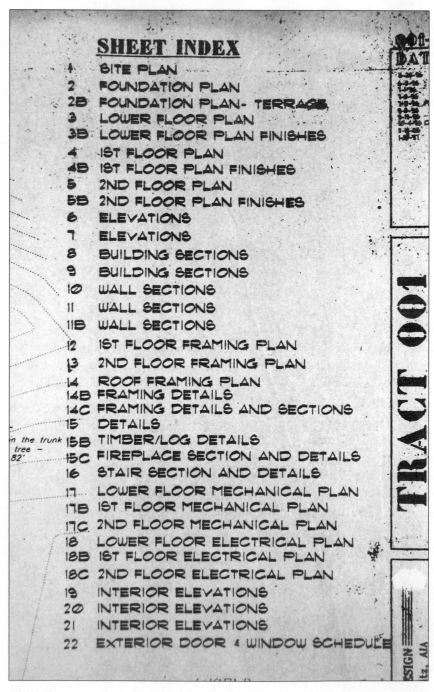

Figure 2-30
Multiple rows of notes should be lettered and spaced properly. Guidelines should either reproduce very lightly or not at all, as seen on this drawing. Evidence of their existence is in the letters' consistent height.

Compass

A **compass** is used for drawing circles (see Figure 2-31) and arcs (see Figure 2-32). It has a point at one end and a lead on the other. There are different compass styles. A drafting (technical) compass has features that distinguish it from a general-purpose compass. One is that it holds a piece of lead, not a whole pencil. Another is that there is a wheel on the center or side that adjusts for maximum accuracy. A drafting compass also has interchangeable points that hold pencils and pens and an extension bar for drawing large circles. General-purpose spring-loaded compasses that hold whole pencils are not accurate enough for drafting work.

To use a compass, place the point where the center of the circle or arc is to be, adjust the distance between the point and the lead to the desired radius, and swing the lead. Placing tape at an often-used compass point location will help keep the paper from tearing at that spot.

The lead should be beveled on a sandpaper block and placed in the compass with the beveled part facing out and extending slightly past the tip of the point.

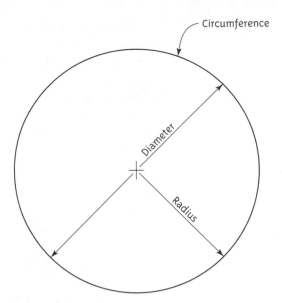

Figure 2-31
Radius is the distance from midpoint to circumference. Diameter is the distance of a straight line passing through the middle, touching opposite points on the circumference. Circumference is the circle's boundary.

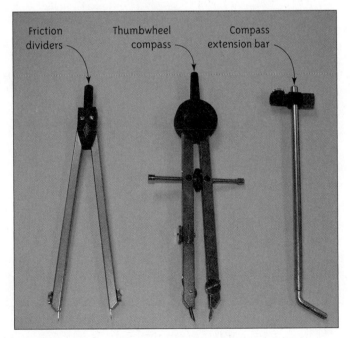

Figure 2-32
Compass, dividers and attachments.

Dividers

The **dividers** resembles a compass, but has points on both legs. It is used for transferring distances where numerical measurement isn't needed. To use the dividers, place its points at the endpoints of the line whose length you want to measure. Then, holding the dividers in that position, move it to where you wish to redraw that line. To double a distance, simply "walk" the dividers once. Walk it twice to triple it.

The most common styles of dividers are the wheel and proportional types. The former has a wheel in the center or on the side and is adjusted to the length desired. It provides the greatest accuracy. The legs on a proportional can be easily moved into place. It is more convenient to use than the wheel type. You can also obtain a tool that combines the compass and dividers functions by enabling the drafter to switch out leads and points.

Irregular and Flexible Curves

Irregular curves, also called French curves, are available in different styles and sizes (Figure 2-33). They allow the drafter to hard-line arcs. Use the irregular curve by finding a place on it that matches the arc to be hard-lined. Use the irregular curve to trace as much of the arc possible with that portion of the irregular curve, and then find a different section on the curve that matches more of the arc being drawn. Overlap the connecting arcs a bit to ensure continuity.

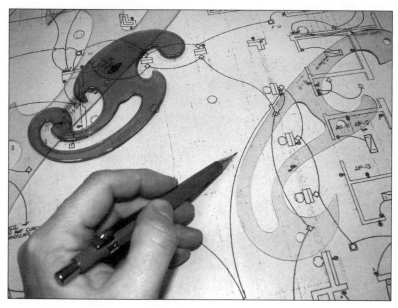

Figure 2-33
An irregular curve is used to hard-line curves that won't fit a circle template or compass.

Figure 2-34
A flexible curve helps turn a freehand sketch into a hard-line drawing.

The **flexible curve** is a piece of rubber that can be shaped to any curve (Figure 2-34). It is used for any curved line that cannot be drawn with a compass or circle template.

Technical Pens

These can be classified as *renewable* and *nonrenewable* (Figures 2-35 and 2-36). Whichever type you choose, a minimum of four pen sizes is needed for effective drafting. Pens of both types are sold in sets or individually (see Figures 2-35 and 2-36).

Renewable pens Renewable pens are multipiece tools that can be taken apart and filled with ink. When the ink runs out, they are refilled. Renewable pens come in different sizes; the sizes relevant to architectural drafting are 000, 00, 0, 1, 2, 3, and 4 (with 000 drawing the smallest lines and 4 drawing the thickest). Smaller and bigger sizes exist, but those are more appropriate for artists than architectural

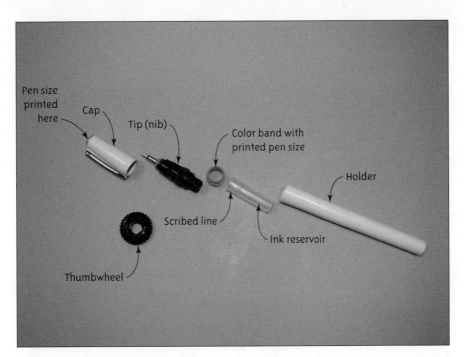

Figure 2-35
A renewable technical pen's individual parts.

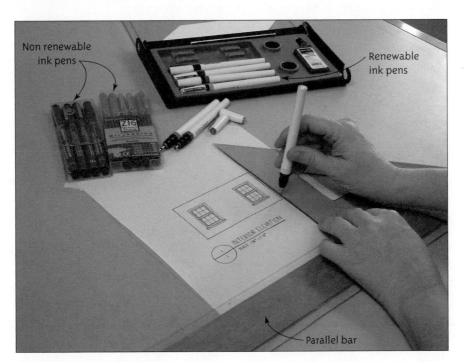

Figure 2-36
Nonrenewable and renewable technical pens for inking drawings.

drafters. Pen tips may be tungsten, jewel, or stainless steel; cost-wise, the last is the most practical.

When the pen tip is shaken, you should hear a rattling noise inside the pen, which means the tip is unclogged and functional. Take the pen apart to fill the ink reservoir. Don't fill it past the line scribed near its top. Shake the pen gently to start the ink flowing; a light tapping on the board with the cap on works, too. The pen must be cleaned after each use or the ink will clog it and render it useless. If it is difficult to unscrew the tip from the rest of the pen, using a thumbwheel (a small tool that is included with most sets of pens) will apply some extra torque to it. To clean, rinse the tip upside down under water and blot it dry. Once in awhile an ultrasonic cleaner (a machine similar to a jewelry cleaner) and special cleaning solution should be used. If there is no sound when the pen tip is shaken, or if the pen leaks around the tip, try an ultrasonic cleaning; it will usually fix the problem.

Nonrenewable pens Nonrenewable pens are glorified felt-tips. When empty, they are thrown out. They provide a line quality good enough for most manual drafting and are often preferred for their low cost and maintenance. Their sizes are typically notated as 01, 03, 05, and 07, with 01 being the smallest and 07 the largest.

INKING TIPS

1. Hold the pen vertically so it is perpendicular to the board. This allows ink to flow smoothly through the tip. Angling the pen can result in inconsistent line widths.

2. Starting at the top of a drawing, draw all horizontal lines, then, starting at the left of a drawing, draw all vertical lines. This allows ink to dry while simultaneously allowing you to continue drawing.

3. To avoid smears, use a triangle with an ink lip. A small space between the tools and paper can also help. Some drafters affix coins or strips of tape to the underside of their triangles and templates to achieve this. Don't press down hard on templates, triangles, or the parallel bar while inking, as that can cause ink to bleed underneath.

4. Lift tools up and away from wet lines instead of sliding them away. This helps prevent smearing of wet ink.

5. Move pens across the paper at a constant speed. Slowing at the end of a line will cause a small pool of ink to form.

6. Special erasers are formulated for drafting film, but any eraser will work if you wet it.

Templates

Templates are laser-cut pieces of plastic used for drawing the same item multiple times (Figure 2-37). There are hundreds of templates available for furniture, fixtures, circles, arrows, electrical symbols, and other images. Templates can be bought in sets or individually. Large furniture and appliance companies often have templates of their products at different scales (Figure 2-37). Some templates have ink risers (small bumps) on one side to facilitate inking work. When used, the risers face the paper.

Drafting Media

Drafting is done on different types of paper and film (Figure 2-38).

Tracing paper Also called, among other things, *flimsy or trash*, this is thin, semitransparent paper used for sketching and designing. It is available in yellow or white and in different sizes of rolls and tablets. Tracing paper is useful because it can be layered, enabling the drafter to copy some elements while redesigning others. It provides an inexpensive medium for quick sketching and brainstorming and is meant to be thrown away afterwards.

Vellum Vellum is a semi-opaque, high-quality cotton paper used for ink or pencil work. It is available in different sizes of rolls and sheets. Its fine, white background makes it suitable for presentation drawings; however, it also tears easily and is not easily ink-erasable. Vellum is "toothed" on the drawing side to diffuse light. The toothed side is slightly darker than the

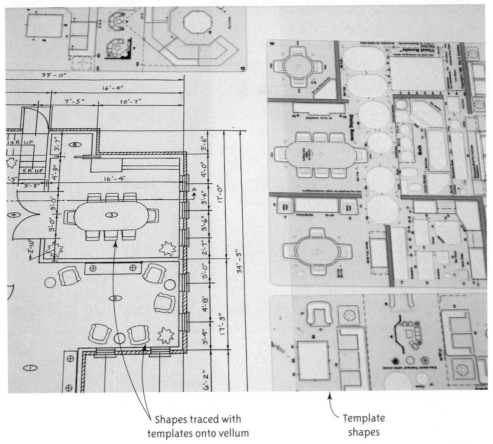

Shapes traced with
templates onto vellum

Template
shapes

Figure 2-37
Furniture templates range from simple rectangles to delineated pieces. Some even have arrangements with proper clearances.
This template set is from Visual Results, Brookfield, WI.

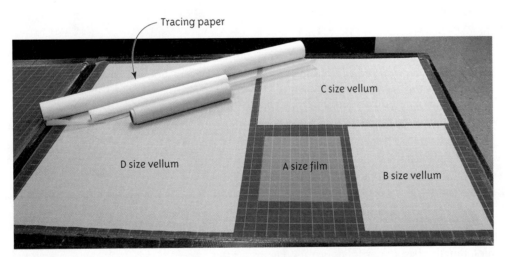

Tracing paper

C size vellum

D size vellum

A size film

B size vellum

Figure 2-38
Illustration board, tracing paper, vellum, and film are used for sketching and presentation work.

opposite side. Vellum may be single- or double-sided. Double-sided vellum is toothed on both sides. The most popular brand is Clearprint®.

Drafting Film Polyester film and plastic sheets are used for ink or plastic lead work, with Mylar® being the brand name for popular film. Plastic is stronger and more resistant to tears than vellum, and easily ink-erasable. Ink and plastic lead are most suitable for these media, because graphite pencil smears. Film cannot be used as a final presentation material due to its transparency, but its high translucency makes for nice sharp copies. It is available in different-size rolls, sheets, and gauges (thicknesses) and, like vellum, can be single- or double-sided. On single-sided film, the matte (dull) side is drawn on, not the shiny opposite side.

Both vellum and film come in ANSI standard sizes. There are two standards: architectural and engineering (Table 2-04). Both have letter designations, but their sizes vary slightly.

ARCHITECTURAL	ENGINEERING
A size—8½" × 11" (letter size)	**A** size—9" × 12"
B size—11" × 17"	**B** size—12" × 18"
C size—17" × 22"	**C** size—18" × 24"
D size—22" × 34"	**D** size—24" × 36"
E size—36" × 48"	**E** size—36" × 48"

Table 2-04
Standard Sizes for Paper used for Architectural and Engineering Projects.

Outside the United States, many countries use ISO metric measurement standards. Its sheet sizes are A4 (210 mm × 297 mm), A3 (297 mm × 420 mm), A1 (94 mm × 841 mm), and A0 (841 mm × 1189 mm).

Board Presentation work is sometimes done on coldpress board. This has a porous surface that will take an ink line. (Hotpress board also exists, but its surface is too slick to draw on.) Board is available in different thicknesses and sizes; a board thick enough to be propped up is desirable. It is not ink-erasable.

Construction Calculator

There are calculators available for all building needs. Some just have the standard math functions of addition, subtraction, multiplication and division. Others will convert numbers to various linear, square, and cubic formats including feet-and-inches format, decimal feet, decimal inches, yards, meters, metric units, and board feet. Construction calculators can manipulate fractions and convert the denominator to any desired accuracy. Some calculators will figure the pitch, rise, and run of roof rafters and stair stringer angles, as well as calculating stair riser, tread number, riser height, and tread width.

Storage and Care of Tools

Storage Tools can be kept in anything that keeps them organized and safe during transport. Compartmentalized plastic boxes work well. Drawings are best stored in flat files, lateral files, round tubes, or portfolios (see also Figure 2-39).

Care of Tools If they are properly cared for, drafting tools will last many years. Keep your tools clean because dirt will rub off them onto the drawing media. Use a "green" cleaner, not a household one, because the latter will remove the painted lines from instruments and templates and leave a film. Clean the underside of your parallel bar and wipe down your drafting board as needed, as dirt on them will embed into the paper. Never use plastic tools as cutting edges! Use a steel straightedge for any cutting. Use tools for their intended purpose only. Chipped triangles and nicked parallel bars are useless as drafting instruments.

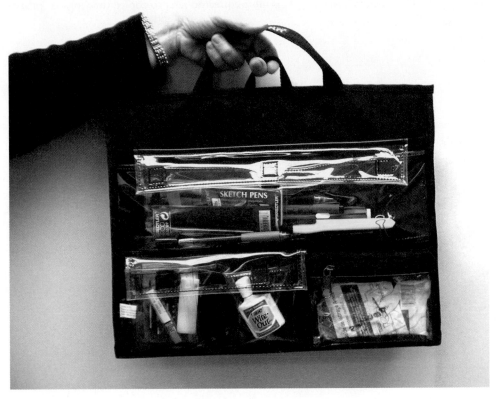

Figure 2-39
A scrapbooking bag is a creative way to hold tools.

Copying Processes

Sets of architectural drawings are often called "**blueprints**," but this is a misnomer. Blueprinting is an old-style, photographic, ammonia-based copying method. A transparent original was needed through which a lamp could shine. This produced copies that had white lines on a dark blue background.

Blueprinting was replaced with blue-line prints, which are the result of a process called diazo printing. Diazo printing, like blueprinting, requires a transparent or semitransparent original, but produces dark blue lines on a light blue background. This diazo-printing process can also produce black-prints; brown-prints (so called because of the paper color); sepia-prints, which are second originals (semitransparent copies).

Today, drafting on transparent media is not strictly necessary because most reprographics shops have photocopiers that make a digitized scan of the original and run black-line **digital copies** onto whatever media is wanted. Hence, the original can be opaque.

Other copying alternatives are multicolor offset prints (the drawing is transferred or "offset" from a plate to a rubber blanket, and then transferred to the paper); reduced-size prints; and photostatic (projection copier) copies that are at a different scale from the original.

Know that each time a photocopy of a drawing is made the lines become slightly stretched; hence, photocopies are not 100 percent accurate representations of the originals. (Scanned black-line digital copies, however, have very little distortion.) Also know that some print shops will not accept originals that have been amended with correction fluid, layout tape, rub-on letters, taped-on tracing paper, or other stick-ons, as these substances may clog or damage the copying machine.

Some trades often refer to sets of drawings as "the plans," but this is technically incorrect, because a "plan" is a specific drawing. It is more correct to refer to copied drawings as "prints" or "bluelines."

Summary

More tools than the ones discussed in this chapter exist; this chapter merely covered the essentials. Browse through an online drafting supply store to learn about specialty products. For instance, you might wish to purchase a chamois pounce to help keep your paper clean when doing pencil work. Start with the basics. As you progress in your drafting you will discover which tools serve you best.

A tape measure is a handy tool to carry at all times. Measuring common objects and clearances will help your internalize the size of the built environment. Record doorknob heights, countertop widths, room lengths, and the distance between restaurant tables. Keep a journal of measurements, information, sketches, and ideas in a pocket notebook.

Suggested Activities

1. On a sheet of vellum, draw lines with different hard and soft leads and observe the differences among them.

2. On a sheet of drafting film, draw lines with different sizes of ink pens and observe the differences among them.

3. Draw the following angles with a drafting machine: 35°, 62°, 95°, and 180°.

4. Measure this textbook with the 1:1 metric scale. Record length, width, and depth in centimeters and millimeters.

5. Disassemble and reassemble a mechanical pencil.

6. Using a tape measure, record the length, width, and depth of three pieces of furniture in your classroom. Record their dimensions in a pocket notebook.

7. Draw different size guidelines on a size "A" sheet of vellum, and then fill them in with letters.

Questions

1. What is the architect's scale used for?

2. What is a dividers used for?

3. What are templates used for?

4. Describe the proper way to draw a vertical line.

5. What number should the Ames lettering guide be set to for guidelines 1/8" tall?

6. What is the difference between vellum, film, and tracing paper?

7. Describe the difference between object and construction lines.

Internet Resources

http://about.com; http://www.ehow.com
Instruction on many subjects, including how to use some drafting tools.

http://www.ansi.org
American National Standards Institute—includes ANSI's history and mission.

http://www.ccaha.org/
Conservation Center of Art and Historic Artifacts—history of architectural drawing reproduction techniques.

http://www.hearlihy.com; http://www.suppliesnet.com
Large selections of drafting supplies.

http://www.leadholder.com/
History of the drafting pencil.

http://www.sciencemadesimple.com/
Imperial-to-metric conversions.

CHAPTER 3

Drafting Conventions

OBJECTIVE

- This chapter discusses line types, symbols, letters, and notes found on architectural drawings.

KEY TERMS

border lines	hidden object lines	note
center line	ID label	plan north
construction lines	leader line	poché
cutting plane	lettering	rendering
cylindrical break line	line	section lining
dimension lines	line quality	section symbols
elevation symbols	line type	short break line
enlargement box	line weight	title blocks
extension lines	local note	visible object lines
general note	long break line	
grid system	match lines	

Drafting involves breaking down complex, difficult ideas into simple, representational lines (see Figure 3-1). This process parallels the way our brains interpret and simplify the huge amount of visual information our eyes receive. A drafter's goal is to make a drawing as readable as a book to people trained in interpreting drawings. These people have different backgrounds; they include designers, contractors, subcontractors, owners, vendors, lenders and others. Making drawings readable to members of these groups is facilitated by following industry conventions and standards, the most common being the Uniform Drawing System (UDS) devised jointly by the American Institute of Architects and the Construction Specifications Institute (see Figure 3-20). However, these conventions can only be discussed at a general level, because each company has its own set of drafting standards. A protocol you learn in one office may be different in another. However, such protocols are never so different that you would not be able to infer what a differently-drafted label or symbol means as long as you understand the label or symbol in the first place.

Whenever you begin a new job, study drawings of the office's past projects to learn that office's specific conventions. Companies often provide a handbook of drafting graphic standards for their designers to follow.

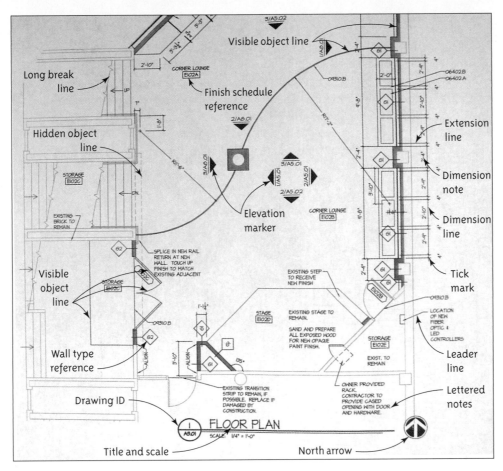

Figure 3-1
A hand-drawn floor plan of a campus building at Johnson County Community College, Overland Park, KS. *Drawing courtesy of PGAV Architects, Westwood, KS.*

The Line

The **line** is the fundamental tool of graphic communication. Different line weights (thicknesses) and line types communicate different ideas. Each line means something, so lines should never be randomly drawn. Because construction drawings are abstract, you should use as few lines as possible to describe an object.

A line is a thin (relative to its length), geometric object. It can also be thought of as an extended point. A line can be long, short, straight, curved, hard, freehand, horizontal, vertical, diagonal, thick, thin, or patterned.

Line Quality

Excellent **line quality** is critical. Each line must be drawn well. With the exception of construction lines, all lines are the color black. Lines should be of a consistent width from end to end and should be the proper width for their importance in the drawing. They should look clear, strong, and dark, and be drawn continuously from one end to the other, rather than as a series of short, overlapped pieces. Light, fuzzy, smudgy, or half-erased lines are weak and won't copy well. All lines must be joined at corners, never falling short; some offices prefer corner lines to overlap. Make a test copy of each drawing before using it for its intended purpose, as a copy will show weak lines more clearly than will the original media.

Line Weight

Line weight (also called hierarchy) refers to thickness. Thicker lines engage the reader's eye first; therefore features that should be read first are drawn with more prominent lines. This practice evolved because most people can interpret a drawing with multiple line thicknesses more easily than a drawing with just one line (see Figure 3-2). Line hierarchy helps clarify questions such as where the edge of a curved object is; which lines are structural and which are textural; and what distinguishes the hardness of a wall from the softness of fabric (Figure 3-2). Illustrators spend their careers addressing these questions and improving their drawings' readability by adjusting line weights and colors.

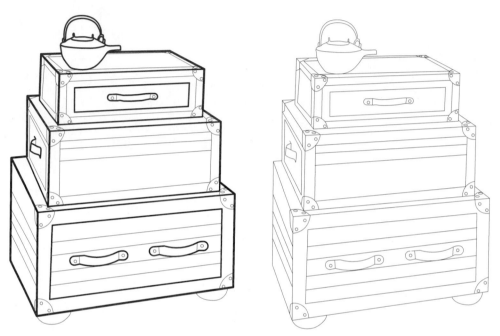

Figure 3-2
Differing line weights direct the eye to a drawing's prominent features first, making it easier to read.

Line Type

Line type (also called the alphabet of lines) is a pattern that a line takes to represent a specific concept. Different types mean different things. Following is a list of architectural line types and suggested pen sizes.

Visible Object Visible object lines define a physical item's outline. Walls, doors, porches, patios, cabinet edges, and floor tiles are examples of things represented with visible object lines. (Lines that represent appliances, toilets, and sinks are sometimes called fixture lines.) Object lines visible in the viewing plane are drawn with a continuous (nonpatterned) line.

Object lines have a hierarchy of relative weight. For instance, walls are thicker than doors and doors are thicker than their swings (Figure 3-3). So if walls are drawn with a 0.6 mm pen, doors might be drawn with a 0.5 mm pen, and arcs with a 0.3 mm pen. If walls are drawn with a 0.5 mm pen, doors might be drawn with a 0.35 mm pen, and arcs with a 0.25 mm one. Line weight also varies with the drawing scale. Larger scale drawings need thicker lines.

Although lines drawn with hard leads look thinner than leads drawn with soft leads, you can assume that a 0.5 mm drafting pencil will draw a 0.5 mm line. When using a drafting pencil, vary line weights by selecting different sizes (e.g., 0.3 mm for the thinnest lines and 0.9 for the thickest). When using the larger 2 mm drafting pencil, vary line weights by beveling one

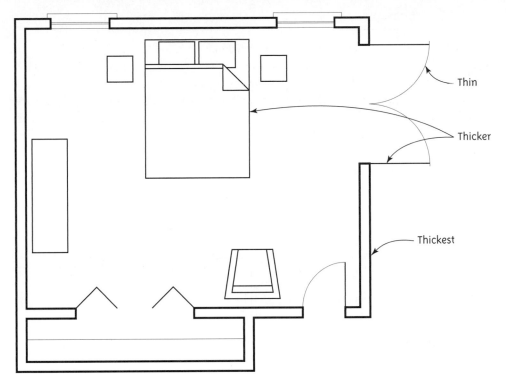

Thin

Thicker

Thickest

Figure 3-3
Line weights differ for the walls, doors, arcs, and furniture.

side of the lead on a sandpaper block and rotating the point as you draw. When inking, vary line weights by choosing different size pen tips. Avoid drawing over one line multiple times to widen it, as this often results in poor line quality. Recommended pen: 0.25–0.70 mm.

Hidden Object **Hidden object lines** define an item that is not visible in the reader's current view but needs to be acknowledged anyhow (see Figure 3-4a). For instance, floor plans show features below a horizontal **cutting plane** four feet above the floor. Features above the four-foot level, such as wall-hung cabinets, are shown with a hidden line. Hidden lines may also show items destined for removal, such as in a demolition plan. Draw hidden lines short and evenly sized (1/8"−3/8" long), with shorter, evenly spaced distances (1/16"−1/8") between them. Some drafters draw hidden lines using different lengths depending on the lines' importance. For example, an overhead beam might be drawn with a longer hidden line than a high closet shelf.

Hidden lines should touch each other at corners and touch any solid lines they intersect, begin or end at (Figure 3-4b). The reason hidden lines must touch solid lines is because when they don't, the drawing describes something else or simply shows poor draftsmanship. Figure 3-5 is a drawing of a cylinder. The left image shows the hidden and visible lines touching; this describes a cylinder open on both ends. The right image, with its gap between the hidden and visible lines, may be poor drafting, or may be describing a "blind hole," which is a hollow cylinder sealed at both ends. Recommended pen: 0.25 mm.

– –

Figure 3-4a
Hidden lines represent features that cannot be seen in the view.

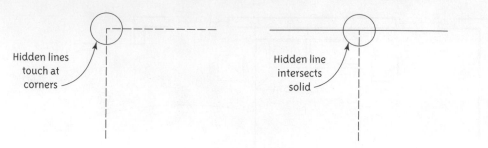

Hidden lines
touch at
corners

Hidden line
intersects
solid

Figure 3-4b
Proper intersection of hidden and solid lines.

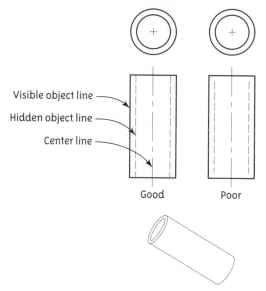

Visible object line

Hidden object line

Center line

Good   Poor

Figure 3-5
Proper termination of hidden lines in a drawing of an open hole.

Center A **center line** is drawn through the center of a feature (Figure 3-6). The notation "CL" is sometimes placed at one end. Recommended pen: 0.25–0.35 mm.

Long break A **long break line** is used to end a feature when drawing it in its entirety is not necessary (Figure 3-7). The long break is used after the feature's defining characteristics have been shown. It can be drawn vertically or diagonally, with a sharp or smooth jag. Recommended pen: 0.25–0.35 mm.

Figure 3-6
Center line. The CL notation at the top is optional.

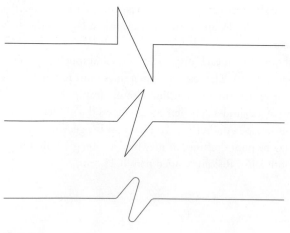

Figure 3-7
Long break lines.

Short break The **short break line** serves the same purpose as the long break (Figure 3-8). It is not used as commonly as the long break. Recommended pen: 0.50 mm.

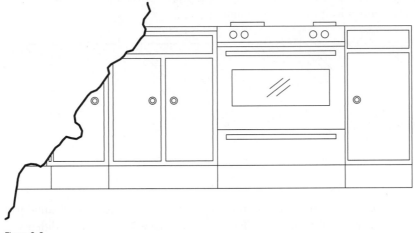

Figure 3-8
Short break line.

Cylindrical break Also called an s-break, the **cylindrical break line** is a freehand line placed through a cylindrical object such as a pipe or a column (Figure 3-9). It is more common in engineering drawings than in architectural drawings. Recommended pen: 0.35 mm.

Cutting plane The **cutting plane line** is drawn on the floor plan and shows where an object is sliced to create a section view (Figure 3-10). When it is drawn through the entire plan and has arrows placed at both ends, it indicates a section drawing. When it is drawn through a small portion of the plan and has just one arrow, it indicates a detail drawing. Either of the line types shown in Figure 3-10 may be used. Sometimes only the two ends of a cutting plane line are shown to avoid obscuring other lines on the plan. Recommended pen: 0.50–0.80 mm.

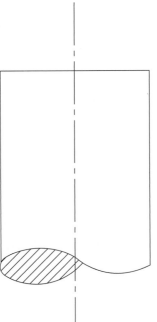

Figure 3-9
Cylindrical "s" break line.

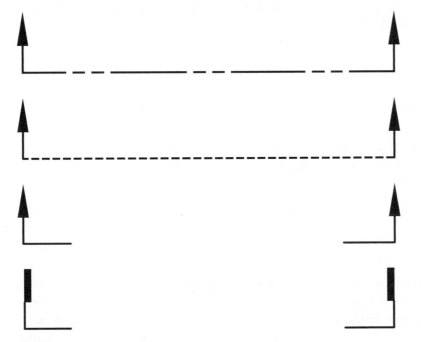

Figure 3-10
Different styles of cutting plane (section) lines and arrowheads. Arrowheads are opaque and three times as long as they are wide.

Section Lines Also called hatching, **section lines are** angled line patterns that indicate an object has been sliced (Figure 3-11). Forty-five degrees is a commonly used angle. Never draw section lines parallel or perpendicular to object lines. Spacing between section lines should be consistent. Recommended pen: 0.18–0.25 mm.

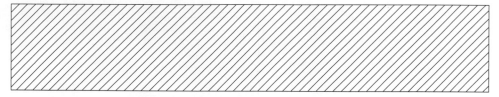

Figure 3-11
Section lines, also called hatching, are angled lines that represent where a material has been cut.

Poché Pronounced "po-shay," **poché** is a French word for a repetitive, textural pattern used to describe the material of which an object is made (Figure 3-12). Poché is different from **rendering,** which is decorative detail on an object. Dozens of pochés are discussed in Chapter 8. Figure 3-12 shows three. Recommended pen: 0.25 mm.

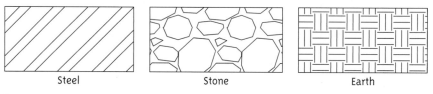

Steel   Stone   Earth

Figure 3-12
Poché symbols.

Match **Match lines** show where to align a large drawing that spans two or more sheets of paper (Figure 3-13). Match lines on a reduced-size floor plan show where the plan is continued on other sheets. Like hidden lines, they are dashed, but match lines are thicker and longer. Recommended pen: 0.50–0.80 mm.

Figure 3-13
Match line.

Border **Border lines** are thick lines that go around the perimeter of a sheet (Figure 3-14). They also can be used to underline titles. Recommended pen: 0.70–2.00 mm.

Figure 3-14
Border line.

Extension and Dimension **Extension lines** can emanate from the endpoints or center of an object (Figure 3-15). **Dimension lines** run perpendicular to them and contain the dimension note. These lines should be dark enough to reproduce on a copy, but thinner than the object lines so as not to be confused with them. On a pencil drawing, choose a hard (light) pencil or a dark, but thinner, pencil (or pen). Recommended pen: 0.25 mm.

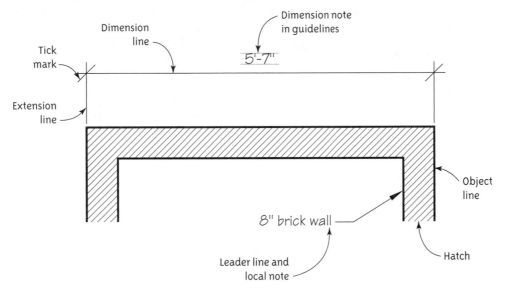

Figure 3-15
Extension, dimension, leader, and hatch lines.

Leader A **leader line** has an arrow or slash mark at one end and a **local note** at the other end, which describes the feature the leader line points to. A local note differs from a **general note**, which is placed elsewhere on the sheet and applies to everything on the sheet. Recommended pen: 0.25 mm.

Construction Also called projection lines, layout lines, or guide lines, **construction lines** are thin lines that help create object lines but are not part of the object itself. They may help position a drawing, regulate letter height, or project features from one picture plane to another. They are drawn lightly so they will reproduce very lightly (or not at all) on a copy. A 4H lead is used. These lines are not inked.

Symbols

Drawings contain symbols whose purpose is to identify specific features and concepts and reference those features to other drawings. Different construction disciplines (e.g., mechanical and electrical) have their own symbol sets, which are discussed in later chapters. The following are common symbols found on architectural drawings.

Section and Elevation Symbols

Section symbols and **elevation symbols,** sometimes referred to as call-outs, are symbols that indicate where, exactly, on the floor plan a section or elevation drawing is made.

Figure 3-16 contains both a section symbol and an elevation symbol.

The section symbol in Figure 3-17 consists of a circle, a line that cuts through the door opening, and numbers and letters. The reader is told that the drawing is drawing 6 on sheet A6.2. (*A* refers to architectural, *6* is the group number, and *2* refers to the group.) The arrow points in the direction of the cut; the material behind the arrow is ignored; and the material in front of the arrow is the subject of the drawing. The elevation symbol does not have a cut line because an elevation is not a cut drawing. Instead, it is a view of the object as seen from the front. The elevation drawing referenced by this symbol is drawing 5 on sheet A6.2.

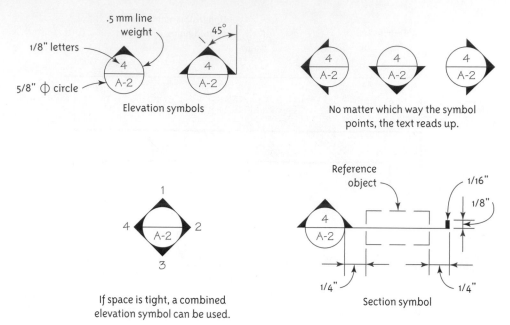

Elevation symbols

.5 mm line weight

1/8" letters

5/8" ⌀ circle

45°

No matter which way the symbol points, the text reads up.

If space is tight, a combined elevation symbol can be used.

Reference object

1/16"

1/8"

1/4"

1/4"

Section symbol

Figure 3-16
Proper orientation of lettering inside section and elevation symbols.

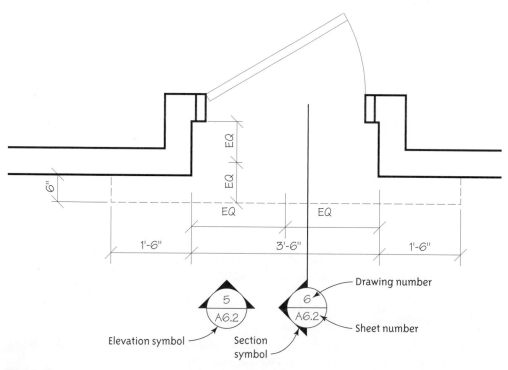

Elevation symbol

Section symbol

Drawing number

Sheet number

Figure 3-17
Section and elevation markers. In the section view the reader looks to the left. In the elevation, the reader looks up, toward the door.

Some elevation and section symbols include an additional number for cross-referencing. Note that the two-sided section symbol in Figure 3-18 has two numbers on its bottom. The first, A-2, states what sheet the section symbol is on. The second, A-3, states what sheet the section drawing referenced by the cut is on. The number 4 refers to which drawing on sheet A-3 the section refers to.

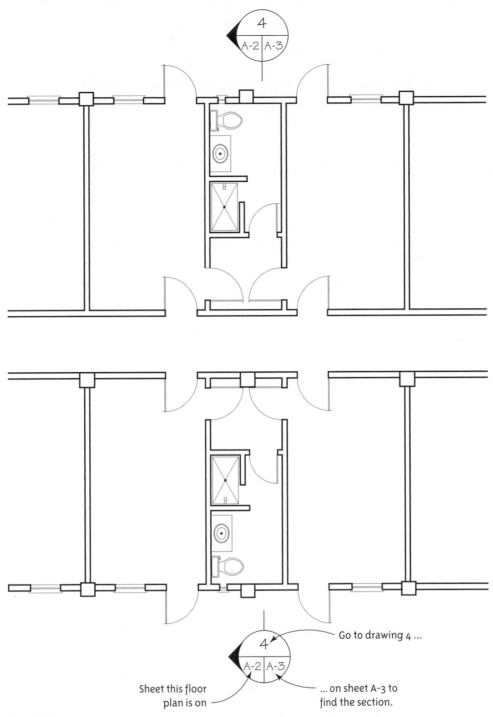

Figure 3-18
Cross-referenced section symbol.

ID Label

This goes under each drawing. Every figure on a sheet, whether it is a plan, section, or schedule, needs an identifier. It usually consists of a 3/4" diameter circle with an attached horizontal line. The drawing and sheet numbers are in the circle, the title is above the line, and the scale is below it. The circle is usually placed to the line's left. Office and regional practices vary; you may find that some drawing sets place it on the right on floor plans and on the left on all other drawings. Figure 3-19 shows the ID label on the section drawing that is referenced in Figure 3-18. Identical notation is found on the section drawing's **ID label.**

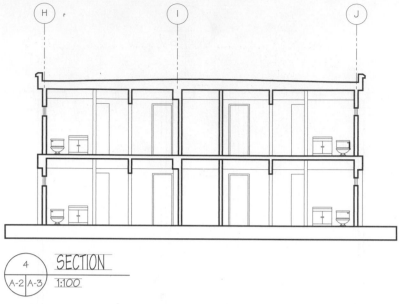

Figure 3-19
An ID label under the section referenced in Figure 3-18.

Miscellaneous IDs

Finishes, wall types, and the structural grid are identified by placing a name and/or number inside a shape. Ovals, rectangles, and circles are the most common shapes used for this purpose. These IDs usually reference a feature to a schedule. Any other repeating feature that the designer needs to call out is likewise assigned a symbol.

Enlargement Box

An area where a close-up is needed is encircled by heavy dashed lines (called an **enlargement box**) and has an attached ID label (see Figure 3-20). The label states where to find the close-up. The close-up should be drawn with the same orientation as it has on the plan.

Grid System

Drawings are laid out on a structural **grid system** that shows the location of columns, load-bearing walls, and other structural elements. **Grid lines** are used mostly for reference to schedules and for dimensioning. Vertical markers are numbered from left to right and horizontal markers, which are placed on the drawing's right, are numbered from bottom to top.

Sheet and Drawing Organization

Identification

Sheets and drawings must be labeled in a manner that makes their content and referencing clear. The American Institute of Architects (AIA) and Construction Specifications Institute (CSI) have developed protocols for this. Figure 3-21 shows the AIA's system, which is used by most architects.

Figures 3-22a and 3-22b show the Uniform Drawing System (UDS), which was developed by the CSI. It is more elaborate than the AIA's and is often used for large, complex construction and public works projects. It consists of a four-part identifier: a discipline designator, hyphen, group number and sheet number. The discipline designator identifies the type of drawings

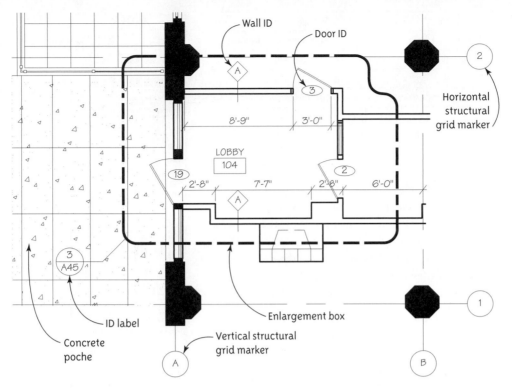

Figure 3-20
Enlargement box.

found on the sheet. The group number describes a specific type of drawing within the discipline. The sequence number states the sheet's number within the set. The group number will always be the same, no matter how many drawings are in the group. This enables drawings to be added or removed without disrupting the set's alphanumeric order.

For less complex projects, only the discipline designator and sheet number may be used. Also, although the discipline designator always comes first, the order of the other elements may be slightly different, as shown in Figure 3-23.

A Architectural (plans, elevations, sections, details)		**K** Dietary (food service)	
C Civil (roads, water, sewage removal, foundations)		**L** Landscape (site work, drainage, ground contouring, planting)	
D Interior design (furniture, color schemes)		**M** Mechanical (heating, ventilation, air conditioning)	
E Electrical (lighting, wiring, phone, security systems)		**P** Plumbing (pipes, fixtures)	
F Fire protection (sprinklers, standpipes)		**S** Structural (framing, load transfer systems)	
G Graphics (signage)		**T** Transportation/conveyance systems	

Figure 3-21
AIA discipline designators.

G General		**S** Structural		**M** Mechanical	
H Hazardous materials		**A** Architectural		**E** Electrical	
V Survey/mapping		**I** Interiors		**T** Telecommunications	
B Geotechnical		**Q** Equipment		**R** Resource	
W Civil works		**F** Fire protection		**X** Other disciplines	
C Civil		**P** Plumbing		**Z** Contractor/shop drawings	
L Landscape		**D** Process		**O** Operations	

Figure 3-22a
UDS coding designators.

0 General	5 Details
1 Plans	6 Schedules and diagrams
2 Elevations	7 User-defined
3 Sections	8 User-defined
4 Large-scale views	9 3-D pictorials

Figure 3-22b
UDS group designators.

Sheet Size

Identification labels, title blocks, dimensions, north arrows, schedules, specifications notes, and room for future users to add their own notes on the **as-builts** (printed sets that have handwritten notes on them acknowledging field changes) add to the size requirements of a drawing. This must be factored into sheet size and layout. All sheets in a set should be the same size.

Title Blocks

Title blocks are square or rectangular boxes. They are typically placed vertically along the right side, in which case they are called title strips (Figure 3-23) or in the sheet's lower right-hand corner (Figure 3-24). Drawings are stapled on the left. Different construction disciplines often use different formats. However, title blocks are always put on each sheet and contain information about the project in general and the sheet in particular. Typical information includes the project title, sheet title, client name and address, architect/consultant contact information, revision dates, professional seal, scale, and date.

Sheet Composition

A simple project might consist of one centered drawing and ID label and a title block, as in the student assignment in Figure 3-25. Steps for centering one drawing are shown in Figures 3-27 and 3-28. In a commercial project sheets are often gridded into modules (with reproducible or nonreproducible lines) with one drawing placed in each module (Figure 3-26). This is

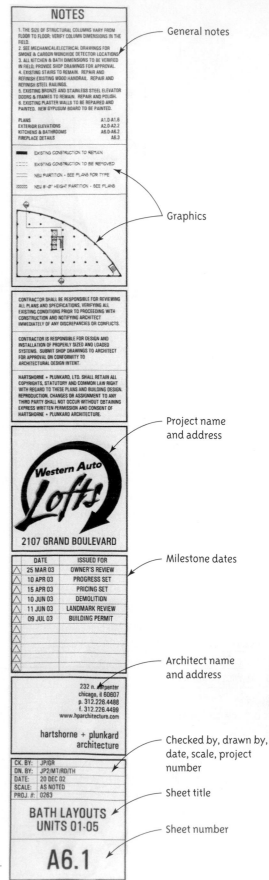

Figure 3-23
Title block on a professional drawing. It runs lengthwise on the right side of the sheet and contains detailed information about the project. *Courtesy of h+p architecture, Chicago, IL.*

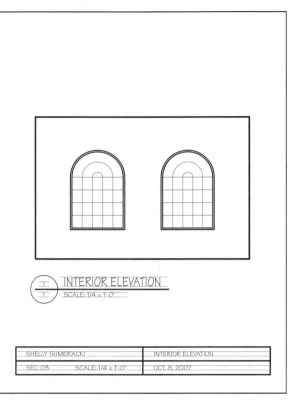

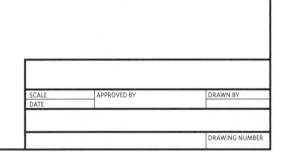

Figure 3-24
Generic title block. It goes in the lower right corner of the sheet.

Figure 3-25
Student drawing of an interior elevation. The sheet is size A (8 ½" × 11"). The title block is 1/2" from the bottom. Whether the sheet is oriented vertically or horizontally, a long title block like this should be placed on the short end. The picture is centered in the space above the title block.

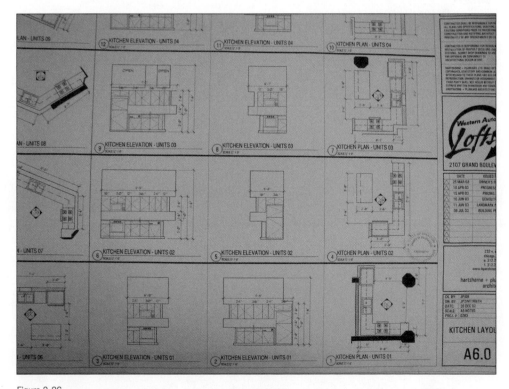

Figure 3-26
A commercial project laid out in gridded-box format. *Drawing courtesy of h+p architecture, Chicago, IL.*

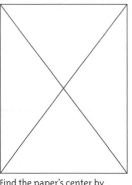 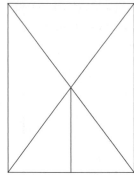 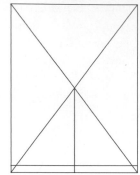

1. Find the paper's center by connecting its corners. Where the diagonals intersect is its center.

2. Project a line down from the intersection to the paper's edge.

3. Measure 1/2" up from the edge to locate the title block's bottom.

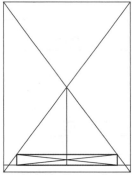 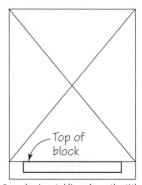 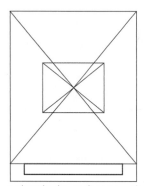

4. Find title block's center by connecting its corners. Align its center with the paper's center.

5. Draw horizontal lines from the title block's top to the paper's edges, then repeat Step #1 to find the center of the paper's remaining space.

Top of block

6. Align the drawing's center with the paper's center, and trace.

Figure 3-27
Steps for centering a drawing on a sheet.

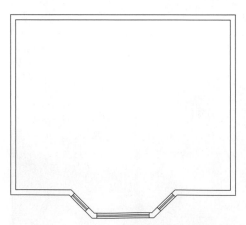 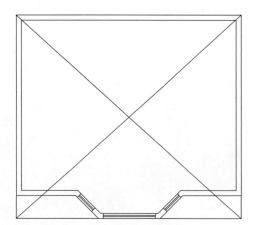

To center an asymmetrical object, block it into a square or rectangle, then find the intersection of the diagonals.

Figure 3-28
Steps for finding the center of an asymmetrical drawing.

so the structural grid or dimension lines on separate drawings won't overlap. Drawings are usually numbered from top to bottom and from left to right. On large, commercial projects, they are sometimes numbered from bottom to top and from right to left with any unused space at the left (stapled) side.

Multiple drawings on one sheet should be related, because this enhances comprehension. The most common combinations are plan/plan, elevation/section, elevation/plan, elevation/ detail, and detail/detail. Avoid completing a partially filled sheet with drawings that don't reference or relate to the others.

On simple projects where there is just one drawing on one sheet, centering it makes a nice presentation. Figures 3-27 and 3-28 show steps centering a drawing and a title block.

Drawing Orientation

When possible, all plans should be drawn parallel to the edges of the paper, with north at the top of the sheet. This is called **plan north** and enables the drafter to give simple names to interior and exterior elevations. If plan north isn't compass (magnetic) north, an additional arrow is added to the drawing that points that way (Figure 3-29).

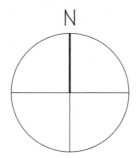
North arrow when plan north and compass north are the same

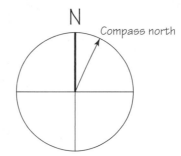
North arrow when plan north and compass north are different

Figure 3-29
True north vs. compass north.

Lettering

Lettering is an aspect of note making. Notes are an integral part of a drawing and lettering's purpose is to make notes easy to read. Lettering is not the same as printing in capital letters, which doesn't provide the level of legibility needed for a set of construction drawings. Printed letters can get especially blurred when the drawings are printed half-size.

Look critically at the letters in Figure 3-30 to see what differentiates lettering from merely printing in capitals. Height is consistent, which is achieved by drawing guidelines (covered in Chapter 2). Letters are the entire height of the guidelines; they don't fall short. They may extend a bit above or below them; this is the case with certain styles and with fractions (fractions are drawn slightly larger than whole numbers). There are no gaps between the pencil lines. Spacing is consistent between words and within words. All same-type letters should look identical. Vertical strokes should be vertical, horizontal strokes should be horizontal, and angled strokes should all angle to the same degree. Numbers must match the style of letters.

Style

There are many different lettering styles, but only a few are suitable for architectural drafting. Avoid fancy styles, as they do not facilitate reading and are difficult to execute. While personalized styles were once a drafter's hallmark, CAD has virtually eliminated this practice. Designers now choose among different computer fonts; "City Blueprint" is one popular choice. For manual drafting, a simple block lettering style with horizontal strokes slightly angled is favored by many designers.

1/8" USE THIS SIZE FOR GENERAL NOTE-MAKING. USE A STRAIGHT-EDGE FOR THE VERTICALS AND FREEHAND THE HORIZONTALS.

1/4" ABCDEFGHIJKLMNOPQRSTUVWXYZ THIS IS A NICE
STYLE TO EMULATE. NUMBERS SHOULD BE THE SAME STYLE AND HEIGHT AS
LETTERS. FRACTIONS CAN BE SLIGHTLY LARGER. 1 2 3 4 5 6 7 8 9 0

1/4" THIS IS A SERVICIBLE, BASIC BLOCK LETTER. 1 2 3 4 5 6 7 8 9 0
ABCDEFGHIJKLMNOPQRSTUVWXYZ

1/2" USE THIS SIZE FOR TITLES AND
LABELS. DRAW LETTERS THE FULL
GUIDELINE HEIGHT. 0 1 2 3 4 5 6 7 8

Figure 3-30
Architectural lettering. Two styles are shown: the basic block and a variant of the basic block (slanted horizontals). Either is appropriate for manual drafting.

Technique

A drafting pencil and a vertical straightedge (such as that of a small triangle or lettering guide) can be efficiently used to create letters (Figure 3-31). Slide the triangle or guide along the parallel bar to create vertical strokes and freehand the others. While using a straightedge is slow going at first, with practice you'll draw letters almost as fast as you handwrite. Practice drawing words and sentences, rather than the alphabet, because words and phrases will incorporate spacing. Make the strokes quickly; slow, labored-over strokes look shaky. Rotate the pencil while drawing to obtain slightly different line weights. If smudging is a problem, use a harder lead or keep a clean sheet of paper under your hand to protect the letters already drawn. When inking, use a 0.40 mm pen. Most notes are made with 1/8" tall letters. Drawing titles are done with 1/4" letters. Only one lettering style should be used throughout a project.

Figure 3-31
A small triangle or the Ames Lettering Guide provides a straightedge for lettering.

Mechanicals

Transfer letters (rub-ons), templates, and lettering wheels are alternatives to hand lettering (Figure 3-32). Rub-ons are bought in sheets, positioned in place, and rubbed with a burnishing tool. Templates are letter outlines that are positioned and traced. Lettering wheels are keyboard machines that print typed notes onto transparent tape. While these all may produce a nicer product than manual lettering does, they are also expensive and time consuming. Hence, they are best used as a supplement to produce larger letters, such as those in titles, which may be more difficult to draw.

Figure 3-32
Using mechanical aids in lieu of hand lettering. Pictured are rub-ons, templates, and a tool for drawing guidelines.

Notes

Notes are lettered comments placed on the drawings. There are general notes, which apply to the whole sheet or set, and local notes, which have a leader line pointing to the feature discussed (Figure 3-33). Notes describe specific features and also clarify intent, such as "slope floor to drain." Here are some tips for their placement.

1. Notes can be placed between dimension lines and object lines. Arrange notes at the same time as dimension lines so they don't overlap. Avoid long leader lines between notes and the items they point to.

2. Group the notes around the construction they refer to. For instance, in a wall section, place all notes that refer to the floor slab in one general area close to that slab. Notes that apply to other areas should be placed elsewhere. Place notes as close to the items they point to as possible; for example, the topmost note directed to the floor slab should describe the topmost material in the floor slab construction, and so on. This will avoid crossed leader lines.

3. Align notes at the left; an even margin improves readability.

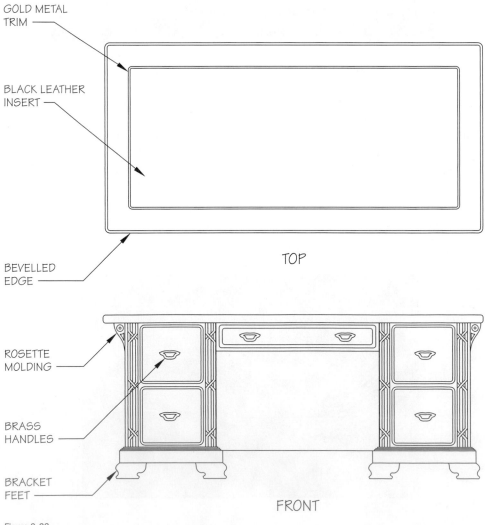

Figure 3-33
Well-placed notes and leader lines on a furniture drawing.

4. Draw leader lines at an angle to avoid confusion of leader lines with the drawing's vertical and horizontal lines. They may also be drawn with an irregular curve.

5. Start leader lines at the beginning or end of the note, not in the center. They should touch the material they're referring to.

6. Notes should be short and to the point. Nomenclature should be consistent throughout the set of drawings.

Summary

The complexity of architectural drawings is managed by using industry standards. Drafters must understand lines, symbols, and the concepts behind those symbols. No matter what part a person plays in the project, it's necessary to understand everyone else's part. If an interior designer has control of a project, such as a residential bath or kitchen redesign, he or she needs to understand the content of consultant submissions to check their overall compliance with the design intent.

Suggested Activities

Obtain sets of working drawings from different sources. Then:

1. Find and identify different line types and symbols on each set

2. Compare how the same symbols are drawn on different sets.

3. Find five objects (e.g., doors, cabinets) on each set and describe the line types used to draw them.

4. Write a list of the similarities and differences between section and elevation drawings.

5. Find and read the general notes. Discuss their relationship to the drawings.

Questions

1. What is the difference between a visible object line and a hidden object line?

2. Why are different line weights used on a drawing?

3. What is the purpose of general notes, local notes, and dimensions?

4. How should drawings be arranged on a sheet?

5. What is the purpose of a title block?

6. Explain how an elevation symbol and a drawing ID label are related to each other.

Further Reading and Internet Resources

Sutherland, M. (1989). *Lettering for Architects and Designers* (2nd ed.) New York: Van Nostrand Rheinhold

http://www.csinet.org

Construction Specification Institute website. The Uniform Drawing System (UDS) conventions are among the information you can find here

CHAPTER 4

Sketching and Drafting Orthographically— Theory

OBJECTIVE

- This chapter discusses how to sketch and draft orthographic views of furniture pieces.

KEY TERMS

elevations
foreshortened
multiview drawings
orthographic drawing

orthographic projection
perspective
picture plane
plans

projection line
sketching
true length

All objects have three physical dimensions: height, depth, and width (or length). An **orthographic drawing** is one that shows two of an object's three dimensions. Interior designers frequently draw orthographic views of rooms and furniture (Figure 4-1). These drawings are created via **orthographic projection**, a drawing technique that deconstructs a 3-D object into multiple 2-D views (Figure 4-2). They are also called **multiview drawings** because multiple views are typically needed to describe the object.

Visualize an object suspended in a glass cube. Each side of the cube is a **picture plane**, a flat surface onto which the suspended object is projected. Think of the object as exploding apart with each side projected (thrown straight ahead) onto the picture plane facing it. **Elevations** are orthographic views projected onto vertical planes. **Plans** are orthographic views projected onto the horizontal plane. These views are flat and depthless, like the planes they're projected onto. This is because only two of the object's three dimensions are shown.

Ortho is a prefix with Greek origins. It means *straight, perpendicular, vertical,* or *correct. Orthodoxy* (meaning *correct teaching*) and *orthodontist* are modern usage examples. In drafting, *ortho* means at a 90-degree angle to the picture plane. To project orthographically means to project perpendicular to something.

Why Orthographic Drawings Are Used for Construction Purposes

Orthographic drawing technique is used because it produces drawings that document space and objects in a way that is measurable and proportionately accurate, unlike the way we see it. Because of the way our brains perceive depth, we see things in **perspective**; objects appear to be smaller as they get farther away, and sets of parallel lines appear to converge to a common point.

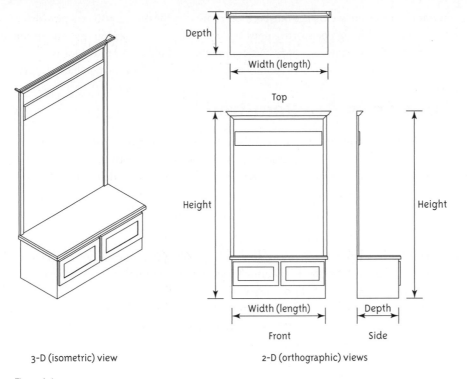

Depth

Width (length)

Top

Height

Height

Width (length)

Depth

Front

Side

3-D (isometric) view

2-D (orthographic) views

Figure 4-1
3-D and 2-D views of a hall bench. The multiple 2-D views show its size and shape. The front view shows height and width. The top shows width and depth. The side view shows depth and height.

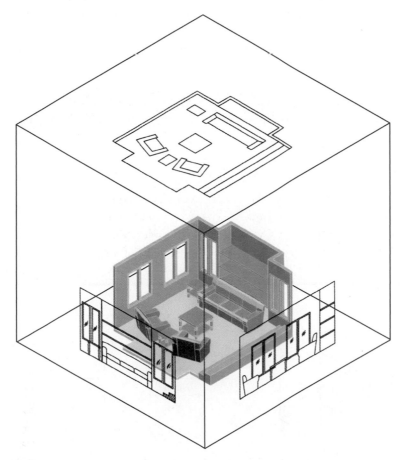

Figure 4-2
Orthographic drawings are made by deconstructing a 3-D object into multiple 2-D views.

Furthermore, each time we change our position relative to the object, we see a new shape and new proportions. Although perspective drawings are good for visualization, they are unsuitable to build from. Orthographic drawings best communicate the details needed for construction, because they provide views that correspond to the subject's actual, or measurable, proportions.

How Many Drawings Are Required?

The number of drawings required depends on the project. If is the goal is to design a very simple object, one view plus a note might suffice. However, most projects require more than this. Buildings can require hundreds of drawings. The number needed is the number that completely describes the project.

Sketching

Sketching is freehand drawing. It is used to think a problem through and solve it. Hence, it is helpful to sketch an object before hard-lining it. Attempting to problem-solve while simultaneously hard-lining slows down both processes.

Sketching is not synonymous with sloppy work. Sketched lines should be neat, long, and continuous, not short and overlapped. Tip the pencil in the direction of movement. A dark pencil and eraser work well, although using a light lead first and darkening later also works. Keep your eye on the point where the line ends to help draw lines straight.

Orthographic Sketching

Visualize a cube with its sides unglued, lying flat on a table, each plane containing a view of the object (Figure 4-3). When the views are drawn, they will be carefully placed within each plane so the object edges will align. Object lines that are parallel to the picture plane show up **true length**, which means actual length. Object lines that are skewed to the picture plane show up **foreshortened** length, which means shorter than actual length. Object lines that are perpendicular to the picture plane show up as points. An object (or a feature on an object) whose boundaries consist entirely of true length lines shows up true shape and size. An object whose boundaries contain foreshortened lines does not show up true shape and size, and hence is not measurable.

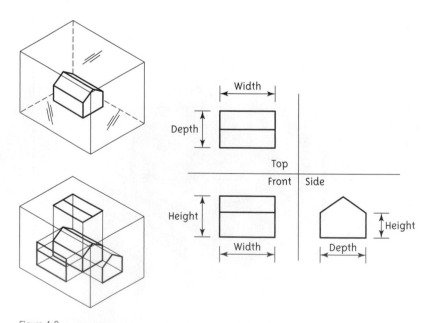

Figure 4-3
3-D view of a house suspended inside a "glass box" of picture planes, and the corresponding 2-D views.

Square Table

Figure 4-4 is a perspective drawing of a square table. Let's sketch two orthographic views of it (Figure 4-5).

The top view is sketched from a birds-eye view; that is, the viewer is looking straight down on it. Its overall dimensions are measured and an outline is drawn. Details are measured and added. All edges are parallel to the top picture plane, so all lines are true length, making the view true in shape and size. Note that it is oriented so lines can be projected down to create the frontal view directly below.

Projection lines are construction lines along which points and features are transferred from one plane to another. They are drawn lightly and at 90 degrees (perpendicular) to a picture plane, never at any other angle.

The front view is sketched with the viewer standing on the floor, looking straight at the table. The projection lines from the top view define the table's width. All edges are parallel to the frontal picture plane, so this view is also true in shape and size.

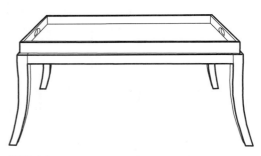

Figure 4-4
3-D (perspective) drawing of a square table.

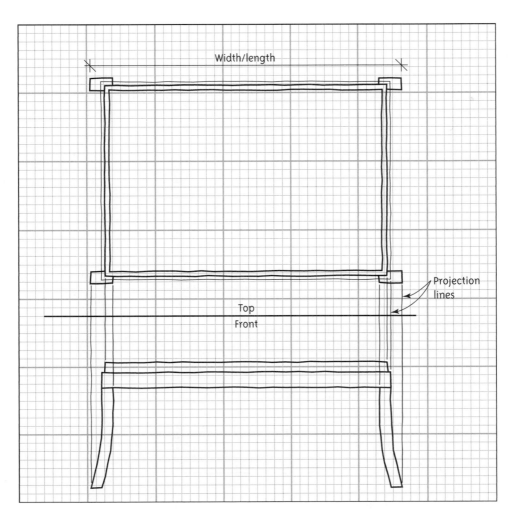

Figure 4-5
Top and front orthographic views of the square table from Figure 4-4. All lines are parallel to the picture planes they are in, making the views true in shape and size.

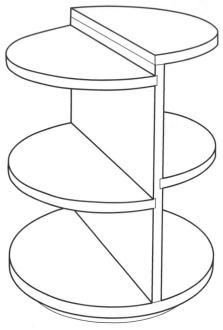

Figure 4-6
3-D (perspective) drawing of a round table.

Round Table

Figure 4-6 is a perspective drawing of a round table. Let's sketch two orthographic views of it.

The table top is measured and sketched in the top view of Figure 4-7. Because its circular lines are parallel to the top picture plane it is true in shape and size. The hidden line represents the round platform underneath the table, which cannot be seen from the top. Projection lines are drawn down to the front view, which define the table's width. However, since the shelves' curved lines are not parallel to the frontal picture plane, they appear **foreshortened**, meaning shorter than their true length. Since a shape with boundaries of foreshortened lines does not appear true in shape and size, we cannot see that the shelves are round.

Once the problem of constructing these tables orthographically has been solved, drafting them with straight-edged tools becomes easier, because the sketches, which is where the problem-solving occurred, serve as a reference. Precise, hard-lined drawings, are needed for construction, and for them, straight-edged tools are brought out.

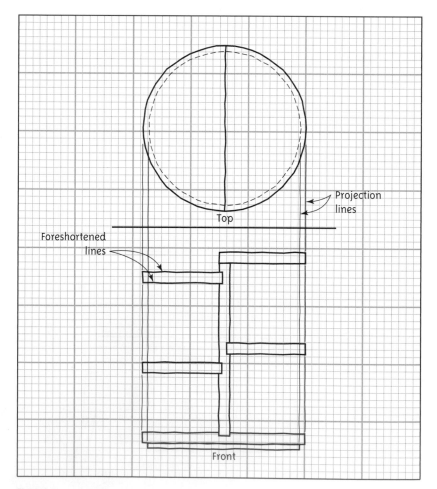

Figure 4-7
Top and front orthographic views of the round table from Figure 4-6. The shelves in the front view are foreshortened, because they are not parallel to the frontal picture plane.

Orthographic Drafting

Cabinet

Figure 4-8 shows an isometric drawing of a cabinet. Let's draft three orthographic views of it.

Top This image is drawn from a birds-eye view, looking straight down. Measure the cabinet's length and width, and then draw a rectangle with those dimensions. Note the orientation. It is oriented so lines can be projected down to the frontal plane to create a frontal view. Add a door and knob at the front.

Front This view is drawn standing on the floor, looking directly at it. Projection lines from the top plane define the cabinet's width, so there is no need to measure it again. Measure and mark the cabinet's height. Draw a rectangle with these measurements. Then measure and mark the individual drawers and knobs.

Side This is drawn standing on the floor, facing one side of the cabinet and looking directly at it. Projecting lines from the front view will define the cabinet's height. Its width must be measured again and marked off.

Notice the line weights in Figure 4-9. The strongest, most predominant feature is drawn with the heaviest lines. Here, that feature is the outline. The eye is led to the outline, reading it first, and then is led to the thinner drawer and knob lines.

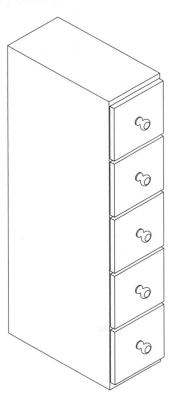

Figure 4-8
3-D (isometric) drawing of a cabinet.

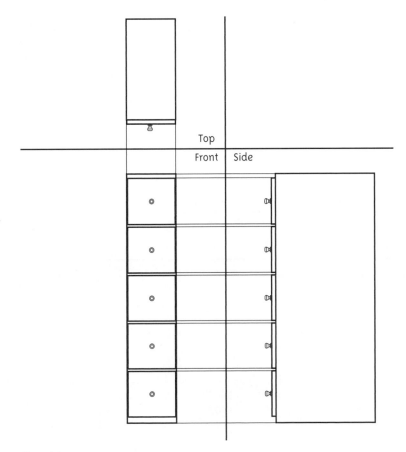

Figure 4-9
Top, front, and side orthographic views of the cabinet from Figure 4-8.

Chest of Drawers

Let's construct orthographic views of a more detailed item, the chest of drawers shown in Figure 4-10.

Figure 4-10
Perspective drawing of a chest of drawers.

Top Measure and mark the chest's length and width, the top overhang, and the plate rail. Draw the outline first, then add detail. Note the hidden (dashed) lines. Hidden lines represent features that can't be seen in a view but must still be acknowledged. In this case they represent the chest's perimeter, which can't be seen because of the lid overhang, and the legs (Figure 4-11). Hidden lines for the drawers are left out for clarity.

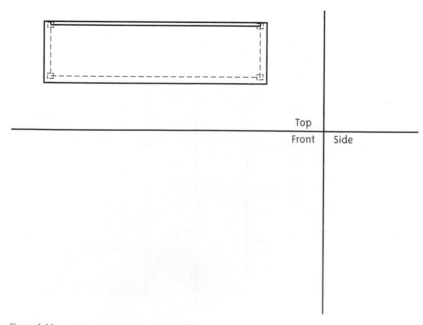

Top

Front | Side

Figure 4-11
Top view of the chest of drawers from Figure 4-10.

Front Project the chest, plate rail, and overhang's width down from the top view. Measure and mark the chest's height, then all measure and mark the heights of all the details (here, the apron rail, handles, and plate rail). Look at the overhang. You can see its thickness and how much it projects at the sides, but because of the lack of depth you can't see how much it projects at the front (or if it overhangs at all). It's only in the side view that the front overhang is visible. There are no hidden features in the front view, so there are no hidden lines (see Figure 4-12).

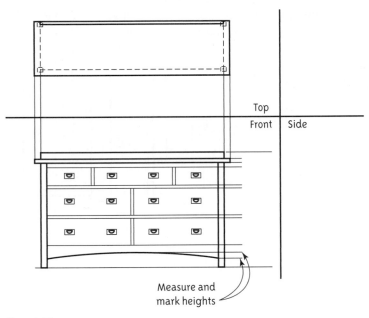

Figure 4-12
Front view of the chest of drawers from Figure 4-10 projected from the top view.

Side Project the heights of the chest, plate rail, and drawers horizontally over from the front view. Note that in Figure 4-13 the drawers are indicated with hidden lines because they can't be seen.

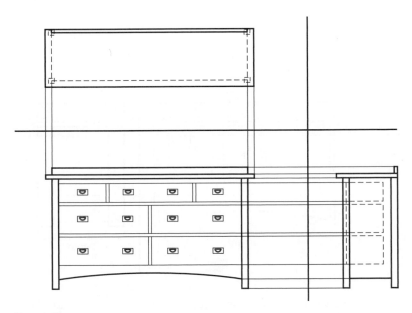

Figure 4-13
Side view of the chest of drawers from Figure 4-10 projected from the side view.

Drawing a Third View

At times a drafter might have two views of an object and be asked to create a third. Although a third view of any object can be created by measuring the width and depth in the available views with a scale or dividers, constructing it orthographically is an easier option if the object is detailed.

If you have top and side views, creating a front view is straightforward; simply project information down from the top view and horizontally from the side view. But if you only have front and side views, how do you construct the top (or, if you only have top and front views, how do you construct the side)? Figure 4-14 shows front and side views of a hope chest. Figures 4-15 and 4-16 show how to construct the top view.

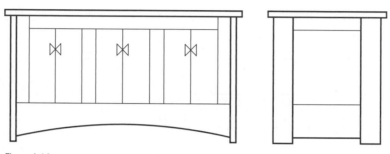

Figure 4-14
Front and side views of a hope chest.

1. Draw lines that delineate the picture planes. Since no partial top view already exists, these lines can be drawn anywhere between the views. Project lines straight up from the front view to the top.

2. Information from the side view must now be transferred to the top. To do this, the quadrant directly above the side view must be bisected with a 45-degree angle. (When a 90-degree angle is bisected, the result is two 45-degree lines. This is why a 45-degree line is used.)

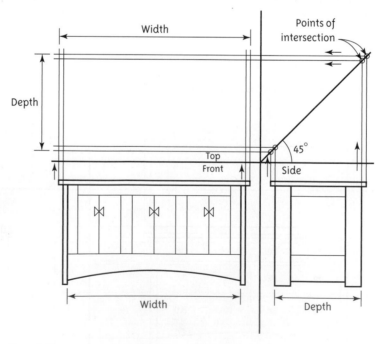

Figure 4-15
Construction of the third view starts with drawing cross-lines to represent the picture planes, and a 45° angle line at their intersection. Lines are then projected from the front view up to the top, and from the side view to the angle, then horizontally to the top.

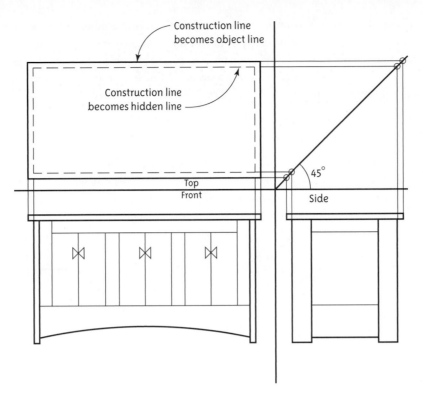

Figure 4-16
Darken the object lines and turn the appropriate construction lines into hidden lines.

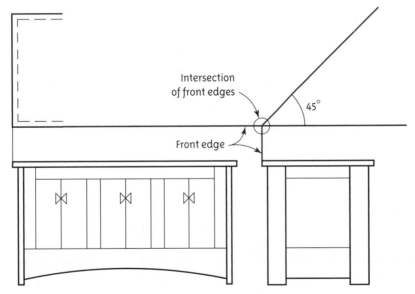

Figure 4-17
When partial views exist in all three picture planes, connect the front edge in two views. Where those edges intersect, draw the 45-degree line.

3. Project lines from the side view up to the angle and, where they intersect, project lines horizontally to the top view. Alternatively, the width and depth from the existing views could simply be measured with dividers and transferred to the top view.

4. All the lines needed to draw the top view now exist. Darken the perimeter, which represents the chest lid. Darken the hidden line, which represents the edge of the chest.

Note that if a partial view of the top exists in addition to the other two views, the intersecting lines that represent the picture planes and the 45-degree angle line cannot be placed randomly between the views. Instead, they must be placed precisely. As shown in Figure 4-17, a line from the chest's front edge in the top view and a line from the chest's front edge in the side view must intersect; place the 45-degree angle at that intersection.

Corner Cabinet

Figures 4-18 and 4-19 present an isometric view and two orthographic views of a corner cabinet. The top view shows the cabinet's true shape and size, because all edges are parallel to the picture plane. But in the front view, the sides of the cabinet skew away, hence they appear foreshortened. Only the middle section appears in its true shape and size.

Figure 4-18
3-D (oblique) view of a corner cabinet.

Top

Front

Figure 4-19
Top and front orthographic views of the cabinet from Figure 4-18.
The angled sides are foreshortened. The middle section has the
thickest line weights because it is closest to the viewer.

Summary

Orthographic drawing is the method by which architectural drawings are created. Views are created by projecting points and lines from one plane to another. In theory these drawings are called top, front, and side views; architecturally, these drawings are usually floor plans, interior elevations, and sections. The benefit of orthographic drawings is that the originals are measurable and accurate portrayals of what the item looks like.

Suggested Activities

1. Measure and draw top, front, and side views of small objects in your classroom. Remember to include dashed lines for hidden features.

2. Connect Lego Duplo® blocks in random arrangements and sketch their top, front, and side views to strengthen your understanding of orthographic drawing (Figure 4-20). Use differently shaped blocks for true length and foreshortened lines. Sketch 3-D pictures of the arrangements and apply different colors to the top, front, and side views to clarify what features go in which view.

3. Obtain photos of furniture and their measurements from catalogs. Using this information, draw orthographic views of the furniture.

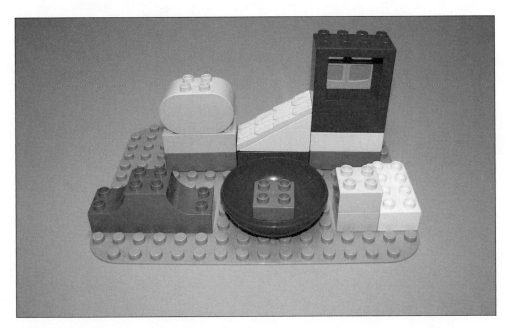

Figure 4-20
Lego blocks are useful for orthographic sketching practice.

Questions

1. What is orthographic projection?

2. What is the advantage of orthographic views over perspective views?

3. How must a line be oriented to a picture plane to show its true length?

4. How must a line be oriented to a picture plane to appear foreshortened?

5. How are features not visible in a view portrayed?

CHAPTER 5

Design Preliminaries

OBJECTIVE

- This chapter discusses the types of information that must be gathered, researched, and analyzed before commencing the design process. Specifically, it discusses programming, taking a visual inventory, sketching, building codes, and accessibility.

KEY TERMS

accessibility	means of egress	sketch
building code	model codes	station point
cone of vision	occupancy	sustainability
environmental	perspective sketching	synthesis
foreshortened	profile gauge	thumbnail sketch
half-diagonal rule	program	universal design
isometric sketch	programming	vanishing point
maintenance	site visit	visual inventory

Competent design requires preliminary research. Whether the design is residential or commercial, information must be gathered, analyzed, and synthesized, a process called **programming**. Research must be done to determine the spaces required; their functions and square footage; how many people will use them and how; which spaces should be adjacent, separated, public, private, or secured; the traffic flow of goods, services and people; and the furnishings and equipment needed. Relevant building codes and regional limitations, context, cost, maintenance, energy usage, and accessibility all must be considered. Aesthetic and functional requirements (mechanical, electrical, and plumbing systems) also need to be evaluated.

Analysis for a commercial building may involve additional considerations of cost life cycle, organizational and image goals, environmental concerns, and sustainability. Input from all parties that will use the building should be solicited. In a residential project this input will come from the owners, but in a commercial project insight may also come from occupants, and operations and maintenance personnel.

Other Considerations

Environmental concerns include any of the interior environment's physical conditions that affect occupant health and safety, such as air quality and circulation, temperature control, ergonomic layout, and physical circulation plan. **Maintenance** concerns include the ability of the products and materials to be kept in good condition, as well as the work required to sustain that condition over the material's lifespan. **Sustainability** concerns include using resources in a manner that does not deplete them and will have the least long-term effect on the environment.

A **program** is a written statement that describes these issues. It lays out the problem, scope, goals, requirements, and constraints. Adequate preparation at the conceptual stage ensures that the design will be based on sound decisions, will be within budget, and will not require later redesigning due to overlooked requirements. Additionally, creating a program maximizes the potential for success of the finished product. Without it, much of the time spent on preliminary design may be wasted.

Taking a Visual Inventory—The Site Visit

If the project involves an existing building, a **visual inventory** needs to be done. This comprises written, sketched, measured, and photographic documentation of existing conditions, accomplished via a **site visit**: a trip to the construction area that includes interviews with the project users, and the creation of sketches, photos, and notes. The site visit should answer questions like "What is your reaction to the space? What catches your eye and where is it led? What does the client like and dislike? What needs to be done? Is the space open to other spaces, requiring visual coordination? What are the individual components, overall arrangement of space, predominant shapes, light, and view?"

Typical tasks include measuring floors and walls for carpet and wallpaper and examining windows to determine where to place a drapery pull. You should note finishes and colors, traffic patterns, and what needs to be moved, removed, or replaced. What can't be moved without major cost? What needs to be there? What doesn't? Is there enough storage? Where will carpet be joined? Where are air vents, outlets, light switches, and thermostats? Thermostats need to be noted because furniture can't be placed on top of or in front of them. List all objects and other visual information in the space. Sketch a floor plan and supplement it with photos and notes. Verify that your work is understandable enough for others to read.

Tools

Travel is typically the biggest time and money expense of a site visit. Hence, arriving at the site properly equipped is important. The following tools are useful to record site features (Figure 5-1).

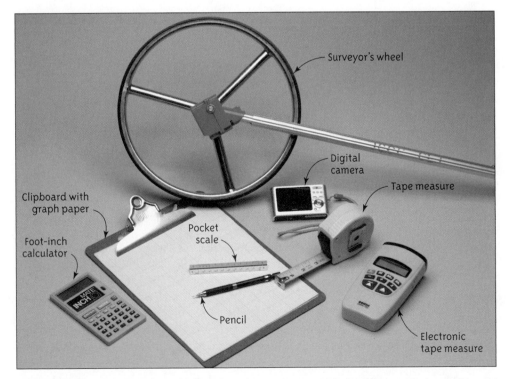

Figure 5-1
Useful tools during a site visit. *Photo courtesy of Bret Gustafson, JCCC*

Tape measure A 25-foot metal tape that is rigid, retractable, and lockable, with a hook at the end for attaching it to a wall (useful when working solo) is best. Models that show increments in inches, feet and inches, and metric are available.

Other measuring tools Surveyor's wheels are measuring devices that click off distances on an odometer when hand-pushed. Electronic tapes are battery-operated and use an ultrasonic beam to record distance on a digital readout. They can even calculate square footage. But they should only supplement a traditional tape since they can be unreliable. Electronic tapes can't record distances when beamed on soft surfaces, (e.g. bulletin boards), because those surfaces scatter the beam, and they only work at distances between 50' and 200'.

Profile gauge A **profile gauge** measures complex moldings in place (Figure 5-2). It consists of a magnetic handle and moving pins that take on the contour of whatever object they're pressed against. When placed against a molding, the pins reproduce its outline or profile. The shape can then be traced from the gauge to a sheet of paper. This tool is usually necessary if a drawing will include ornamental detail beyond an outline.

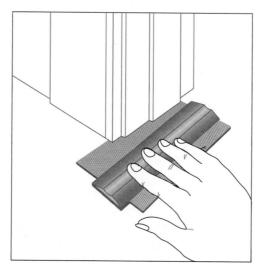

Figure 5-2
A profile gauge has short pins that take the shape of the object they are pressed against.

Digital camera A digital camera with a zoom lens and a high-capacity memory stick works well. Carry it in a soft case with extra batteries. Photograph the site and use the zoom for close-ups. Pictures can be printed and used as an underlay for sketching thumbnail perspectives.

Clipboard An oversized one (11" × 14") is preferable, but an 8−1/2" × 11" clipboard will also do and provides a nice portable surface for sketching and holding paper down.

Other Other useful tools include a plumb bob to check vertical walls, a metal triangle to check square corners, a level to establish horizontal reference (or datum) lines, a flashlight, and a ladder.

Grid Paper or Vellum

Paper printed with four, eight, or twelve squares per inch is available. One-eighth–inch or quarter-inch squares make sketching to those scales, as well as drawing straight lines, easy. Paper printed with faint lines allows penciled lines to show up better. Grid paper can be drawn on directly or used as an underlay for tracing paper. Use the boxes to scale your drawing (Figures 5-3, 5-4, 5-5). A pocket architect's scale is also useful for this purpose.

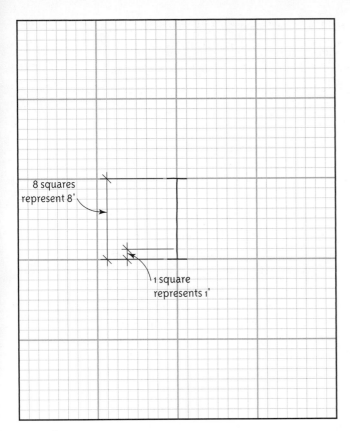

Figure 5-3
This grid can be used as a 1/8"=1'-0" scale.

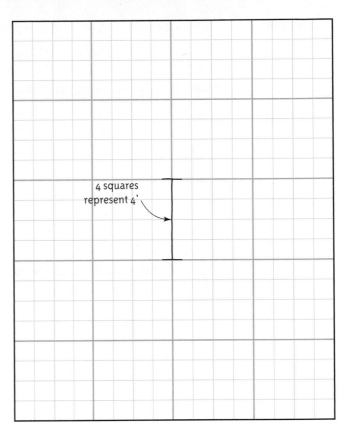

Figure 5-4
This grid can be used as a 1/4"=1'-0" scale.

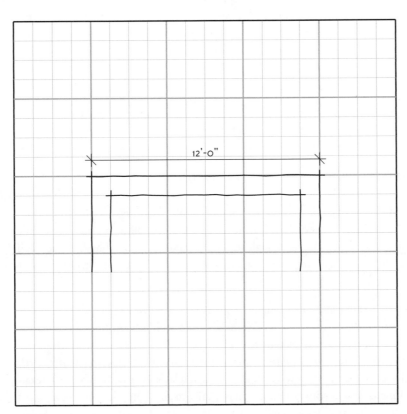

Figure 5-5
Grid paper makes sketching to scale easier. Each square represents a distance. Sketch with long, continuous freehand lines from one end to the other. Overlap at corners.

Calculator

There are calculator models that calculate in decimal units, feet and inches, and metric, as well as converting between scales. A basic four-function calculator is enough for most uses, although one with trigonometry functions is useful for calculating angles.

Pencil and Notebook

It's helpful to carry a dark lead pencil and eraser, and a pocket-size spiral notebook for note-taking.

Sketching

As discussed in Chapter 4, **sketching** is freehand drawing. It is used to brainstorm and record information. Sketched thumbnail drawings enable a designer to problem-solve in 2-D and 3-D manners. Straight-edged tools should not be brought out at this point, as premature hard-lining of a drawing will circumvent programming and problem-solving efforts. Hard-line drawings also imply that the design stage is over, which may make clients and consultants hesitate to suggest changes, assuming too much time and money has already been invested. Sketches should look like what they are: early stages of design. The input they invite results in a better product. Sketching is a particularly important skill for designers, because the lack of it can hinder design solutions; for example, some designers fail to include a desired element because of their inability to sketch it.

If you've watched skillful sketchers at work it may seem they have natural ability. While that is true for some, anyone can sketch. Artistic ability isn't needed because a drafter's sketches serve a different purpose from an artist's. An artist employs color, shade, and shadow to please his or her audience. A drafter simply explains enough for the subject to be built. There are techniques the beginner can study. Sketching ability improves with heightened observational ability and practice. Your goal should be to reach the point where sketching is done automatically, with the seeming effortlessness of experienced drafters.

Sketching Tips

A small percentage of drafters are born with the artistic ability to freehand sketch in perspective. For everyone else, study of the mechanical construction of orthographic and perspective drawings is a good start. In addition, the following techniques will help the beginner.

Hone Observational Skills Seeing and observing are two different things. When we see something, we usually just give it a cursory glance to identify it. We might see a wallpaper pattern and remember its general appearance. However, if asked what the specific fruit in the pattern was, or if it was more purple than red, our recall is less precise because we didn't *observe* it, or didn't look closely enough at it. Observation is a disciplined, learned skill. Learning to look at the built environment with a critical eye will improve your sketching ability and help you put on paper exactly what you see.

Estimate Proportions Knowing an object's actual measurements isn't necessary to sketch it proportionally, but estimating relative heights and widths is. Although the use of grid paper and scales is appropriate, developing the skill of "eyeballing" accurate sizes without such help is invaluable. To accurately draw images smaller than actual size, study apparent-size relationships: the relative sizes of the parts of your subject and their relationship to the whole. For example, if you are sketching a dresser, estimate how wide it is relative to its height. There are techniques to help you do this. One is the pencil trick. Stand parallel to the dresser. To estimate its width, hold your pencil at arm's length. Turn the pencil sideways, align one end with the right or left side of the dresser, and move your thumb to mark the end of the dresser's other side on the pencil's length. Mark that length on your paper. Then turn the pencil vertically with the bottom aligned with the bottom of the dresser and slide your thumb until it aligns with the top. Mark that length on the paper.

Any lines that are not parallel to you should be **foreshortened**. To sketch a foreshortened line, align your pencil with one, and, keeping that angle, bring your arm to your paper. Draw a line parallel to it in the place desired. The accuracy of this trick depends on maintaining the pencil at a constant distance from your eyes (full arm's length is best) and keeping the pencil parallel to the line being sketched.

Point-to-Point Method Another technique for sketching proportionately is the point-to-point method, which consists of drawing and connecting points on an object to form its shape. This is analogous to drawing pictures of star constellations. Constellations are actually three-dimensional; the stars making up a constellation are at light-year–different distances from Earth. But because we cannot perceive this depth from where we stand, we just see a flat group of stars with measurable proportions. When we look at the Big Dipper we are actually seeing the 2-D shape of a 3-D cluster of points. So, if drawing an object's proportions relative to itself or its sur-roundings is difficult, you can try to ignore its 3-D character and focus on its 2-D shape.

Count tiles in a floor or an acoustic ceiling grid, or pace off rooms with long strides, each equaling about one yard, to develop a feel for the size of spaces.

Scale the Picture to the Paper Part of layout and proportion is figuring how large to sketch the subject relative to the paper size. Too-large sketches fall off the paper and too-small sketches are difficult to detail. Establish a scale on the grid paper and then mark the four corners of the object you're drawing. Put those four corners close to the corners of the paper for the largest sketch. If the subject is square or rectangular, position your clipboard accordingly. The clips will keep the paper from falling forward when you use the clipboard horizontally.

Sketching Geometric Figures

Whatever the shape, sketch a box of its approximate size first, and then inscribe the figure inside (Figure 5-6).

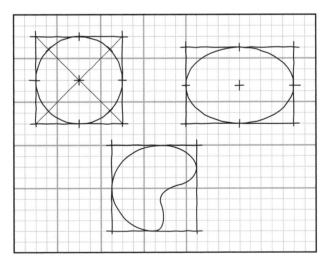

Figure 5-6
Sketch circles, ellipses, and irregular curve shapes by boxing in their approximate size and using the box edges as a guide.

Sketching Irregularly Shaped Items

A trick for sketching an odd-shaped item proportionately is to sketch a square of its approxi-mate size first, and then inscribe the figure inside.

Sketching in Perspective

Perspective drawing is photographic in nature; that is, it shows spaces and objects as we see them, not as they actually are. Lines are drawn to a **vanishing point** (the place where sets of parallel lines appear to converge), objects appear smaller the farther away they are, and the amount of information drawn depends on the viewer's **station point** (location in the room) and **cone of vision** (a 60-degree angle emanating from the viewer's eye). The portion of the room that falls within that cone is sketched. **Perspective sketching** is probably the hardest kind of sketching to do well, and for most people a background in the mechanical drawing of perspective (discussed in Chapter 14) is a required precursor. Perspective grids (see Figure 5-7) can be used; these are pre-drawn grids on which a station point, direction of view, and vanishing points have already been chosen.

Think of a grid as a piece of 3-D paper. It is useful for **thumbnail sketches**—quick, freehand drawings done mainly during the conceptual design phase and to communicate while conferring with clients. A grid's simplicity is deceptive, though, because a sketcher needs some knowledge of perspective drawing to make the most effective use of it. Pre-printed grids are available in books and sets. To use a grid, select one whose view highlights the part of the space you want to draw. Each square is a unit of measurement. Whatever measurement you choose, such as 1' or 2', all grid squares have it, whether they're vertical or horizontal. Figures 5-8, 5-9 and 5-10 show steps for using a one-point grid to draw a small shop.

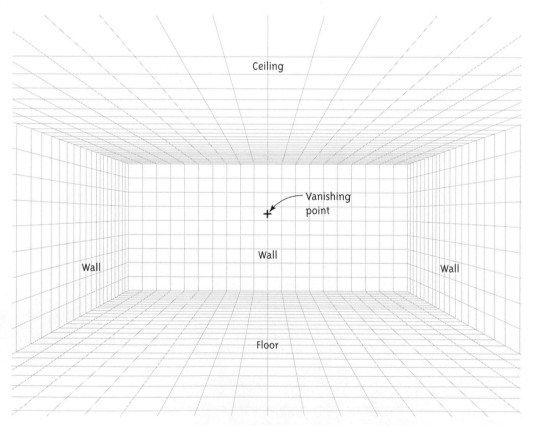

Figure 5-7
One-point perspective grid.

First, draw the room's outline (back wall, floor, and ceiling). Draw the plan view of all furniture by counting grid squares to establish their distance from the walls, then counting grid squares to establish their size.

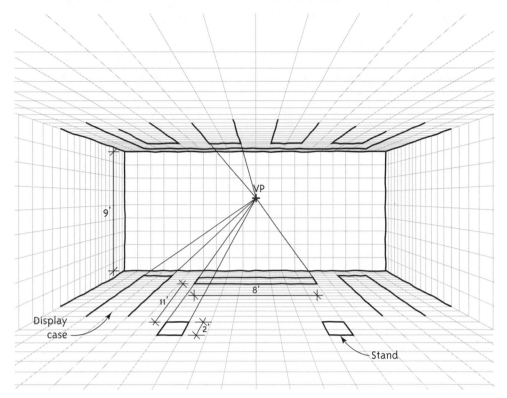

Figure 5-8
Draw the back wall and count squares on the grid to determine the location of furniture and ceiling lights.

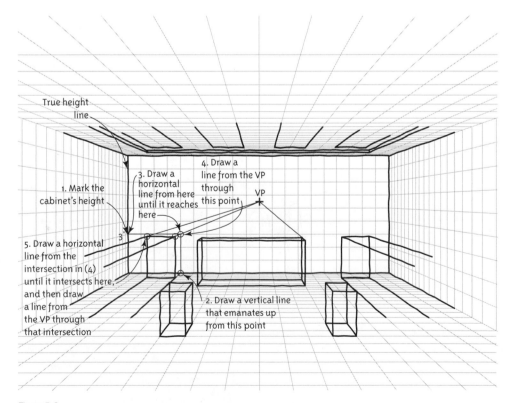

Figure 5-9
Once the locations of objects in plan are established, find heights.

Figure 5-9 shows steps for drawing the left side display cabinet.

1. Mark the cabinet's 3' height on the true height line.
2. Draw a vertical line emanating up from the intersection shown until it intersects a horizontal line that emanates from the 3' height line mark.
3. Draw a horizontal line from the 3' height mark until it intersects the vertical line drawn in Step 2.
4. Draw a line from the VP through the intersection found in Step 3 and bring it forward to intersect a vertical line emanating up from the plan corner shown.
5. Draw a horizontal line from the intersection found in (4) until it intersects a vertical line emanating from the opposite plan corner shown. Then draw a line from the VP through that intersection.

Finally, add detail (Figure 5-10) Draw all lines line parallel to the walls to the vanishing point. Draw all lines that are perpendicular to the walls horizontally.

Adding people to a drawing gives it scale and enables you to judge whether wall art, windows, and other items are too high or too low, and to determine whether other elements are in or out of proportion. Chapter 14 discusses how to do this.

Take photos of interior spaces, including close-ups of beams that intersect walls, vaulted ceilings, and other notable architectural features. Print the photos and trace them onto vellum. This will help you learn how objects are drawn in perspective.

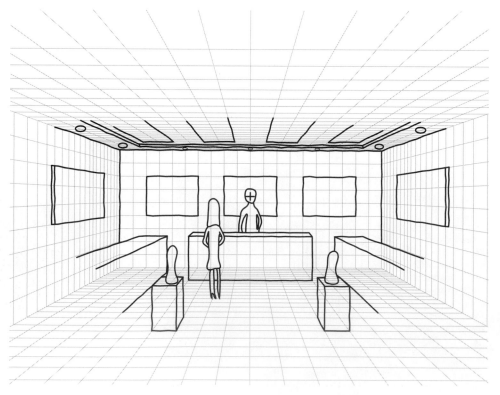

Figure 5-10
Add detail to the sketch.

Sketching in Isometric

An **isometric** drawing is a type of three-dimensional drawing. Horizontal lines are drawn at a 30-degree angle to the horizontal and remain parallel; unlike lines in a perspective drawing, they do not converge to a vanishing point. Although this method results in a picture that isn't as realistic as a perspective drawing, the benefit is that the image has measurable dimensions and is generally easier to sketch. Isometric drawing and other 3-D drawing types are discussed at more length in Chapter 14.

Wall Lockers Figure 5-11 is a perspective drawing of some wall lockers. Figures 5-12, 5-13 and 5-14 are steps showing how to draw them in isometric.

1. Draw a vertical z axis, and then draw x and y axes at a 30-degree angle to the horizontal.

2. Measure the length, depth, and height of the lockers along the x and y axes for the outline.

3. Measure and draw details. Remember that all vertical lines are drawn parallel to the z axis and all horizontal lines are drawn parallel to the x and y axes.

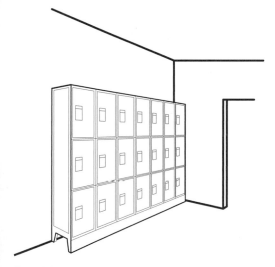

Figure 5-11
Perspective drawing of wall lockers.

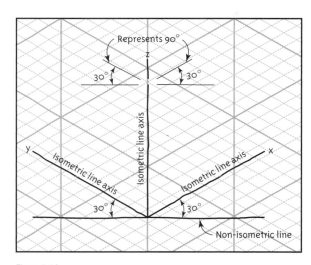

Figure 5-12
The isometric axis.

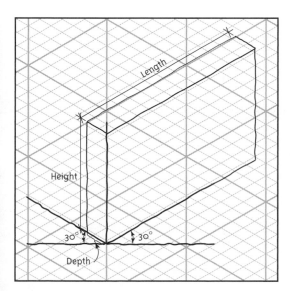

Figure 5-13
Mark the lockers' length, width, and height.

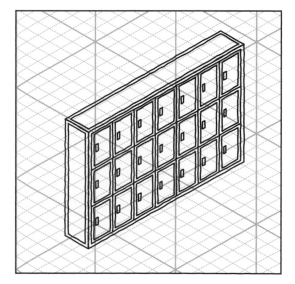

Figure 5-14
Fill in locker detail.

Round Table Figures 5-15 and 5-16 show how to sketch an isometric drawing of a round table from two orthographic views. First, draw an isometric cube of the object's approximate size, then sketch the circle inside. It will appear as an ellipse. Ellipse templates are available for this purpose.

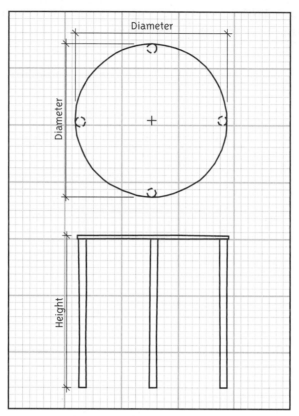

Figure 5-15
Orthographic views of a round table.

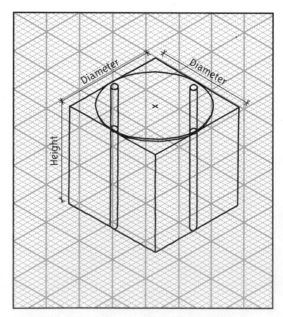

Figure 5-16
Isometric view of a round table. Draw a cube the table's diameter first, and then sketch in an ellipse.

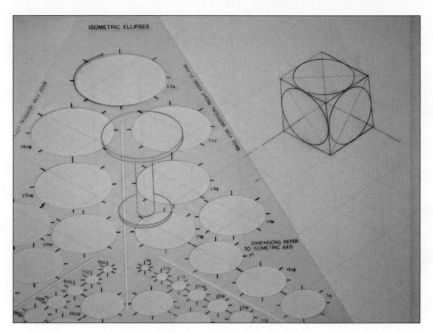

Figure 5-17
Ellipse templates come in different sizes and axis angles.

Sketching the Floor Plan of an Existing Space

Sketching a floor plan may be necessary when an existing plan is not available. Its scale will depend on the amount of detail needed, which in turn will depend on the plan's purpose and the amount of accuracy required. The overall size must be sufficient to clearly show everything while fitting on the clipboard. Obviously, you must work with enough detail and accuracy to permit you to reconstruct the drawing when you get back to the office. When sketching a floor plan, only length and width is recorded. Height isn't needed. Wall attachments such as marker boards, mirrors, and wood trim are not drawn unless the scale is very large.

Let's sketch the drafting lab shown in Figures 5-18 and 5-19. The room is square with one short, angled wall. The photos shown were taken from different corners.

Figure 5-18
View 1 of drafting lab.

Figure 5-19
View 2 of drafting lab.

You, the viewer, are standing with your back against wall AE. Select the corner to your left (marked A) to start the drawing (Figure 5-20).

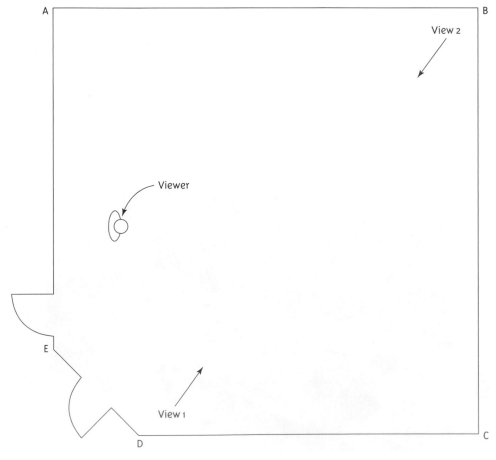

Figure 5-20
Plan of drafting lab.

1. Sketch a line from corner A to corner B (Figure 5-21).

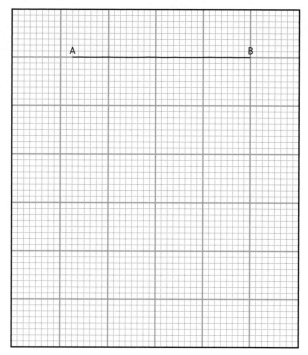

Figure 5-21
Step 1.

2. Sketch a line from corner B to corner C (Figure 5-22).

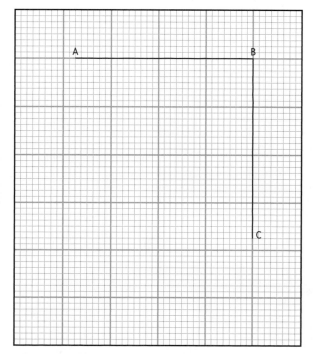

Figure 5-22
Step 2.

3. Sketch a line from corner C to corner D. Note that wall CD is shorter than the others (Figure 5-23).

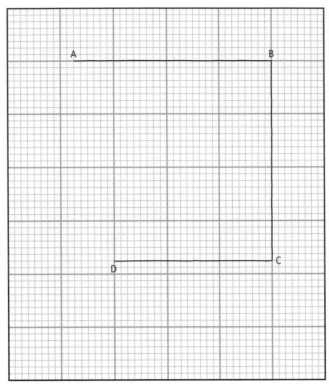

Figure 5-23
Step 3.

4. Sketch wall AE (Figure 5-24).

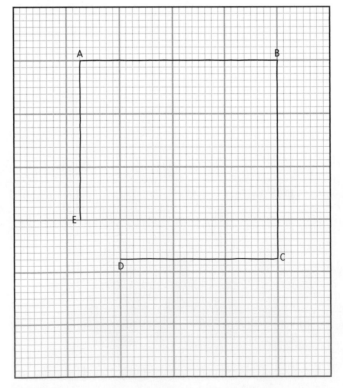

Figure 5-24
Step 4.

5. Complete the outline of the room by connecting corners E and D (Figure 5-25).

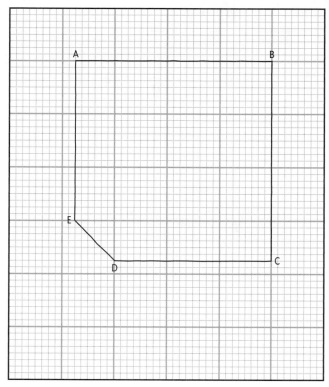

Figure 5-25
Step 5.

6. Add a second line to represent wall thickness; don't draw walls as single lines (Figure 5-26).

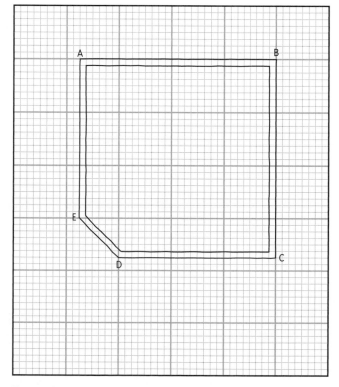

Figure 5-26
Step 6.

7. Add details, starting with the walls. Wall AB has two windows, each starting at the corners. Estimate their lengths and mark lines inside the wall, perpendicular to it. Note that there is a pilaster (a column attached to the wall) in between the windows (Figure 5-27).

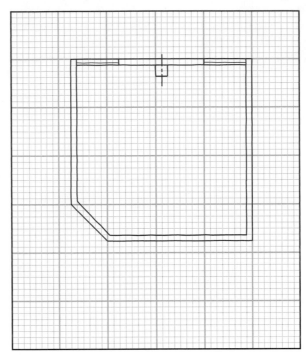

Figure 5-27
Step 7.

8. Turn to face the door on wall DE. (You are still "inside" the room.) The door is hinged on your left and, when opened, it swings away from you. The door on wall AE is also hinged on your left, and when opened, swings away from you. Sketch door symbols indicating this (Figure 5-28).

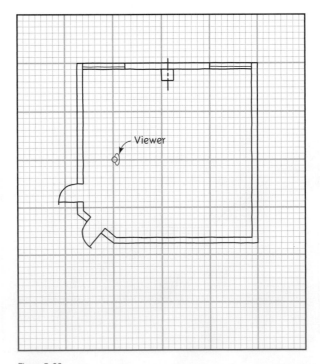

Viewer

Figure 5-28
Step 8.

9. Sketch the furniture. There is a row of four drafting boards against wall AB. There are seven boards in the middle of the room and four more against wall CD. There are small cabinets between the boards. A table, bookcase and projector stand are near wall AE. Sketch these items as squares and rectangles, estimating their sizes (Figure 5-29). Don't draw the equipment on tables and stands, rather, acknowledge their presence with notes.

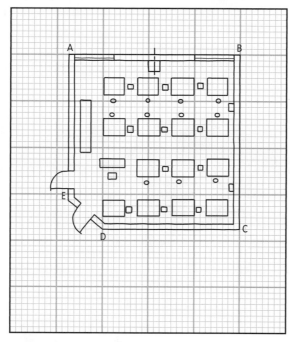

Figure 5-29
Step 9.

10. Draw extension and dimension lines indicating features that need to be measured (Figure 5-30).

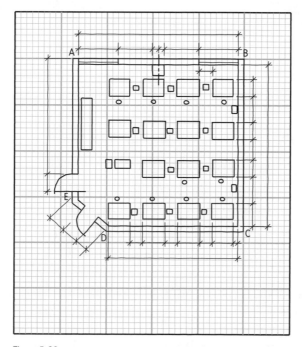

Figure 5-30
Step 10.

Additions to the Sketch If you draw an adjacent room, show the wall between it and the first room as shared; don't draw two separate floor plans. If a room contains a lot of items, you may want to layer plans on multiple sheets with different information on each sheet. Add a north arrow, room names, the name/address of the building, the identity of anyone who helped measure it, and the date. If relevant, you might note building additions and differentiate them from the original. You could indicate them by using different poché styles on the floor plan, such as black for the original walls, diagonal lines for the next oldest, and open for the most recent. Add a legend for your poché.

Another type of "sketching" that some designers and furniture sellers do is laying out floor plans with magnetic piece sets (Figure 5-31).

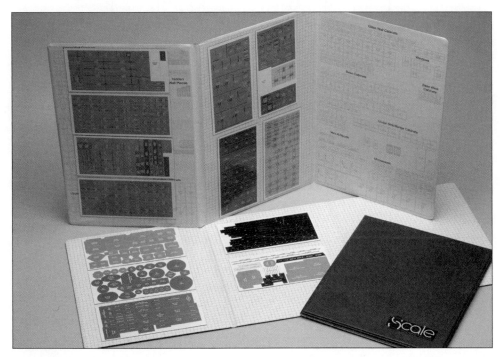

Figure 5-31
Magnetic cut-outs on printed boards, such as these from furnitureplanning.com, enable furniture sellers or residential designers to make quick arrangments. *Photo courtesy of Bret Gustafson, JCCC.*

Measuring an Interior Space

Size and *location* are the two main characteristics that need to be recorded. Specifically, you must record the size of the room and the size and location of everything in it. The amount of accuracy needed varies. If a living room is incorrectly measured by a few inches, the new sofa will probably still fit. If a cabinet or kitchen wall is incorrectly measured by 1/8", custom work may not fit. Even if blueprints are available, careful measuring is still needed, because measurements taken after the interior walls are in place may not be exactly the same as what the drawing dimensions show.

Measuring Floor Plans

Be systematic. Look at the big picture first and then fill in details. Measure the room's overall dimensions first. Choose a corner to reference everything from and commence measuring from it. Work your way around the plan. Lay the tape along the floor's baseboards for greatest accuracy and measure from the walls. If furniture or obstructions prevent this, place the tape on the wall, but ensure that the tape is parallel to the floor.

Horizontal distances must be measured with the tape held level and taut. Tension will affect the accuracy of tape measurements by causing the tape to stretch or sag. Know where the tape's zero point is; it's not always at the end. Placing a small carpenter's level below the tape will help you verify that it's horizontal.

When measuring along the floor from baseboard to baseboard, you must add back in the thickness of the baseboard on each wall measurement. Hold the tape at a level that allows you to get measurements for key features; for instance, don't start so low on the wall that the tape runs below the bottom of the windows.

Next, measure all smaller, permanent features and their length, width and distance from each other. Locate and measure doors, windows, chimneys, closets, columns, base cabinets, countertop overhangs, floor drains, air vents, outlets, switches, thermostats, other utility connections, and any obstructions. Measure each feature's distance from a wall or corner. Draw in any radiators or air ducts in the walls, the location of water supply drainage pipes and vents, gas pipes, and electrical outlets. Note door swing direction. Indicate wall thickness and, when possible, indicate whether the walls are load-bearing or non–load-bearing. Take *running*, or cumulative measurements. Hold the measuring tape at one corner (or datum point) and read all desired points along that line without moving the tape, rather than continually moving the tape and taking each measurement from the last reading. This prevents small errors from accumulating and makes measuring errors apparent.

Measuring Elevations

Measure exterior and interior wall thickness. This can be done by measuring the width of exterior and interior door jambs (correcting for the dimensions of the frame). Measure ceiling height, baseboard, chair rail, and other trim width. Measure the distance from the top of the window trim to the ceiling and from the bottom of the window trim to the floor. Measure any *soffit* dimensions (a soffit is an enclosed area below the ceiling and above the wall cabinets). Measure the distance from the soffit to the floor. Measure the thickness of flooring and wall materials where possible.

Measuring Stairs

Count the number of steps. Measure the length and width of one tread and any nosing (the overhang of the tread on the riser). Measure the riser height and any landings. When measuring for carpet and stair runners, factors like the width of carpet bolts and the placement of joins will add to the square footage of material needed (see Figure 5-32).

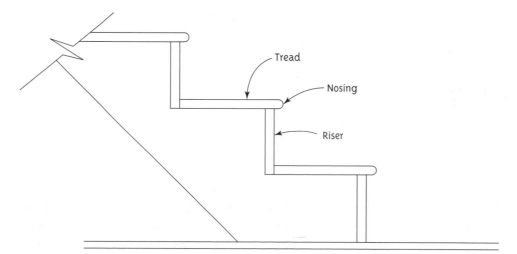

Figure 5-32
Stair parts.

Additional Measurements

If furniture is part of your drawing, record its size and location. Note odd and oversized pieces, existing furniture you'll be working with, and special characteristics of the room.

Measuring Tips

Practice the following for greatest accuracy.

- Don't assume that rooms are square, walls are plumb, or floors are level. Take diagonal measurements and check walls and floors to determine distortion early in the documentation process.
- Check corners for squareness. If your corners are not square, you will need to make some adjustments during the installation of cabinets. There are two ways to check for squareness. The first is with a carpenter's square, a triangular tool placed inside a corner to see if the walls align with it. The second is the 3/4/5 method. Choose a corner and measure 3' along the wall in one direction and 4' in the other direction. Connect the points with a straightedge. If the distance is 5', the corner is square.
- Compare the wall's overall measurements with the sum of its subsections. Measure within 1/4".
- Measure walls with windows this way:

    From wall to wall
    From corner to outside trim of window
    From outside window trim to outside window trim
    From outside trim of window to corner

Photograping Details

Cameras photograph in perspective, so hold the lens parallel to the object to minimize distortion. Measure by hand for verification. Hand rubbings can be useful when drawing some historic elements.

Calculating Square Footage (Area)

The area of a rectangle or square is its length multiplied by its width. The area of a triangle is 1/2 base multiplied by its height. The area of a circle is π (*pi*) r^2 (*pi* $\approx$ 3.12; r^2 is the radius of the circle multiplied by itself). If a shape is irregular, split it into recognizable shapes, calculate each area separately, then add the products together (see Figure 5-33).

Design Constraints

All projects are bound by constraints and requirements. The most common ones are **building codes** and **accessibility** issues. These must be thoroughly researched before designing.

Building Codes

These are rules that govern the materials and methods of construction for residential and commercial design. Their purpose is to provide the public with safe places to live and work. Interior designers must be familiar with building codes because many aspects of design, such as room size, number of exits, lighting, hallway length, interior finish selection, and furniture placement, are code-dictated.

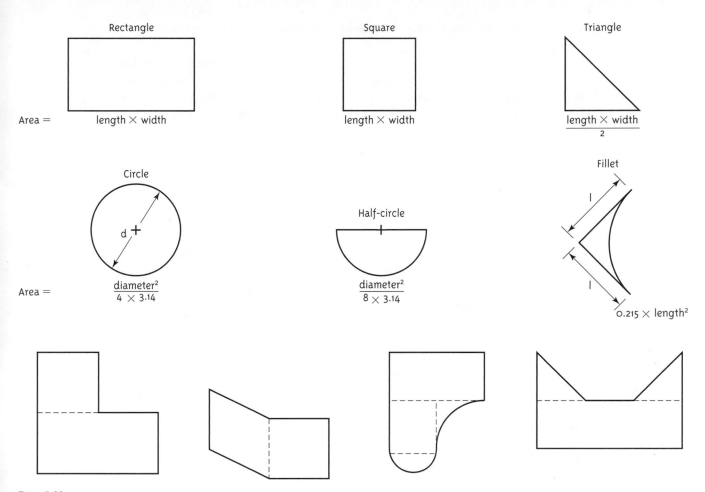

Figure 5-33

Area of common geometric shapes. Calculate areas of odd shapes by breaking them down into common shapes. If an area cannot be broken down into a common shape, draw the figure to scale on a grid and count the number of squares it takes up. Then multiply that number by the grid square's area.

The first building code was instituted by Hammurabi, a 2nd millennium B.C. Babylonian king. This code required a builder to reimburse the owner for damages due to faulty construction. It even called for the builder to be put to death if an improperly constructed home collapsed and killed its owner. However, building construction was largely unregulated for most of history. Early timber-frame builders were hired based on their reputation and worked with little official oversight. Most building codes have been enacted reactively, typically as a result of fires. The earliest building code on record in the United States is from 1625. It regulated roof materials based on their resistance to chimney sparks.

For most of the past 150 years, building codes in the United States were regional documents promulgated by various groups. In 1994 a standards organization emerged called the International Code Council (ICC), which promulgated a single comprehensive set of codes called the International Building Codes (IBC) or "I-codes." This set of standards, which has no regional limitations, contains 14 individual codes. For example, one is the International Residential Code (IRC), applicable to one- and two-family homes of three stories or less.

The I-codes are **model codes**, meaning that they are developed and maintained by a standards organization that is separate from and independent of the authorities that enforce the codes and give them the force of law. A model code does not become law until a governing authority (usually a state government, fire district, or municipality) enacts it.

The National Fire Protection Agency (NFPA) is another standards organization that puts out its own codes, such as the Life Safety Code and the National Electric Code. However, the IBC is the most widely referenced and adopted source of building standards in the United States.

Codes are revised yearly. Revisions apply to new construction; existing buildings don't have to comply unless the revisions are considered essential for public safety. When an existing building is altered or its use changes, it must meet the latest code. Building codes generally don't discuss aesthetics, traffic, or activities inside a building; local codes, ordinances, and zoning laws cover those topics.

An in-depth discussion of codes is beyond this chapter's scope. However, code requirements for occupancy, egress, and accessibility must be understood for competent floor plan design and drafting.

Occupancy

Occupancy is a category that describes a building's use. **Occupant load** is the number of people a space is expected to hold. Occupancy and occupant load need to be determined at the outset of design. The IRC has ten occupancy categories: assembly, business, educational, factory, high-hazard, institutional, mercantile, residential, storage, and utility. Buildings may be "mixed occupancy," meaning they have more than one type of occupant. Spaces within a building are also categorized; categories include occupiable, dwelling, sleeping, living, and habitable. (Much of this categorization is a function of whether or not windows are present.)

Codes address risk factors that characterize the people and activities in the space. For instance, a nightclub has dim light and loud music; offices tend to contain upholstered furniture, paper, and other flammables; nursing homes have occupants with limited mobility; prisons have occupants with restricted mobility; and theatres have many occupants in one space. General safety requirements, such as lighted signage, maximum room area size, dead-end corridor length, fire sprinklers, and multiple exits, ensure that all types of buildings will provide their users with the identical level of safety.

Means of Egress

Means of egress refers to occupants' ability to get to exits, especially in an emergency. Egress has three components: exit access, exit, and exit discharge (Figures 5-34 and 5-35). *Exit access* is the portion of the building that leads to an exit; it includes halls and stairs. An exit is the fully enclosed, fire-protected space between the exit access and the exit discharge. *Exit discharge* is the area between the termination of an exit and a public way. A *public way* is the area outside a building between the exit discharge and a public street, and is usually the final destination of a means of egress. Another type of final destination is an *area of refuge,* which is a space protected from fire and smoke where people unable to use stairs or an elevator can wait for help.

Codes dictate the number of exits and their sizes. For instance, hallways must be at least 36" wide in residences, 44" in commercial buildings, and up to 48" wide for handicapped accessibility. Egress doors must have a minimum clear width of 32" and be 80" high. Note that a single 36" wide door yields 34" of clear width. Therefore, if 40" of exit width is required, a 44" door will be sufficient but a 36" one will not. Dead-end corridors in exits cannot be longer than 20'.

Depending on their location, doors must be of a specific type, size, and swing. Usually a swing door is required, and it must swing in the direction of exit travel to prevent crowds from piling up against a door that opens inward. Sometimes revolving or power-operated sliding doors are permitted as part of the means of egress (usually in a mercantile-occupancy building). These doors rarely meet full egress requirements, however; for instance, revolving doors are sometimes credited with meeting 50 percent of the exit requirements (if they are credited at all), so adjacent swing doors must be included.

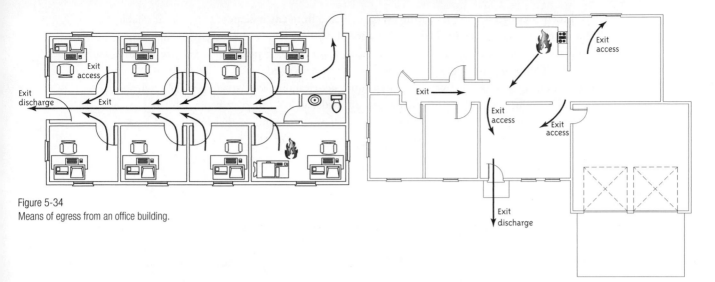

Figure 5-34
Means of egress from an office building.

Figure 5-35
Means of egress from a residence.

Exits must be located as far from each other as possible. When two or more exits are required, at least two of the exits must be a certain minimum distance apart. The **half-diagonal rule** requires this minimum distance between two exits to be at least one-half of the longest diagonal distance within the building or room that the exits serve (see Figure 5-36).

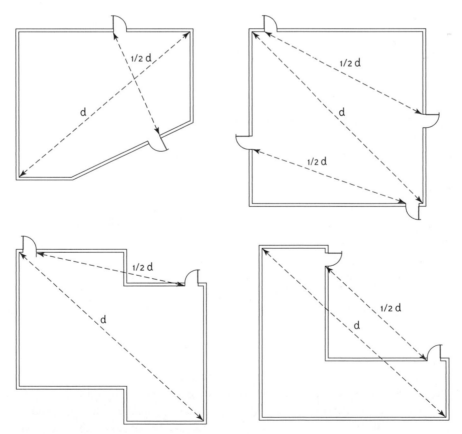

Figure 5-36
The half-diagonal rule for exit placement. D is the diagonal or maximum distance, permitted. 1/2D is half that distance. Some codes allow the minimum distance to be one-third instead of one-half if the building has an automatic sprinkler system.

Emergency egress is required from every room and from any basement that contains habitable space. Egress may be made through a door or window that opens directly to the outside. Windows may be used as emergency exits; codes discuss their placement and size for this. The sill of all bedroom emergency escape windows must be within 44" of the floor. Windows used for emergency egress must have a minimum net clearance of 3 square feet. The net clear opening area must be at least 5.7 square feet, 20" wide, and 24" high. When a habitable space is in a basement, the room must be provided with a rescue window that has a window well (which must include an integral ladder if the window is over 44" deep). Tools must not be needed to open any egress windows.

Accessibility

This term refers to the accommodation of people with disabilities in buildings. Disability is a broad term that encompasses physical and psychological issues. Accessible design is a challenge; what works for one disabled person will not work for another. What is certain, though, is that the difference between reluctant incorporation and creative, compassionate integration is huge. Between 43 and 60 million Americans are estimated to be disabled in some manner. The Americans with Disabilities Act, enacted in 1990, acknowledged this. The government-written Americans with Disabilities Act Accessibility Guidelines (ADAAG) sprang from this act. They cover all design issues that affect accessibility, such as parking, entering, elevators, and communications systems. Figures 5-37 and 5-38 show examples of accessibility incorporated into design.

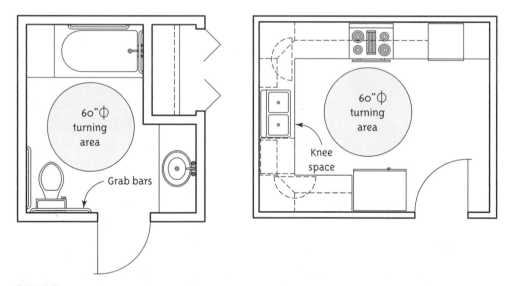

Figure 5-37
Characteristics of accessible kitchens and baths include a 60" diameter circle for wheelchair turning, knee space under sinks, and grab bars.

Accessible design is not the same as **universal design**, which is the philosophy of making buildings and products usable to as many people as possible, regardless of disability. Young, old, tall, short, abled, disabled, or temporarily injured, all should be able to use the design without adaptation, extra cost, or specialized accessibility features. Popular examples of universal design are the close-captioned TV shows and curbs that slant into the street instead of dropping straight down 6".

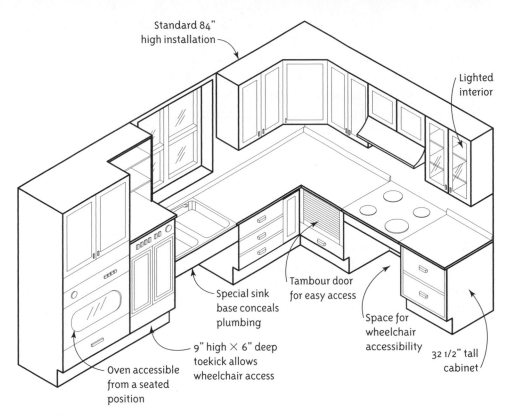

Standard 84"
high installation

Lighted
interior

Oven accessible
from a seated
position

Special sink
base conceals
plumbing

9" high × 6" deep
toekick allows
wheelchair access

Tambour door
for easy access

Space for
wheelchair
accessibility

32 1/2" tall
cabinet

Figure 5-38
Features of an accessible kitchen.

Summary

Information must be gathered before you begin the design process. The project must be studied, and information on goals, existing conditions, and constraints must be researched. This provides a foundation on which to build the design. When all the information has been gathered and analyzed, the project is ready for **synthesis:** the process of putting together the parts to form a whole. Synthesis is discussed in Chapter 8.

Suggested Activities

1. Sketch and measure your drafting lab on grid paper.
2. Sketch orthographic and isometric views of easily measurable objects (such as a chair or computer) on grid paper.
3. Document, sketch, and measure two rooms in your home. Ask a classmate to hard-line a drawing of the rooms from your notes and sketches.
4. Draw a one-point perspective on a grid of a room of your choice.
5. Overlay vellum on a photograph of a room and trace. Make design changes.
6. Draw a bubble diagram for a residence.
7. Obtain a copy of the 2006 International Residential Code for One- and Two-Family Dwellings and browse through it.

Questions

1. What is a program?
2. What does taking a visual inventory mean?
3. What is the purpose of sketching?
4. Name three tools used for measuring and inventorying a room.
5. Why do building codes exist?
6. What is the difference between accessible and universal design?

Further Reading and Internet Resources

Guthrie, P. (2004). *Interior Designer's Portable Handbook*. New York: McGraw-Hill.

International Building Code (2006). Whittier, CA: International Code Council.

International Residential Code for One and Two Family Dwellings (2006). Whittier, CA: International Code Council.

Kumlin, R. (1995). *Architectural Programming: Creative Techniques for Design Professionals*. New York: McGraw-Hill.

Null, R., & Cherry, K. (1996). *Universal Design: Creative Solutions for ADA Compliance*. Belmont, CA: Professional Publications, Inc.

Pable, J. (2004). *Sketching at the Speed of Thought*. New York: Fairchild Books.

Bienfang Designer Grid sets, available at art and drafting supply stores

http://www.aarp.org/
The American Association of Retired Persons (AARP) website contains lots of information on universal design.

http://www.access-board.gov/
This site contains information on ADAAG (Americans with Disabilities Act Guidelines).

http://www.cs.wright.edu/
The Wright State University website contains a page on designing an accessible kitchen.

http://www.furnitureplanning.com
This is a source for magnetic furniture sets.

http://www.iccsafe.org/
The website of the International Code Council, the organization that oversees the International Building Code, has information and code help.

http://www.osha.gov
The Occupational Safety and Hazards Act (OSHA) website lets users search for information on means of egress and exit route requirements.

http://www.usdoj.gov/
The U. S. Department of Justice website has links to Americans with Disabilities Act information, including legal standards and technical information.

http://www.wbdg.org/
The Whole Building Design website contains discussions of the building design process and different building space types.

CHAPTER 6

Space Planning

OBJECTIVE

- This chapter discusses planning, sizes, and clearances for rooms in a single-family house.

KEY TERMS

anthropometry	galley	proxemics
blind cabinets	GFI	range
bubble diagram	grab bar	receptacles
chase	half-bath	single-wall space
clearances	islands	planning
cooktop	L-shape	synthesis
ensuites	lavatory	three-quarter bath
ergonomics	lazy Susan	U-shape
exhaust fan	medicine cabinet	vanity
fittings	mudrooms	warming drawer
foyer	orientation	water closet
full bath	peninsula	work triangle
G-shape		

What is Space Planning?

Space planning is the process of designing a space to make it functional for the occupants (Figure 6-1). It integrates design concepts with programming needs. Spatial and occupancy requirements, layouts, knowledge of architectural features, furniture sizes and **clearances,** or distances between objects, human proportions and behavior, and what is safe, functional, and aesthetically appropriate are all considered. In Chapter 5 the process of analysis was discussed. **Synthesis,** the process of putting together the parts to form a whole, is the next step. It is the actual design process. Information that was researched and gathered must be synthesized into a workable design.

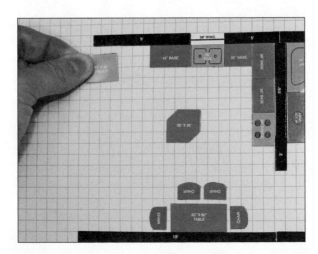

Figure 6-1
Careful space planning makes a design functional and pleasant for the occupants. While decorating makes a space fit fashion criteria, design is about functionality.

The Bubble Diagram

Once a problem has been defined, the designer may draw a **bubble diagram,** also called a *schematic* or *relationship* diagram. A bubble diagram is a visual means of organizing thoughts and developing a concept. It consists of labeled circles that represent spaces and functions that occur inside. Arrows show *circulation,* which is movement within a room and from one room to another. Short lines between the bubbles may represent visual or auditory inaccessibility. Such diagrams facilitate seeing the spaces and how they relate to each other. Freehand sketching is the most effective technique to use at this stage; therefore, the only supplies needed are a pencil, colored markers, and tracing paper.

During the design process, multiple pieces of tracing paper are layered over each other as the plan is developed (Figure 6-2). Expect to explore different options and layouts before settling on the final one, because design is a slow, thoughtful, and exciting process of discovery.

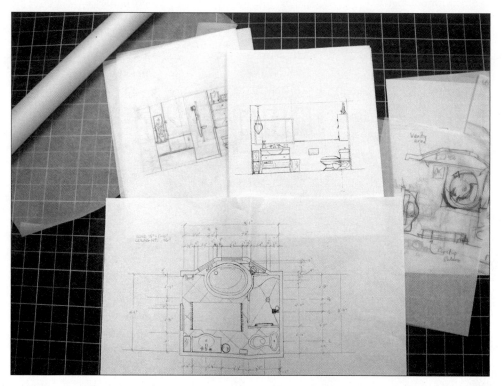

Figure 6-2
Development sketches. *Courtesy of Rhiannon Craven, Craven Design.*

As the bubble diagram is refined, it starts to resemble a floor plan. Rooms, windows, and doors are drawn in place. Once scale is introduced, you can draw the transition from bubble diagram to floor plan on graph paper Figure 6-3 shows the documentation of a site's features. Figures 6-4 through 6-8 show the genesis of a floor plan design from bubble diagram to rough floor plan.

There are different approaches to laying out a floor plan. The addition approach starts with one major room and continues by laying out all the other rooms that radiate from it. The subtraction approach starts with a shape for the whole building and then divides the interior. It is used when the footprint must be a certain size or shape. With either approach, the designer must know how many people will be in the room, how much furniture, how big the furniture is, and the *clearance,* or distance between pieces, in order to know how big to make the room. Good floor plans contain doors that do not open into each other, closets of suitable depth, and sufficient room between furniture, fixtures, and walls. Design occurs from the inside out; if

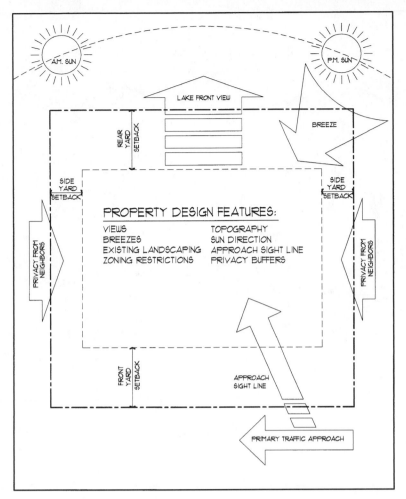

Figure 6-3
Property design features. *Courtesy of Jay Colestock, Colestock and Muir Architects, Boca Raton, FL.*

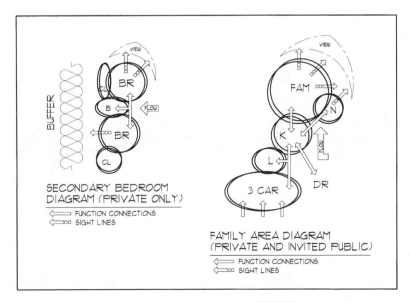

Figure 6-4
Diagramming the bedroom and family areas. *Courtesy of Jay Colestock, Colestock and Muir Architects, Boca Raton, FL.*

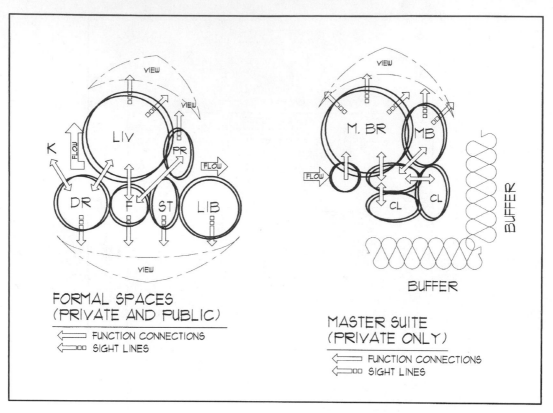

Figure 6-5
Diagramming the formal spaces and master suite. *Courtesy of Jay Colestock, Colestock and Muir Architects, Boca Raton, FL.*

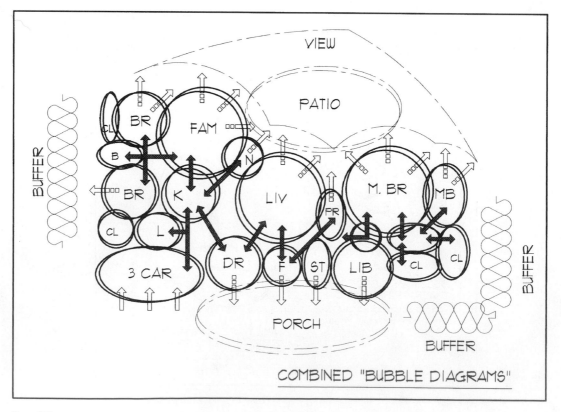

Figure 6-6
Pulling all the bubble diagrams together. *Courtesy of Jay Colestock, Colestock and Muir Architects, Boca Raton, FL.*

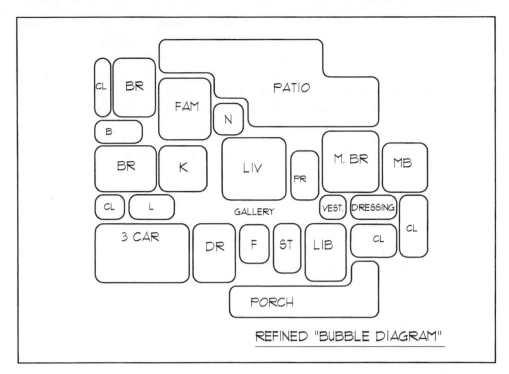

Figure 6-7
Refining the bubble diagram. *Courtesy of Jay Colestock, Colestock and Muir Architects, Boca Raton, FL.*

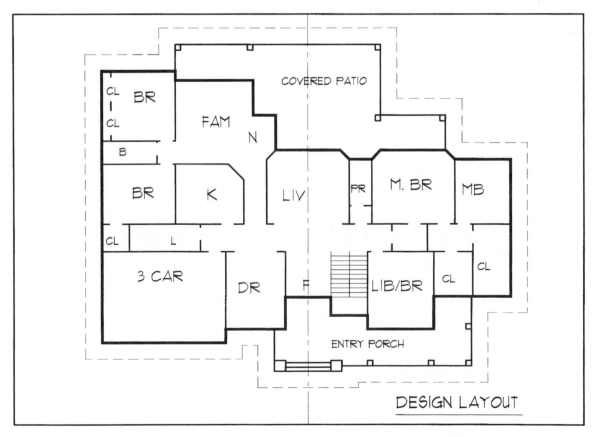

Figure 6-8
The design layout, or rough floor plan, takes shape. *Courtesy of Jay Colestock, Colestock and Muir Architects, Boca Raton, FL.*

rooms are designed in random sizes, fixtures, and furnishings must later be forced to fit inside. Templates are helpful at this stage because they let the designer quickly trace furniture and fixtures, allowing him or her to calculate the approximate room dimensions required. Most large furniture manufacturers and stores have manual drafting templates for this purpose, as well as interactive design planning tools on their Websites.

Here are some tips for the synthesis phase of laying out a floor plan.

1. Decide the **orientation,** or compass location, that will take best advantage of the site's views, features, and breezes. The north side is usually shaded and the south receives heavy sun, so in cool climates a living room placed on the south will benefit from winter sun. In a hot climate, a northern orientation will keep the room cooler. Important spaces such as the living room and master bedrooms should face the best views; closets and garages should face the worst. Similarly, important rooms should be positioned to take best advantage of cooling summer breezes, while less important ones can be positioned to block heavy winter winds.

2. Group areas into three zones: public (living room, front entrance, bathroom); private (bedrooms and bathrooms); and dining (dining area and kitchen). Then, carefully plot the circulation among them. Successful circulation requires convenient routes among rooms or areas that have the most connecting traffic or whose functions are dependent on each other, such as the kitchen and the dining area. This is usually done by making those rooms adjacent (with a common doorway) or near to each other.

    Circulation areas are halls, stairs, and travel routes through rooms. Circulation paths through rooms should be avoided; for instance, you shouldn't have to cross through the living room to get from a bedroom to the bathroom. When crossing a room is unavoidable, try to route traffic across a room's corner or along its side.

3. Make the kitchen physically and visually accessible to the outside to provide occupants with a view and to facilitate grocery loading and garbage disposing.

4. Plan bathroom placement effectively. House size and design dictates the number and type of bathrooms needed. Bathrooms should be convenient to bedrooms; this requirement is more important than plumbing convenience. Do not site the bathroom near the kitchen simply so the plumbing can share a wall. The bathroom should be accessible via a hall, not another room (Figure 6-9). A one-bathroom house should be designed so the bathroom is convenient to all rooms.

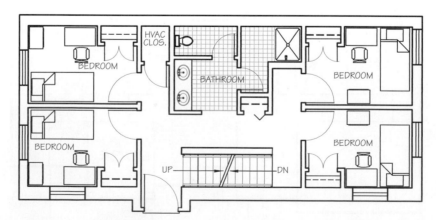

Figure 6-9
Group bedrooms around the bathroom.

5. Keep halls short and to a minimum. They should not occupy more than 10 percent of the plan's total area. One measure of a home's efficiency is the percentage of area occupied by rooms as opposed to stairs, furnaces, halls, walls, closets, and entryways. Seventy-five percent is considered efficient; less than 70 percent is not.

6. Determine the types and placement of doors. When locating doors, consider swing direction and furniture placement. Swing doors are easiest to operate, but sliding, folding, and accordion doors are useful in tight spaces. Doors opening into rooms from a hall should swing into the room, against the wall. They should be able to completely open without being blocked by appliances or furniture.

7. Place windows high on walls that face unattractive views to hide the view while admitting light.

8. There is no such thing as an "average" or "ideal" room size; however, the Federal Housing Authority (FHA) suggests minimum sizes (Figures 6-10). Figures 6-11 through 6-14 illustrate different design layouts.

ROOM SIZES (INTERIOR USABLE SPACE)

ROOM	MINIMUM	AREA 中	AVERAGE	AREA 中
LIVING ROOM	12' × 16'	192	15' × 21'	315
DINING ROOM	10' × 12'	120	13' × 14'	182
DINING AREA (in kitchen)	7' × 9'	63	10' × 10'	100
MASTER BEDROOM	11'-6" × 12'-6"	143.75	13' × 15'	195
BEDROOM	9' × 11'	99	11' × 13'	143
KITCHEN	7' × 10'	70	9' × 13'	117
BATHROOM	5' × 7'	35	6'-6" × 9'	58.5
UTILITY (NO BASEMENT)	7' × 8'	56	9' × 12'	108
HALL WIDTH	3'		3' -6'	18
CLOSET	2' × 2'	4	3' × 5'	15
GARAGE (SINGLE)	9'-6" × 19'	180.5	13' × 21'	273
GARAGE (DOUBLE)	18' × 19'	342	21' × 21'	441
RECREATION ROOM	10' × 15'	150	12' × 20'	180
MUD ROOM	4' × 4'	16	4' × 9'	36

Figure 6-10
Room sizes chart.

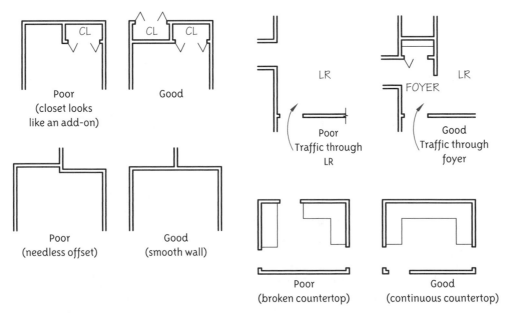

Figure 6-11
Good and bad design layouts.

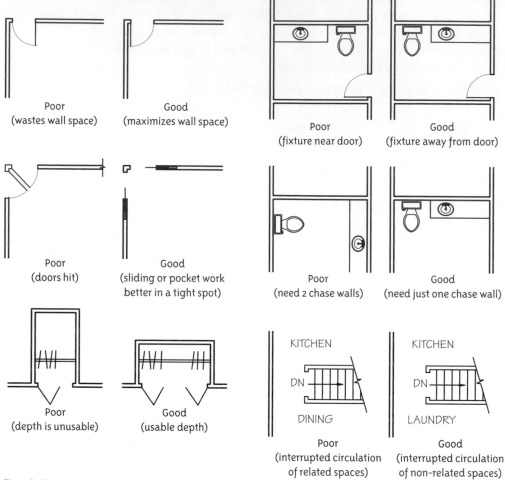

Poor
(wastes wall space)

Good
(maximizes wall space)

Poor
(fixture near door)

Good
(fixture away from door)

Poor
(doors hit)

Good
(sliding or pocket work
better in a tight spot)

Poor
(need 2 chase walls)

Good
(need just one chase wall)

Poor
(depth is unusable)

Good
(usable depth)

KITCHEN

DN

DINING

KITCHEN

DN

LAUNDRY

Poor
(interrupted circulation
of related spaces)

Good
(interrupted circulation
of non-related spaces)

Figure 6-12
Good and bad design layouts.

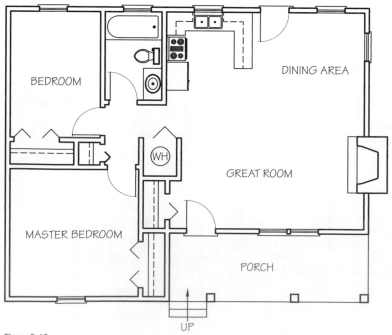

BEDROOM

DINING AREA

WH

GREAT ROOM

MASTER BEDROOM

PORCH

UP

Figure 6-13
This floor plan has good traffic circulation.

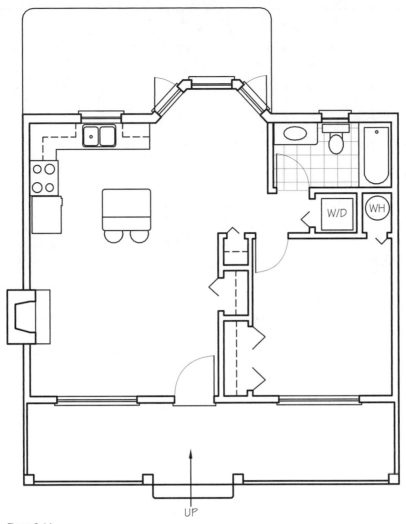

Figure 6-14
This small plan makes efficient use of space.

Anthropometry, Proxemics, and Ergonomics

Three areas that impact the synthesis process are anthropometry, proxemics, and ergonomics.

Anthropometry Anthropometry is the study of human sizes and proportions and their variations and reaches Figures 6-15 through 6-21 show common dimensions in inches. The word is derived from the Greek *anthropos,* meaning "man," and *metric,* meaning "measure." The word literally means "measurement of humans." Statistical data on height, weight, limb, and body segment size enables designers to create clothing, furniture, toys, cars, and buildings, among other things, that are sized to fit the user.

Anthropometric data is regularly updated to account for changing lifestyles, nutrition, and ethnic make-up. Different age, gender, and ethnic populations have different size and proportion norms; designers must know the specific market targeted. The anthropometric range for products designed for professional male basketball players is different than that for products designed for Southeast Asians or for a general, mixed population. Designers must also know how much of the population they want to accommodate. For instance, when designing overhead luggage racks for public transportation, accommodation of 90 percent of the rider population is reasonable. However, placement and operation of emergency latches or buttons should accommodate 99 percent of the rider population. Adjustable devices are designed to accommodate a broad swath of users.

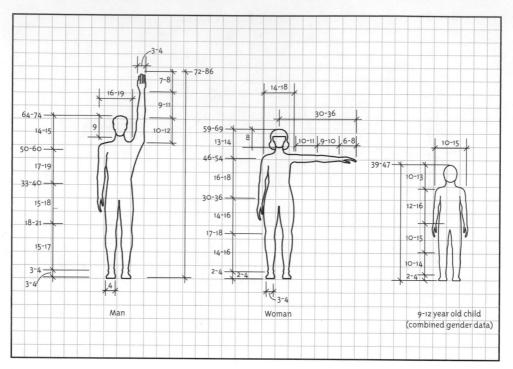

Figure 6-15
Anthropomorphic data: Body sizes.

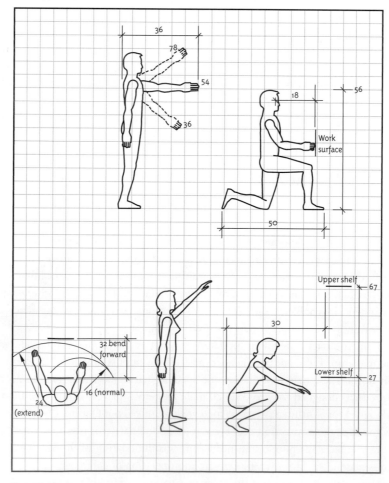

Figure 6-16
Anthropomorphic data: Reaches.

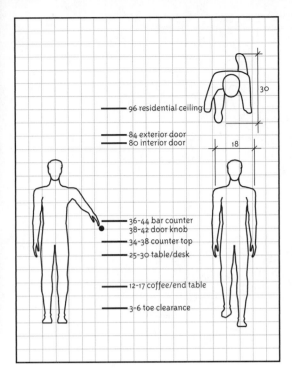

Figure 6-17
Anthropomorphic data: Heights of common items.

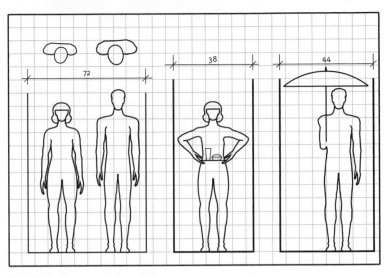

Figure 6-18
Anthropomorphic data: Space needed when holding common items.

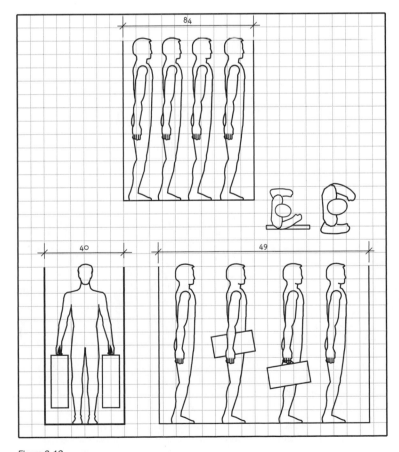

Figure 6-19
Anthropomorphic data: Space needed when standing in line.

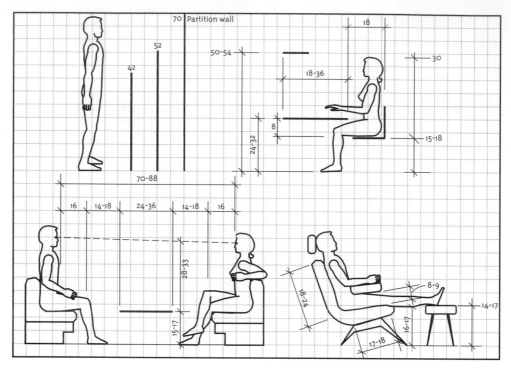

Figure 6-20
Anthropomorphic data: Workplace dimensions.

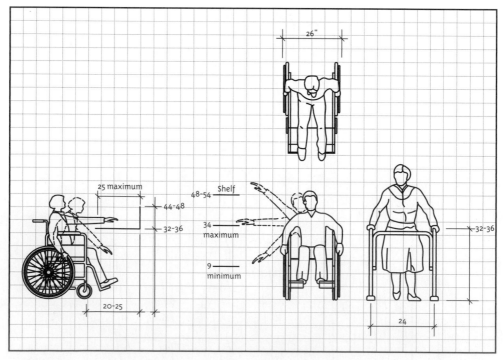

Figure 6-21
Anthropomorphic data: Accessibility. Wheelchair users have an eye level that is typically 15 to 16 inches below that of most standing people. This seated position influences reach, controls, and access. A standing person occupies 25" × 15" of floor space, requires 16" to 26" aisle width, and can turn on the spot; a wheelchair user occupies up to 57" × 25" of floor space, requires an aisle clearance of 32," and has a turning circle between 59" and 67" in diameter.

Proxemics **Proxemics** describes set, measurable distances among people as they interact. It encompasses spatial relationships, body language, boundaries, colors, physical territory, and personal territory. We perceive appropriate distances for different types of messages and we establish distances for personal interaction, nonverbally defining this as our personal space (the "bubble" we keep between ourselves and the person ahead of us in line) (Figure 6-22). These perceptions vary with culture, gender, age, social situation, and individual preference.

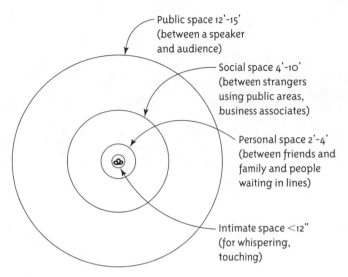

Figure 6-22
American proxemic spaces. Personal space varies culturally and ethnically.

Proxemic data documents our use of space and how differences in its use affect our emotions, such as making us feel comfortable or nervous. Desks, chairs, and partitions can serve as physical or implied barriers. Designers use proxemic data to arrange floor and furniture plans; hierarchies, status, attitudes, personal space, and boundaries are built into them. For instance, an extra chair can be a status symbol to an employee because it indicates that an office's occupant has a professional reason to entertain visitors. Proxemic data can suggest how to place the chair and what relation it should have to the desk in order to engender a particular emotion in the chair's owner (such as control, protection, hierarchy, or equality).

Proxemic data is also used to determine which rooms are public and which are private. Some rooms are used for public gatherings, others for close friends and relatives; some are even considered off-limits to certain family members. The proxemics of the furniture can define whether the space feels cozy or formal. For instance, rooms with a linear or curved seating alignment are not conducive to small, intimate gatherings, because they do not facilitate face-to-face contact.

Ergonomics The *erg* is a measurement of work done; here it relates to the efficiency of a design. The suffix "-onomics" is related to the effective application of something. Together, they mean "effective management." **Ergonomics** is the applied science of using design and anthropometric data to make the work environment more comfortable. Ergonomic designs take into account the typical user's body size, shape, and movements with relation to a product, such as a piece of furniture, tool, or household good. Ergonomics seeks to reduce operator fatigue by making products easier and more comfortable to use than those of non-ergonomic design.

Residential Rooms

Now that we have discussed some general space planning principles, we shall present some specific design principles for the typical rooms in a house.

The Kitchen

Figure 6-23
An elegant kitchen. Notice the stone walls and wood floor; the floor plan would show them pochéd as such. *Courtesy of Bosch Home Appliances.*

Kitchens vary greatly in size and style, from the elaborate (Figure 23) to the simple, but the same planning basics apply to all of them. One hundred square feet of usable floor space is the minimum needed; more than 160 square feet creates a space too large to manage well. Efficient kitchens are organized around three activity centers: cooking, clean-up, and storage. The cooking center comprises the **range, cooktop,** and microwave. The cleaning center comprises the sink, dishwasher, trash compactor, and garbage disposal. Storage consists of the refrigerator, pantry, and cabinets.

Each center needs storage and counter space. There should be at least 50 square feet of cabinet storage, which is provided by 6 linear feet of base cabinets with wall cabinets above. At least 11 square feet of drawer space are needed, which can be provided by four drawers. Centers should not be subjected to pass-through traffic, and work should flow in one direction from center to center.

Work Triangle The **work triangle** is the area between the stove, sink, and refrigerator (Figure 6-24). Kitchen efficiency is measured by the amount of walking needed between them, so these appliances should be in close proximity. When the lengths of the triangle's three sides are added together, the total length should be between 12' and 26'. Each side should be between 4' and 9'. There should be 4' to 6' between the sink and range, 4' to 7' between the refrigerator and the sink, and 4' to 9' between the range and the refrigerator. The dishwasher should be within the triangle and next to the sink. The work triangle should not be subject to pass-through traffic or intersected by an island (Figure 6-25).

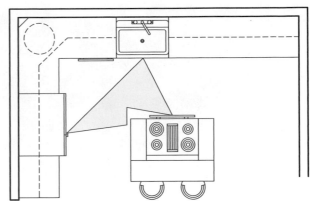

Figure 6-25
Poor layout. An island should not intersect the work triangle.

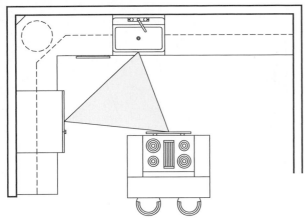

Figure 6-24
The work triangle. Total perimeter should be between 12' and 26'.

Kitchen Styles There are five kitchen styles: **U-shape, G-shape** (peninsula), **L-shape, galley** (corridor), and **single wall**. The arrangement chosen depends on room dimensions and door and window location.

- U-shape: This is an efficient style for large and small kitchens (Figures 6-26, 6-27). It surrounds the cook is surrounded with lots of countertop and storage space and gives him or her easy access to it all. The designer may remove one of the walls to open up the kitchen while allowing the counter to be both a food preparation zone and a seating area. The legs of the U should be at least 10' apart.

Figure 6-26
U-shaped kitchen.

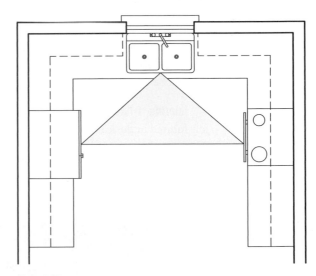

Figure 6-27
U-shaped kitchen.

- G-shape: Also called **peninsula,** this is a modified U-shape kitchen (Figures 6-28, 6-29). It offers the benefits of the U-shape plus additional storage, food preparation, and appliance areas. The peninsula is usually a "pass-through" that provides bar seating and opens up the kitchen into another space. This design usually only has one opening, which means less pass-through traffic.

- L-shape: This is an open plan that maximizes a small space's counter and storage areas and accommodates an island (Figures 6-30, 6-31). It can be adapted to almost any space but is often placed in a corner of a house. The open area of the "L" provides space for an eating area.

Figure 6-28
G-shaped (peninsula) kitchen. *Courtesy of Bosch Home Appliances.*

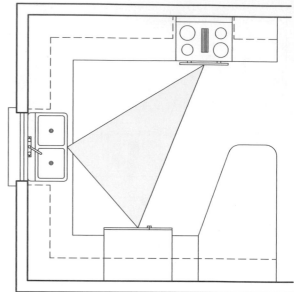

Figure 6-29
G-shaped (peninsula) kitchen.

- Galley: This plan maximizes a small space; hence its name, because it is used on boats (Figure 6-32). In a galley kitchen, long countertops uninterrupted by corners provide the most efficiency possible. It is best used when there are doorways on opposite ends of the kitchen; if there is only one doorway, traffic will become an issue. There is little storage space, but roll-out shelves under the counters can offset that and a built-in rolling trash cabinet can keep floor space clear. The corridor in between the counters should be between 4' and 6' wide and the whole room should be at least 8' wide.

- Single wall: This plan is used in confined spaces, such as small apartments or lofts, where the floor plan allows no other arrangement (Figure 6-33). Its efficiency is in its use of space, because its layout doesn't produce a work triangle. However, it can be improved upon by adding an island.

Islands Islands are stand-alone work centers, usually located in the kitchen's center (Figures 6-34, 6-35). They may have a simple countertop for work space, or have multiple levels containing cabinets, wine racks, appliances, cooktop, sink, and eating space. An island

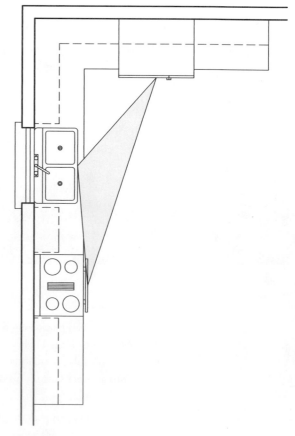

Figure 6-31
L-shaped kitchen.

Figure 6-30
L-shaped kitchen. *Courtesy of Bosch Home Appliances.*

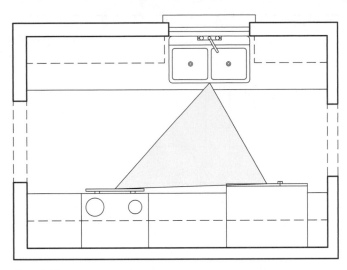

Figure 6-32
Galley kitchen.

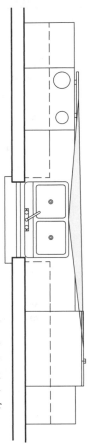

Figure 6-33
Single wall kitchen.

should not intersect the work triangle; in fact, a well-placed island can deflect traffic away from it. Not all kitchen designs are conducive to an island, however. They do not fit in a G-shape or galley design due to its limited space. Islands work well in single-wall and L-shape kitchens, as those usually have less space for storage or appliances. In a U-shape, islands work when the space between the U's two legs is at least 10'. An island should be at least 2' 6" by 3' (larger if the island houses an appliance), and accessible from all sides, with at least 36" of clearance; 42" or more are needed if cabinet or appliance doors open into the clearance.

Cabinets Although there are many cabinet manufacturers, cabinet sizes are standardized (Figures 6-36, 6-37). Wall cabinets are 12" to 13" deep and 12" to 36" high. Base cabinets are 24" deep and 34" high; when the countertop is added, the base cabinet is a standard 36" high. Base cabinets can also be custom-made taller or shorter. Widths are multiples of 3" and range from 9" to 48". They can be made with revolving shelves, pull-out drawers, stackable spice racks, wine racks, partitions for lid and tray storage and trash can concealment.

Figure 6-34
Kitchen island. *Courtesy of Hy-Lite Products, Beaumont, CA.*

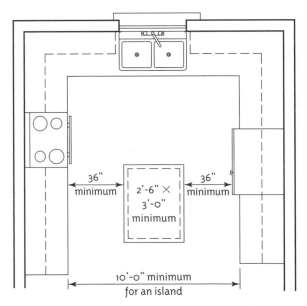

Figure 6-35
Island in a U kitchen.

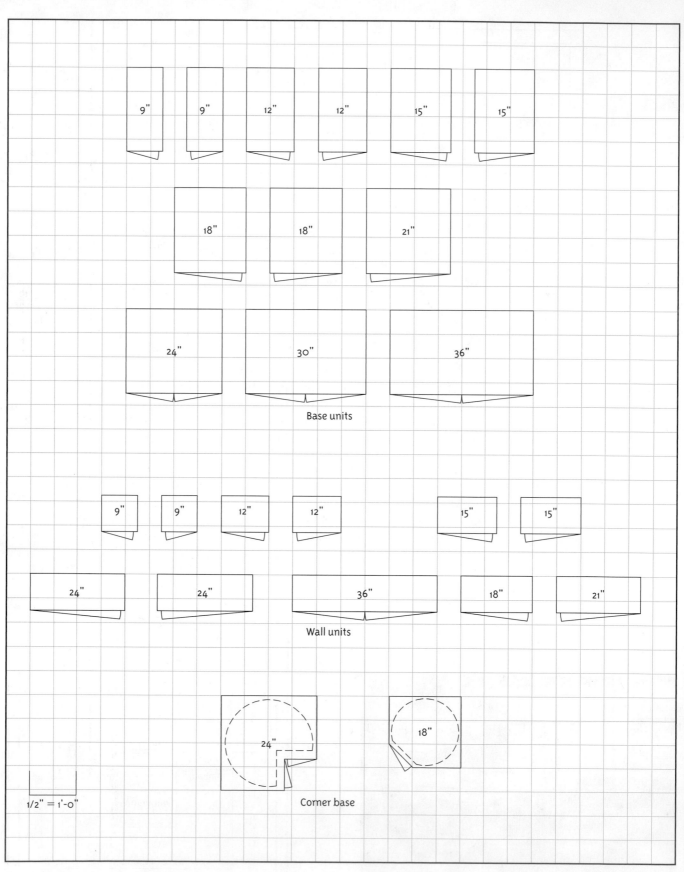

Base units

Wall units

1/2" = 1'-0"

Corner base

Figure 6-36
Cabinet dimensions on a ½"=1'-0" grid.

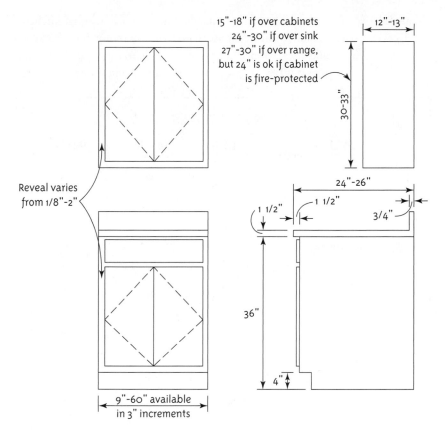

15"-18" if over cabinets
24"-30" if over sink
27"-30" if over range,
but 24" is ok if cabinet
is fire-protected

12"-13"

30-33"

Reveal varies
from 1/8"-2"

24"-26"

1 1/2"   1 1/2"

3/4"

36"

4"

9"-60" available
in 3" increments

Figure 6-37
Cabinet elevation dimensions.

Corner cabinets These are base and wall cabinets made to fit in corners. Contents can be hard to access, but adding a **lazy Susan** (rotating platform) (Figure 6-38) or slide-out drawers or using a diagonally-shaped cabinet can make the contents easier to reach. **Blind cabinets** (Figure 6-39) are cabinets in corners that have just one door; the remainder of the storage is dead (inaccessible) space.

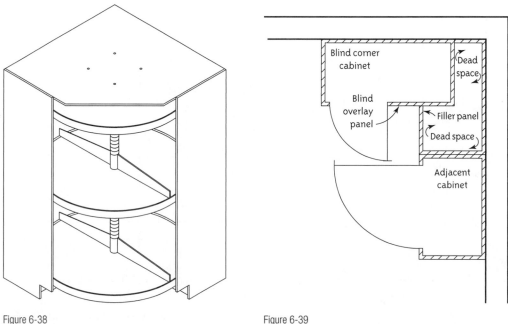

Figure 6-38
Multi-tiered lazy Susan (rotating platform).

Figure 6-39
Blind cabinet.

Blind corner
cabinet

Dead
space

Blind
overlay
panel

Filler panel

Dead space

Adjacent
cabinet

Appliances Any or all of these appliances may be found in a residential kitchen: range, range hood, cooktop, refrigerator, dishwasher, microwave, and warming drawers. All are available in dozens of models and sizes, so specific manufacturer literature must be consulted for precise measurements. Figures 6-40, 6-41 give generic maximum and minimum sizes.

KITCHEN APPLIANCES SIZES

APPLIANCE	WIDTH	HEIGHT	DEPTH
REFRIGERATOR	24"	56"	29"
	30"	68"	30"
	31"	63"	24"
	34"	70"	29"
	36"	66"	29"
FREE-STANDING RANGE	20"	30"	24"
	21"	36"	25"
	30"	26"	36"
	40"	36"	27"
DOUBLE-OVEN RANGE	30"	61"	26"
	30"	67"	27"
	30"	71"	27"
DROP-IN RANGE	23"	23"	22"
	24"	23"	22"
	30"	24"	25"
BUILT-IN COOKTOP	12"	2"	18"
	24"	3"	22"
	48"	3"	22"
RANGE HOOD	24"	5"	12"
	30"	6"	17"
	66"	7"	26"
	72"	8"	28"
SINGLE-COMPARTMENT SINK	24"		21"
	30"		20"
DOUBLE-COMPARTMENT SINK	32"		21"
	36"		20"
	42"		21"
DISHWASHER	24"	34"	24"
	18"	34"	24"
MICROWAVE	20"	17"	12"
	20"	13"	11"
	17"	12"	15"
	29"	16"	15"
TRASH COMPACTOR	14"	13"	23"
	15"	34"	24"
WARMING DRAWER	24"	12"	24"
	27"	12"	24"
	30"	12"	24"

Figure 6-40
Kitchen appliances sizes.

Range: A **range** is an oven (Figure 6-42). There are different types; the most common is the *free-standing* range, which is self-contained and finished on both sides, sits on the floor, and usually is placed between two base cabinets or at the end of a line of cabinets. Some free-standing ranges have a second, smaller oven mounted above the cooking surface. A *slide-in*

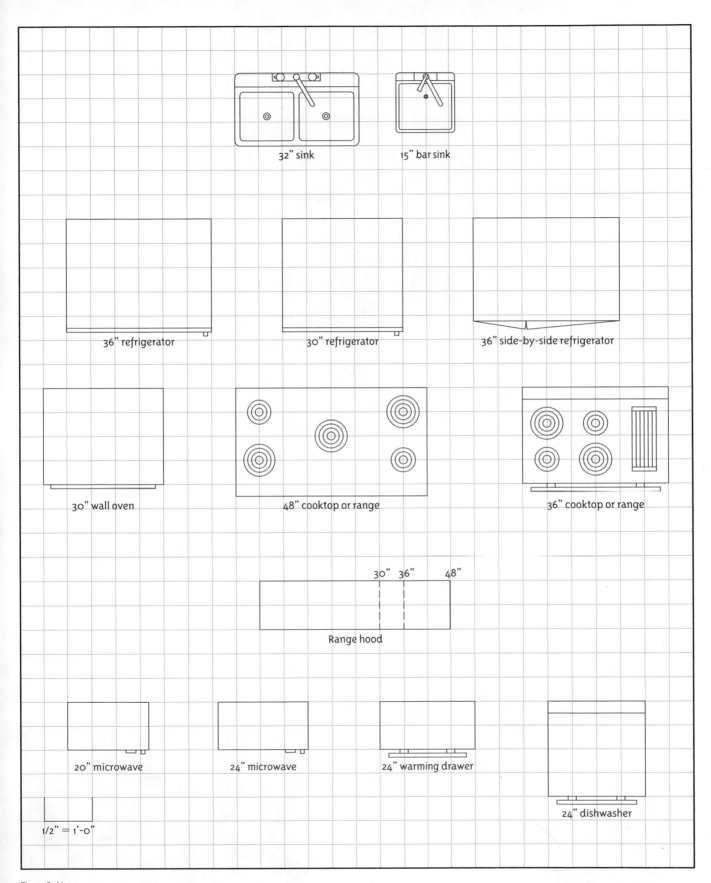

Figure 6-41
Kitchen appliance sizes on a 1/2" = 1'-0" grid.

range is self-contained and rests on the floor; the sides are usually unfinished (end panels are available for visible sides). A *drop-in* range is installed flush with the base cabinets and is supported on a low cabinet base. This type is appropriate for an island. Avoid placing a range next to the refrigerator because this will create an energy drag on both appliances.

Figure 6-42
Built-in range. *Courtesy of Bosch Home Appliances.*

Ranges need to be vented. Some have built-in venting, but an overhead exhaust hood that vents through the roof or wall is most common (Figure 6-43). There are also downdraft systems vented through the floor, typically used for islands or G-shaped kitchens. The **cooktop** (Figure 6-44), which is the appliance that contains the burners, may be on top of the range or be a separate appliance located elsewhere.

Figure 6-43
Range hood. *Courtesy of Bosch Home Appliances.*

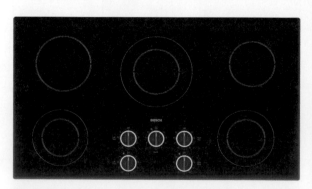

Figure 6-44
Glass cooktop. *Courtesy of Bosch Home Appliances.*

Refrigerator: This cold-storage appliance may have an upper or lower freezer, be a side-by-side model (Figure 6-45), or consist of a completely separate refrigerator and freezer that are the same size and placed side-by side. Built-in refrigerators and freezers are designed to be flush with surrounding cabinets. Top-mounted refrigerator/freezer models are 30" deep; built-ins are 24" deep, and some are taller (up to 19") and wider (up to 5") than top-mounted models.

Figure 6-45
Side-by-side refrigerator. *Courtesy of Bosch Home Appliances.*

Dishwasher: There are different types of this cleaning appliance; the most common is a built-in model designed to be installed under a countertop in a 24" wide space between cabinets. Small-capacity models exist that fit into an 18" wide space. Built-ins can be finished or paneled to match kitchen cabinets. Convertible–portable dishwashers are the same width as built-ins (24") but have finished sides and tops and rolling casters. An under-sink dishwasher's purpose is to save space in small kitchens. It fits under a special double sink. There are even microwave-sized dishwashers that attach to the sink and sit on the countertop.

Microwave oven: This small appliance can be freestanding, built-in, or mounted under wall cabinets. It is available in many different sizes.

Warming drawer: A **warming drawer** keeps plates and cooked food warm. It is a slide-out appliance that fits in the cabinetry. Figure 6-46 shows a microwave and warming drawer built into the wall.

Figure 6-46
Microwave and warming drawer. *Courtesy of Bosch Home Appliances.*

Sink: This plumbing fixture comes in different materials, sizes, and depths. It can be mounted on top of the cabinetry (Figure 6-47) or under it. There is also a wide variety of **fittings,** or faucets and controls, available; Figure 6-48 shows one particular model.

Figure 6-47
Top-mounted, double bowl sink. *Courtesy of American Standard.*

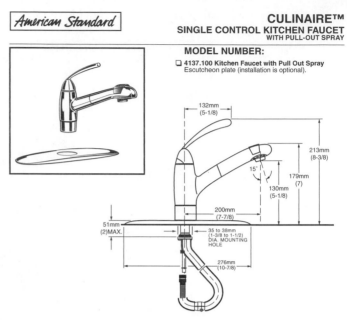

Figure 6-48
Detail drawing for a faucet. *Courtesy of American Standard.*

Eating Area This can be part of the kitchen (Figures 6-49, 6-50), a separate dining room, or an outdoor deck (Figure 6-51). Since eating and entertaining often go hand in hand, it is useful to locate the eating area near other entertaining areas or even combine it with them, perhaps visually dividing them with a screen, furniture, or a level change. While many homes do have the traditional separate dining room, many now have combined living/dining areas or kitchen/dining areas, because families are increasingly using the traditional dining room for other purposes. The minimum recommended dining room size is 120 square feet.

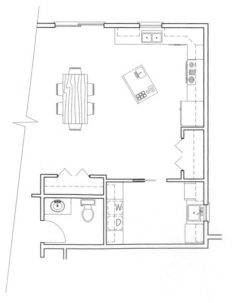

Figure 6-49
An efficient layout for a kitchen, eating space, laundry room, and bathroom.

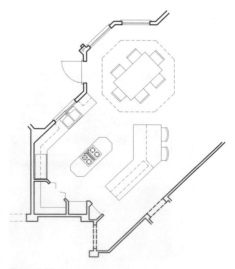

Figure 6-50
This elaborate kitchen nicely incorporates the eating space.

Figure 6-51
Attractive outdoor eating area. *Courtesy of Laneventure, Conover, NC.*

Clearances Clearance standards (Figures 6-52, 6-53, 6-54) vary depending on the source; the Federal Housing Authority (FHA), the National Kitchen and Bath Association (NKBA), and different codes suggest or require different ones.

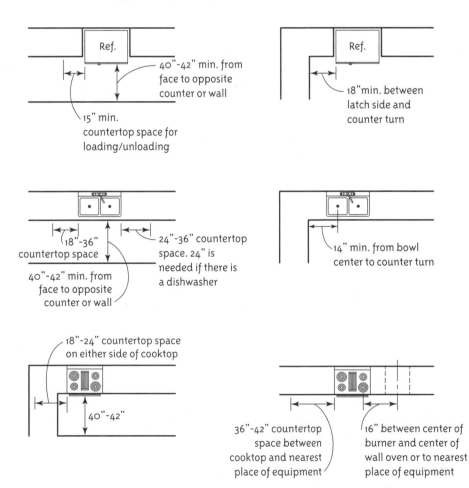

Figure 6-52
General kitchen clearances.

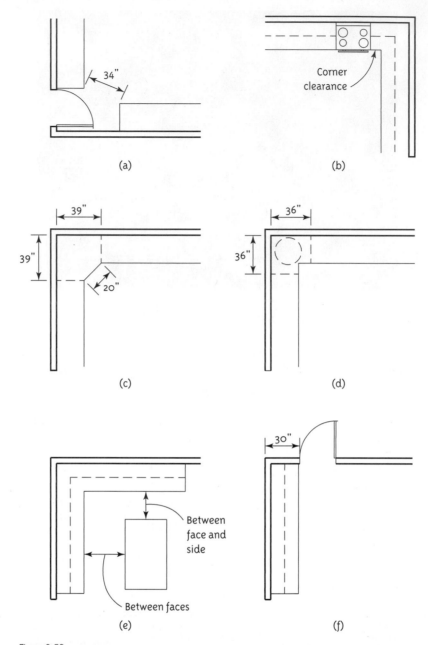

Figure 6-53
Kitchen countertop clearances:
(a) 34 inches is the corner-to-corner clearance space between appliances or base cabinets perpendicular to each other. (b) Corner clearances: 12" between edge of range and corner cabinet; 15" between edge of refrigerator and corner cabinet. 9" to 12" between edge of sink and corner cabinet. (c) A diagonal cabinet of 20" requires 39" on either wall. A diagonal cabinet or appliance of 30" inches requires 40" on either wall. (d) A lazy Susan (round, rotating platform) requires 36" on either side. (e) The distance between base cabinet and island faces is 48" to 60". (f) Doors should not be placed closer than 30" from the corner or windows 12 ¾" from the corner if cabinets are extended to the corner.

Generally, at least 8'-'8" of clear space is needed for a 36" × 48" table and four chairs. A change in countertop height can separate cooking and eating areas; counter height where chairs are used should be 26" and provide at least 21" of knee space depth underneath, although 28" to 30" of knee space is needed for accessibility (an overhanging or extended countertop can help with this). Knee space should be 27" to 30" high. Figures 6-55 and 6-56 show some generic furniture sizes.

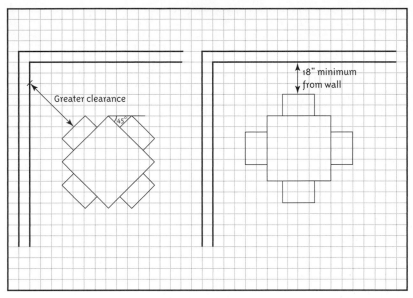

Figure 6-54
Rotating a table 45 degrees to the wall gives a greater clearance than placing the table parallel to it.

DINING ROOM FURNITURE SIZES

FURNITURE	LENGTH	DEPTH	HEIGHT
RECTANGULAR DINING TABLE	42"	30"	29"
	48"	30"	29"
	48"	42"	29"
	60"	40"	28"
	60"	42"	29"
	72"	36"	28"
OVAL DINING TABLE	54"	42"	28"
	60"	42"	28"
	54"	30"	36"
	72"	40"	28"
	72"	48"	28"
	84"	42"	28"
ROUND DINING TABLE	DIAMETER		HEIGHT
	32"		28"
	36"		28"
	42"		28"
	48"		28"
CHINA CABINET OR HUTCH	LENGTH	WIDTH	HEIGHT
	48"	16"	65"
	50"	20"	60"
BUFFET	62"	16"	66"
	36"	16"	31"
	48"	16"	31"
	52"	18"	31"
CORNER CABINET	WIDTH	DEPTH	HEIGHT
	36"	15"	80"
	38"	16"	80"
DINING CHAIRS	17"	19"	29"
	20"	17"	36"
	22"	19"	29"
	24"	21"	31"

Figure 6-55
Dining room furniture sizes.

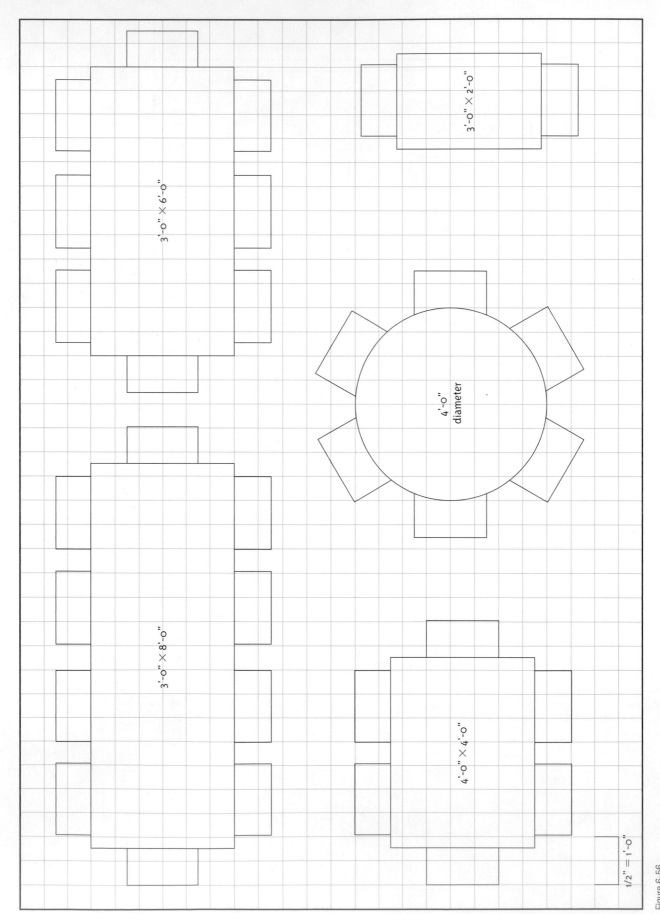

3'-0" × 2'-0"

3'-0" × 6'-0"

4'-0"
diameter

3'-0" × 8'-0"

4'-0" × 4'-0"

1/2" = 1'-0"

Figure 6-56
Kitchen and dining table sizes on a 1/2" = 1'-0" grid. There are no standardized sizes for tables, buffets, or china cabinets.

- To make cabinets accessible, the typical 36" cabinet and cooktop height of 36" will work, but a height of 30" to 33" is better. Knee space must be provided under the sink counter with a protective panel in front of the pipes to prevent scalding (pipes can also be wrapped in insulation).
- Cabinet toe space (the indented area at the bottom to accommodate the foot) 6" deep and 8 to 11" high is needed for wheelchair footrests. The designer should locate the bottom of wall cabinets to make the first shelf reachable from a seated position, typically no more than 17" above the counter or 48" above the floor. Shelves in wall cabinets should be adjustable.
- Wall-mounted pegboards for pots, pans, and utensils are more accessible than cabinets, as are vertical drawers in base cabinets and narrow shelves attached to the backs of cabinet and closet doors.
- For accessibility, place cooktop controls at the front to prevent users from reaching across burners and to alert the visually impaired where the burners are. Place wall ranges so that the top of the open oven door is 2'-7" above the floor.
- Side-by-side refrigerator/freezers are the most accessible, although refrigerators with the freezer on the bottom also work.
- Front-loading dishwashers are the most accessible.
- Tables with pedestal legs and round tables with pedestal bases are wheelchair-friendly. A 4'-0" diameter table accommodates two wheelchair users; a 4'-5" diameter accommodates four wheelchair users. Thirty-six inches between any two points is enough for a wheelchair to pass, but 5' is needed to fully rotate 180 degrees.

Bathrooms

Bathrooms vary widely in size and design. They may be small rooms with just the essential fixtures (Figure 6-57) or elaborate areas that incorporate exercise, dressing, saunas, steam baths, and foot spas (Figures 6-58, 6-59). Hundreds of options for fixtures and features exist, such as radiant heat flooring, towel warmers, bidets, TV sets built into fog-free mirrors, and waterfall tubs that fill with color and water.

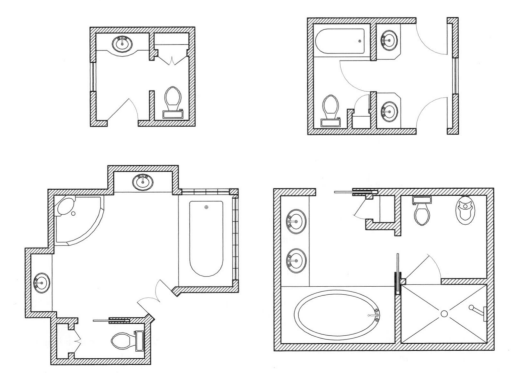

Figure 6-57
Efficient bathroom layouts.

Figure 6-58
Spacious bathroom that incorporates areas for fixtures, grooming and dressing. *Courtesy of Hy-Lite Products, Beaumont, CA.*

Figure 6-59
Bathroom with partitioned area for the water closet. *Courtesy of American Standard.*

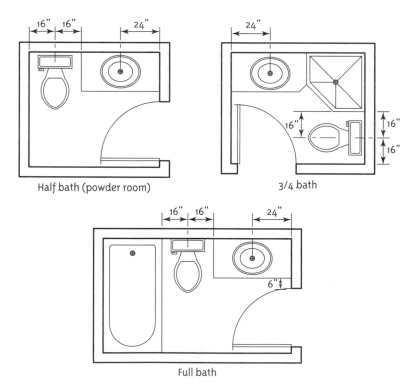

Figure 6-60
Types of bathrooms and NKBA-recommended fixture clearances.

Bathrooms are categorized by the fixtures they contain (Figure 6-60). A **half-bath**, also called a *powder room,* has a water closet and lavatory. A **three-quarter bath** has a water closet, lavatory, and shower. A **full bath** has a lavatory, water closet, and tub or a tub/shower combination. A bathroom's efficiency is a function of its fixture location.

A bathroom needs at least 40 square feet and should have minimum dimensions of 5' × 8'. An accessible bathroom needs at least 60" diameter of clear rotation space between all elements to turn a wheelchair. Master and family bathrooms are typically larger than guest bathrooms, because they require more countertop and storage space. Bathroom doors are typically narrower than bedroom doors, with 2'-6" being common. They can be as narrow as 2-4"; however, accessible bathrooms require doors at least 2'-8" wide. Doors should swing into the bathroom.

Bathroom Fixtures and Accessories The required fixtures in all bathrooms are the **lavatory** (sink), **water closet** (toilet), and a bathub or shower. While some sizes are common, there is no standard size or shape for any of them; products vary widely between manufacturers. Templates with generic sizes are available, but if you need precise dimensions, you must refer to specific catalogs. Bathrooms may also contain vanities, cabinets, medicine cabinets, grab bars, towel racks, and other amenities. Figures 6-61, 6-62, and 6-63 show generic bathroom fixture sizes.

BATHROOM FIXTURE SIZES

FIXTURE	WIDTH	LENGTH	HEIGHT	
STANDARD TUB	30"	60"	12"	
	30"	60"	14"	
	30"	60"	16"	
	30"	66"	16"	
	32"	60"	18"	
	36"	72"	16"	
WHIRLPOOL TUB	37"	42"	12"	
	42"	48"	14"	
	42"	60"	16"	
	42"	72"	16"	
	72"	42"	19"	
	WIDTH	**DEPTH**	**HEIGHT**	
WATER CLOSET (FLOOR MOUNTED, TWO PIECE)	17"	25"	29"	17-½" ACCESSIBLE SEAT HEIGHT
	21"	26"	28"	
	21"	28"	28"	
WATER CLOSET (FLOOR MOUNTED, ONE PIECE)	20"	27"	20"	
	20"	29"	20"	
WATER CLOSET (WALL HUNG, TWO-PIECE)	22"	26"	21"	
WATER CLOSET (FLOOR MOUNTED, ONE-PIECE)	14"	24"	15"	
BIDET	15"	22"	15"	
LAVATORY (WALL HUNG)	19"	17"		
	20"	18"		
	22"	19"		
	24"	20"		
LAVATORY (FLOOR MOUNTED)	27"	20"	35"	
SHOWER	30"	30"	72"	
	36"	36"	73"	
	36"	48"	75"	
UTILITY SINK	21"	20"		
BAR SINK	15"	15"		

Figure 6-61
Bathroom fixture sizes.

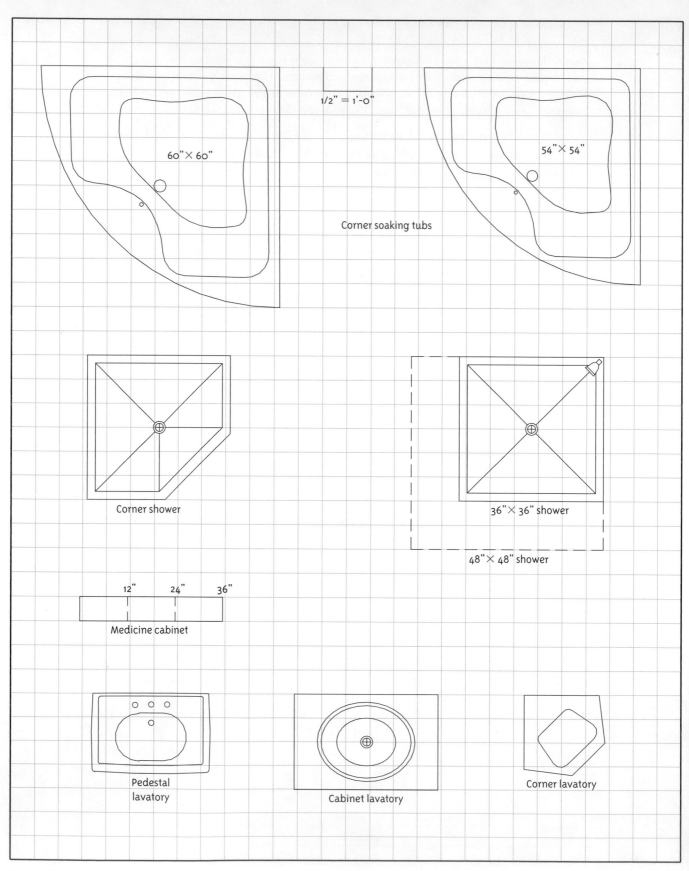

1/2" = 1'-0"

60" × 60"

54" × 54"

Corner soaking tubs

Corner shower

36" × 36" shower

48" × 48" shower

12" 24" 36"

Medicine cabinet

Pedestal lavatory

Cabinet lavatory

Corner lavatory

Figure 6-62
Bathroom fixtures sizes in plan on a 1/2" = 1'-0" grid.

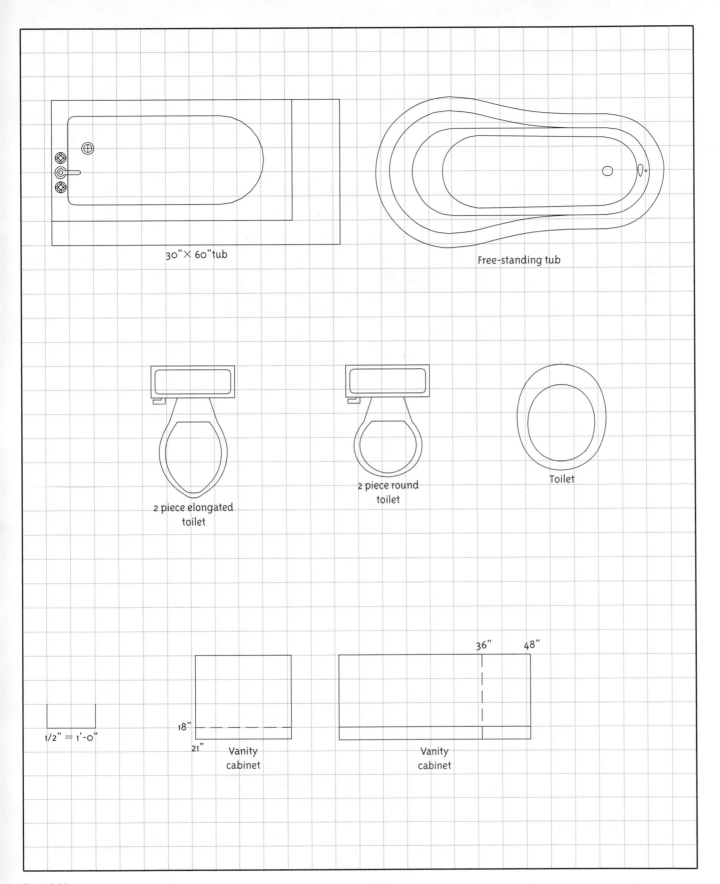

30" × 60" tub

Free-standing tub

2 piece elongated
toilet

2 piece round
toilet

Toilet

36"   48"

1/2" = 1'-0"

18"

21"   Vanity
cabinet

Vanity
cabinet

Figure 6-63
Bathroom fixtures sizes in plan on a 1/2" = 1'-0" grid.

Here is a description of common bathroom components.

- Lavatory: This is a bathroom sink (Figures 6-64, 6-65). It can be wall-hung, corner-hung, console, pedestal, under-counter, over-counter, partially recessed, drop-in, circular, rectangular, or integrally molded with the cabinet. Avoid locating it under a window. A mirror should be placed over the sink.

Figure 6-64
Sinks mounted at different heights for different users. *Courtesy of Kohler Plumbing Products, Kohler, WI.*

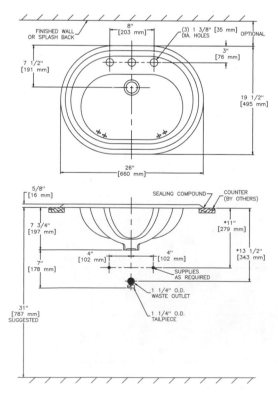

Figure 6-65
Counter-mounted lavatory. *Courtesy of American Standard.*

- Two lavatories are recommended for a shared bathroom; there needs to be enough clearance so that two people can comfortably use them at once. Vessel and pedestal sinks are alternatives to the typical under-mounted-in-a-cabinet sink. A vessel sink sit on top of a counter or **vanity,** so it is higher than most under-mounted sinks, requiring less bending. A pedestal sink's advantage is that it consumes less visual space than a built-in one.

- A minimum clear floor space of 30" × 48" is needed in front of the sink. For accessibility, up to 19" of a 48" floor space can extend under the sink when a knee space is provided. Minimum recommended clear space from the centerline of the sink to any side wall is 16". Wall-mounted and pedestal sinks provide the knee space required for wheelchair users; any exposed pipes should be insulated. For accessibility, a distance of 26" to 30" from its underside to the floor is the minimum needed for wheelchair armrests; 30" to 34" from the rim of the sink to the floor is better.

- Tub: A tub can be built-in, free-standing, clawfooted, small, accessible, or an oversized soaking tub (Figures 6-66, 6-67, 6-68). Often there is a shower above the tub; this eliminates the need for a separate shower. Tubs can have jets or a waterfall and be made of porcelain, acrylic, stainless steel, copper, or teak. There are no code requirements for size, but the built-in 30" × 60" tub is the most common. A small space can hold a corner tub. Rim height is important; the higher the tub rim is, the greater the water coverage possible. Minimum clear floor space alongside the tub should be at least 60" long and 30" wide. Freestanding tubs are becoming popular due to their flexible placement options. Clear space for access to plumbing and any whirlpool motors is also needed.

Figure 6-66
Corner tub. *Courtesy of American Standard.*

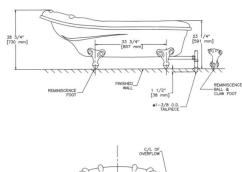

Figure 6-67
Free-standing soaking tub. Also pictured are pedestal lavatories and a two-piece toilet. *Courtesy of American Standard.*

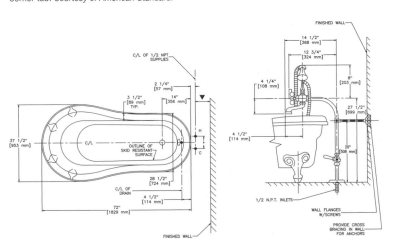

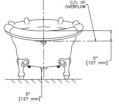

Figure 6-68
Orthographic drawings of the tub in Figure 6-67.

Figure 6-69
One-piece elongated toilet. *Courtesy of Kohler Plumbing Products, Kohler, WI.*

- Water closet: This is the toilet (6-69, 6-70, 6-71). It can be floor-mounted, wall-mounted, one-piece, two-piece (separate tank and stool), round, elongated, or compact-elongated. Most commercial toilet bowls are elongated. The most common residential toilet bowl has historically been round, because a round bowl takes up less space in a small bathroom. However, houses now feature larger bathrooms, and the trend now is to use elongated bowls. Round toilets extend 25" to 28" from the wall and are about 16½" long. Elongated toilets extend 29" to 31" from the wall and are 2" longer (about 18½" long). Both require a clear space 30" wide for installation. Most codes require a clearance of at least 15" from the center of the toilet to any side wall and at least 30" to the center of any other sanitary fixture. (The NKBA recommends 16" and 32", respectively.) There should be at least 24" of clear space in front of the toilet. A clear space 36" wide is needed for accessibility. Wall-mounted toilets are more wheelchair friendly, since a 20" high toilet seat is about the same height as most wheelchair seats. The same requirements hold for a *bidet,* a plumbing fixture used for body cleaning. It is similar to a toilet in size and is placed next to it.

- Universally designed "chair-height" toilet seats 16 1/2" to 19" from the floor are becoming popular with the general (non-elderly or disabled) population, because sitting down and getting up again is easier from these than from the typical 15" high seat.

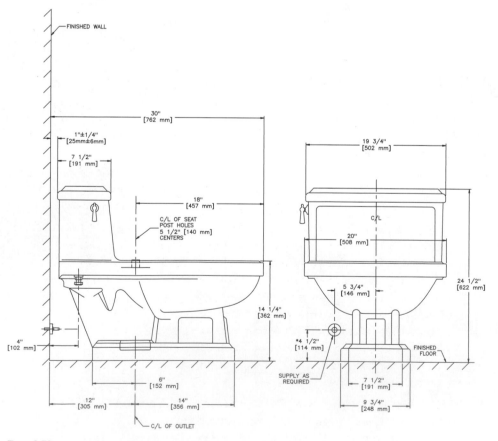

Figure 6-70
Orthographic drawings of a one-piece toilet. *Courtesy of American Standard.*

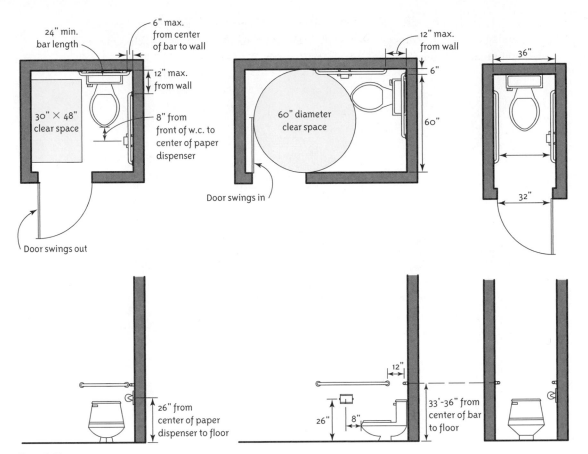

Figure 6-71
Accessible, compartmented water closets.

- In larger bathrooms, the water closet may be partitioned in a separate area from the rest of the room. Where clearance is not an issue, doors should swing into a toilet compartment or stall, not into the room. Avoid locating toilets and bidets under windows and where they can be seen from other rooms when the bathroom door is open.

- Shower stall: This can be installed when there's no room for a bathtub (Figure 6-72). Corner showers are particularly good space-savers. Shower stalls may be made of metal, fiberglass, plastic, ceramic tile, terrazzo, or marble. Many are prefabricated with a built-in soap dish, shelves, a seat, water jets, and electronic temperature controls. Wall-mounted, fold-down seats are available for wheelchair users. Showers can be custom-made to fit any dimension, but anything less than 36" square (interior wall-to-wall dimensions) is considered tiny. Most codes require a minimum width of 32" (the NKBA recommends 34") and an interior circular diameter at least 30" regardless of shape. Stall doors should swing out into the bathroom, not into the stall.

- Fittings: These are pieces and accessories for pipes (Figure 6-73). The ones most relevant to interior designers are faucets and temperature controls. Hundreds of choices are available. Faucet handles should be a maximum of 18" from the front of the

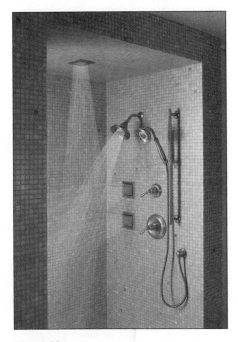

Figure 6-72
Shower and its fittings. *Courtesy of Kohler Plumbing Products, Kohler, WI.*

sink; lever-types are easiest to use. Small sinks use the same faucets as larger models; positioning them in the corner optimizes space, as does using single-control, 4-inch-center and wall-mounted fittings.

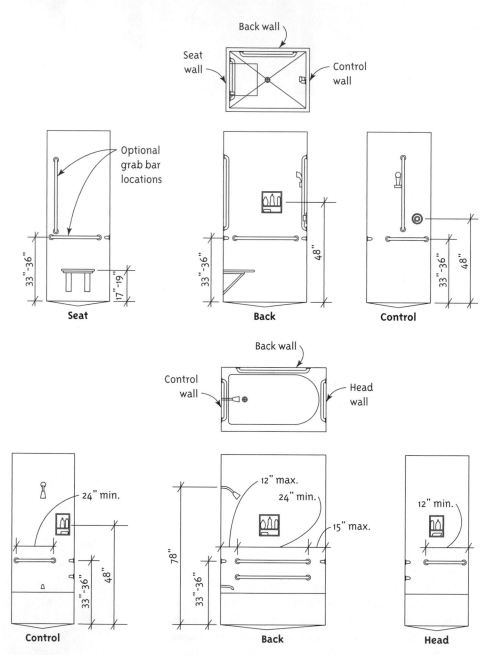

Figure 6-73
Grab bar, soap dish, and shower seat clearances.

- **Grab bars** are bars anchored into the wall over tubs, toilets and in showers to help provide accessibility. Horizontal bars are for pushing up; vertical bars are for pulling up. Grab bars must be well anchored, 1¼" in diameter, 1½" away from the wall, and have an easily grasped profile with no sharp edges.

- Vanity: This is a sink cabinet (Figure 6-74). It provides countertop and storage space. Modular vanities come in three standard depths: 16", 18", and 21". Most are between

29" and 31" tall. Vanities come in 18", 24", 30", 36", 48", and 72" widths. A common vanity size is 18" – 21" deep and 24" wide. Small vanities 16" deep and 18" wide with an integral sink and top are available for tiny bathrooms. Some bathrooms have storage cabinetry in which electric plugs can be installed inside to hide appliance cords. You can also order custom cabinetry with features like extra-deep drawers and different counter-top heights to accommodate tall and short users.

Figure 6-74
Vanity with vessel sinks. *Courtesy of Strasser Woodenworks, Woodinville, WA.*

Accessibility and Bathroom Design

- An accessible bathroom needs at least 60" diameter of clear rotation space between all elements to turn a wheelchair.
- Up to 19" of a 48" floor space can extend under the sink when a knee space is provided. A distance of 26" to 30" from its underside to the floor is the minimum needed for wheelchair armrests; 30" to 34" from the rim of the sink to the floor is better.
- Toilets require a clear space 36" wide is needed for accessibility. Wall-mounted toilets are more wheelchair friendly, since a 20" high toilet seat is about the same height as most wheelchair seats.

Medicine Cabinets **Medicine cabinets** are small bathroom wall cabinets (Figure 6-75). They may be mirrored and lit and have swing or sliding doors. Medicine cabinets should be mounted so the top shelf is no more than 50½" from the floor; lower if mounted over a counter or sink.

Figure 6-75
Medicine cabinet with mirrors and lights. *Courtesy of Strasser Woodenworks, Woodinville, WA.*

General Design Tips Codes and practical considerations affect the location of plumbing fixtures and the stack (vertical) and branch (horizontal) pipes serving them. When placing plumbing fixtures, think three-dimensionally and visualize the spaces above and below them.

- Ventilation must be provided by a window, an **exhaust fan** (an appliance that removes polluted and humid air from a space, allowing room for fresh air to enter), or both. To avoid drafts and contact with moisture, don't place windows near showers or tubs. Windows must also guard privacy, either by their location or by using special types of glass. A fan should be located near the tub and toilet areas. It should vent to the outside, not to an attic space. Electrical switches should not be accessible from the tub and all **receptacles** (plug-in outlets) should be **GFI** (Ground Fault Interrupter) ones, which interrupt the current when water is sensed.

- Avoid locating fixtures on exterior walls, because structural members, cross-bracing, and windows can cause problems with stack pipe installation, and cold temperatures can cause them to freeze. A wider **chase**, or plumbing wall, can solve these problems but cuts into the room's usable square footage. Fixtures may be located below windows only when the bottom seal is at least 48" above the floor. If a fixture is located below a window, the wall beneath the window will also require additional reinforcement to compensate for the holes cut into its studs.

- When laying out walls and plumbing fixtures, know where the beams and joists beneath the floor are (they're usually located on column grid lines), because they may interfere with pipes. Avoid locating fixtures next to columns because the footing under the column will be larger than the column itself. This can cause problems when installing pipes. Space availability within the ceiling directly below fixtures is also an issue; for instance, vaulted or high ceilings in the rooms below can hinder pipe installation.

- Building codes typically forbid plumbing pipes to be located above electrical equipment rooms and food preparation, storage, serving, or dining areas. In commercial construction they cannot be above sensitive areas like hospital operating rooms, intensive care units, or nurseries.

- Utilize common walls and chases by setting fixtures back to back on walls that are larger than the standard 2" × 4" stud. Stack chases on multi-floor homes and anywhere where the vent stack of a first-floor bath must pass through a second-floor wall. Arrange fixtures so that all pipes are concealed in the wall or beneath the flooring. Place the fixture end of a recessed or built-in tub along an inside wall to permit easier inspection of the piping if needed. When possible, place all fixtures so they can be vented through one main vent.

Vents, pipes, and plumbing systems are discussed in more detail in Chapter 13. Chase walls are discussed in Chapter 8.

Living Room/Gathering Centers

Modern homes still contain living rooms (also called family or great rooms) as traditional gathering centers (Figures 6-76–6-81). However, specialty rooms such as bonus rooms, home theatres, music, exercise, game, and recreation rooms are becoming popular for gathering and entertaining. Along with traditional furnishings, these rooms may contain arcade games, air hockey tables, pool tables, specialized game room and home theatre chairs, jukeboxes, countertop games, pianos, electronics, large-screen TVs, wet bars, or exercise equipment. There may be special power, lighting, wiring, and acoustical requirements. With an open floor plan, these may be areas instead of rooms, but the space they require generally ranges between 220 and 400 square feet, with 250 square feet being average.

Figure 6-76
Family Room. *Courtesy of Rowe Furniture, Elliston, VA.*

Figure 6-77
Bonus room. *Courtesy of Sauder Woodworking, Archbold, OH.*

Figure 6-78
Living room. *Courtesy of Willis Construction, Overland Park, KS.*

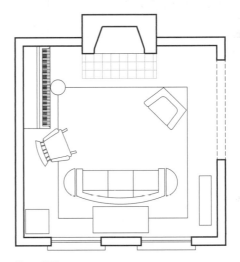

Figure 6-79
An efficient living room layout.

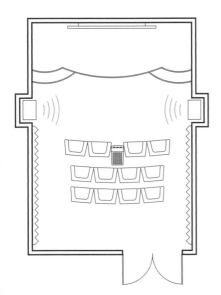

Figure 6-80
Home theatre layout.

Figure 6-81
Home theatres and media centers are popular features in new construction.

Gathering centers should not have through traffic to other rooms. Slightly raising or lowering the room level can help set it apart and discourage through traffic. The main outside entry should not open directly into it; instead, a hall or foyer should act as a buffer. Arrange the major furniture grouping so that a circle of chairs will fall within a 12' diameter. In a living room, it is desirable for the furniture to have a focal point. Figures 6-82, 6-83, and 6-84 give some generic gathering room furniture sizes.

LIVING ROOM FURNITURE SIZES

FURNITURE	WIDTH	DEPTH	HEIGHT	FURNITURE	WIDTH	DEPTH	HEIGHT
SOFA	72"	36"	28"	END TABLE	21"	28"	20"
	76"	35"	35"		26"	20"	21"
	84"	36"	37"		28"	28"	20"
	87"	31"	31"	CORNER TABLE	28"	28"	20"
	91"	32"	30"		30"	30"	15"
SOFA	72"	30"	30"		36"	36"	15"
	90"	30"	30"	OTTOMAN	22"	18"	13"
LOVE SEAT	47"	28"	36"		24"	119"	16"
	54"	30"	36"	SIDE TABLE	20"	20"	15"
	59"	36"	37"		21"	21"	16"
DESK	50"	21"	30"	SOFA TABLE	48"	16"	29"
	55"	26"	29"		44"	26"	30"
	60"	30"	29"	FLAT SCREEN TV	37"	17"	29"
	72"	36"	29"		40"	18"	30"
RECLINER CHAIR	31"	30"	40"		47"	19"	30"
	32"	35"	41"	SHELF UNITS	17"	10"	60"
	36"	37"	41"		24"	10"	60"
ARM CHAIR	30"	30"	25"		36"	10"	36"
	36"	36"	25"		36"	10"	60"
SMALL ARMCHAIR	18"	18"	29"		48"	10"	60"
	21"	22"	32"				
COFFEE TABLE	35"	19"	17"			DIAMETER	
	50"	18"	15"	ROUND COFFEE TABLE		24"	16"
	54"	20"	15"			36"	16"
	57"	19"	15"			48"	16"
	61"	21"	17"				
	66"	20"	15"				

Figure 6-82
Living room furniture sizes.

Home Office

With large numbers of people now telecommuting, the home office (Figures 6-85, 6-86) is increasingly important. Whether it is in a spare bedroom, the basement, or a room specifically designed for it, making it a comfortable, efficient place takes planning. The designer's goal is to make the home office a place that has the support infrastructure of the corporate office without the drawbacks that caused the worker to leave it.

When choosing its location, determine if clients will be visiting; if so, the areas they enter and pass through should be considered. The home office should be located in a quiet area of the house. Make sure the area has sufficient ventilation, overhead lighting, outlets, cable connections, and phone jacks for whatever equipment is needed, and that the wires and cords can be managed. Arrange furniture so that natural light will not cause glare to compromise visibility on the computer screen. Typical home office furnishings and equipment include a desk, chair, computer, printer, fax, scanner, phone, storage, file cabinet, tables, seating, shelves, desk lamp, whiteboard, and specialized furniture for computers. Executive and secretarial desks, credenzas, and file cabinets have standardized sizes.

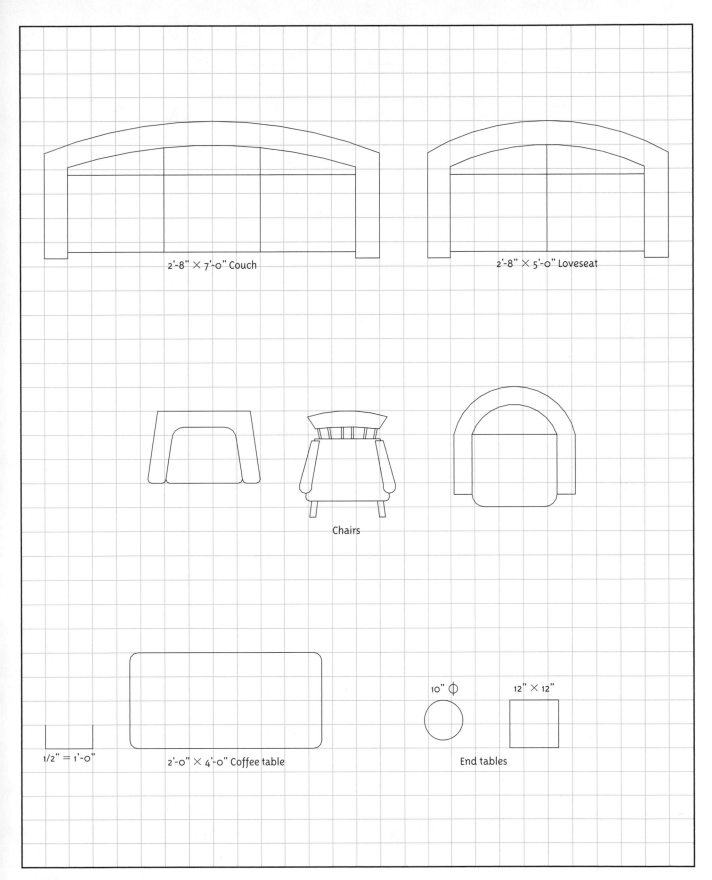

Figure 6-83
Living room furniture on a 1/2" = 1'-0" grid.

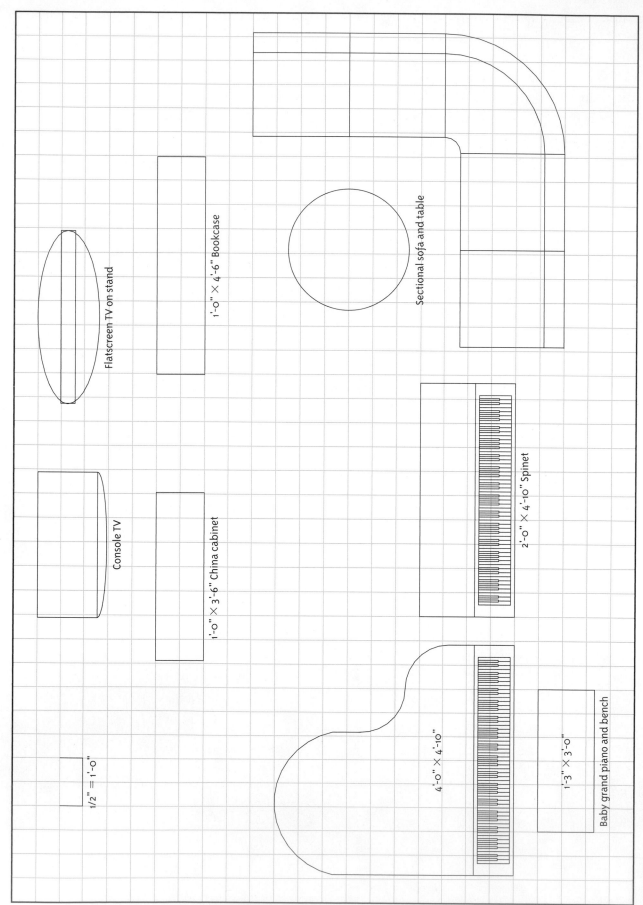

Flatscreen TV on stand

1'-0" × 4'-6" Bookcase

Sectional sofa and table

Console TV

1'-0" × 3'-6" China cabinet

2'-0" × 4'-10" Spinet

1/2" = 1'-0"

4'-0" × 4'-10"

1'-3" × 3'-0"

Baby grand piano and bench

Figure 6-84
Living room furniture on a 1/2" = 1'-0" grid.

Figure 6-85
A home office in a college student's apartment. *Courtesy of Ashley Furniture, Arcadia, WI.*

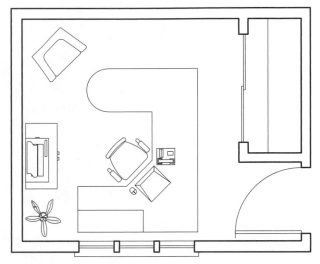

Figure 6-86
An efficient home office layout.

Ergonomics should be researched when designing and drafting workstations. Distance from the computer monitor and seat and keyboard heights are all critical in creating a comfortable workspace. The user should be at a 45-degree angle to other elements in the work area when sitting in the chair, and the chair and the desk should be ergonomic and height and tilt adjustable. The keyboard should be at seated elbow height (25"–32" from the floor), and the screen should be positioned at eye level. The eye/monitor distance should be about 24" for a 14"–17" monitor. That said, generic clearances for distance from monitor, height of chair, etc., are not very useful because ergonomics is all about fitting the furniture to the precise body using it. The most important chair adjustments are seat height from the floor (feet should be able to rest flat on the floor), depth from the front of the seat to the backrest, and lumbar support height. There are no standardized sizes for ergonomic chairs and other furniture; they vary by manufacturer.

Bedroom

One-third of our lives are estimated to be spent sleeping; thoughtful planning will help make the bedroom a nice space instead of just a place to sleep (Figures 6-87, 6-88, 6-89).

Figure 6-87
Bunk beds are space savers in children's bedrooms. *Courtesy of Lea Furniture, Richmond, VA.*

Figure 6-88
Leave clearance on either side of a double or larger bed for easy entrance and linen changing. *Courtesy of Century Furniture, Richmond, VA.*

Figure 6-89
Arrange furniture diagonal to walls for maximum clearance. *Courtesy of roomstogo.com.*

The FHA recommends 100 square feet as a minimum to hold required furniture. However, 175 square feet are necessary to hold a double bed, chest of drawers, dresser, and nightstand. Group bedrooms together in the home's private zone, in an area quiet and away from traffic. Placing them on the second floor helps maintain that privacy; however, it is common to place a master or accessible bedroom on the first floor. Bedrooms should be near bathrooms. **Ensuites** (Figure 6-90) are bedrooms connected to bathrooms.

When planning a bedroom layout, placing the bed is the main factor. A double bed may be used by two people, so it should not be placed in a corner. (Accessibility requires a 36" clearance on at least one side.) A bed that is in a corner should have 18"–24" clearance from the wall

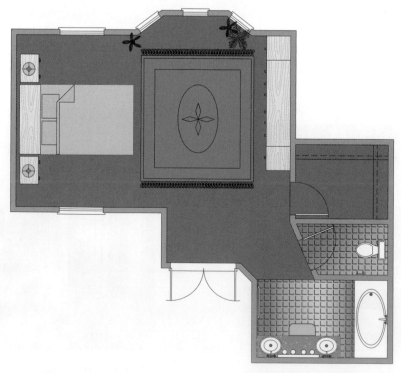

Figure 6-90
Master bedroom ensuite.

so sheets may be easily changed. Bunk beds need clearance around them for access to the upper bunk. Place nightstands, dressers, chests of drawers, armoires, and dressing tables so that there is clearance in front of their opened doors and drawers. If there will be a TV in the bedroom, arrange furniture and outlets for best viewing and cord management. Some bedrooms double as offices or reading areas, so you may need to add shelving. Figures 6-91, 6-92, and 6-93 give generic bedroom furniture sizes.

BEDROOM FURNITURE SIZES

FURNITURE	LENGTH	WIDTH	HEIGHT
SINGLE BED	75"	39"	20"-36"
DOUBLE BED	75"	54"	20"-36"
	80"	54"	20"-36"
	84"	54"	20"-36"
QUEEN-SIZE BED	80"	60"	20"-36"
KING-SIZE BED	80"	72"	20"-36"
	80"	76"	20"-36"
CALIFORNIA KING-SIZE BED	84"	72"	20"-36"
	84"	76"	20"-36"
SOFA BED	87"	31"	
	91"	32"	
	79"	34"	
NIGHT STAND	24"	15"	22"
	22"	16"	22"
	24"	18"	22"
	22"	22"	22"
DRESSER	WIDTH	DEPTH	HEIGHT
	30"	18"	30"
CHEST OF DRAWERS	20"	16"	50"
	26"	16"	37"
	28"	15"	34"
	32"	17"	43"
	36"	18"	45"
ARMOIRE	36"	22"	66"
	48"	22"	66"
	60"	22"	66"
DESK	33"	16"	29"
	40"	20"	30"
	43"	16"	30"
RECLINER	30"	31"	
	32"	35"	
	36"	38"	
CRIB	20"	50"	36"

Figure 6-91
Bedroom furniture sizes. There are industry standards for beds. A twin bed is always 3'-3" wide and is called a 3/3. A full bed is always 4'-'6" wide and is called a 4/6. A queen bed is always 5'0" wide and a king is always 6'6" wide (the latter is called a 6/6). A California king is not standardized and varies by manufacturer. Twin and full beds are always 75" long. They can be special ordered 80" long and those are called 3/3 XL and 4/6 XL. Queen and king beds are always 80" long.

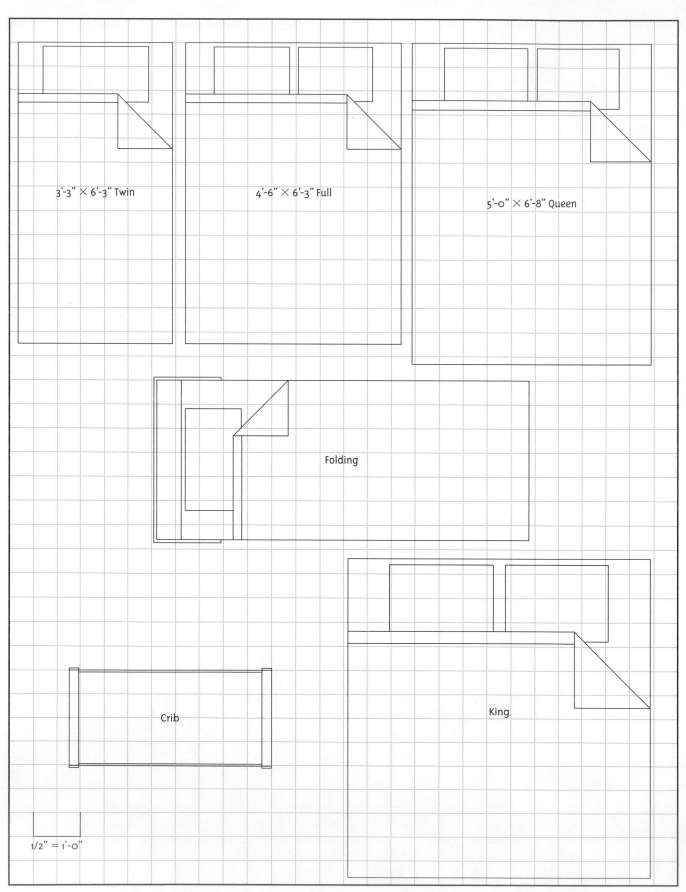

3'-3" × 6'-3" Twin

4'-6" × 6'-3" Full

5'-0" × 6'-8" Queen

Folding

Crib

King

1/2" = 1'-0"

Figure 6-92
Bedroom furniture sizes on a 1/2"=1'-0" grid.

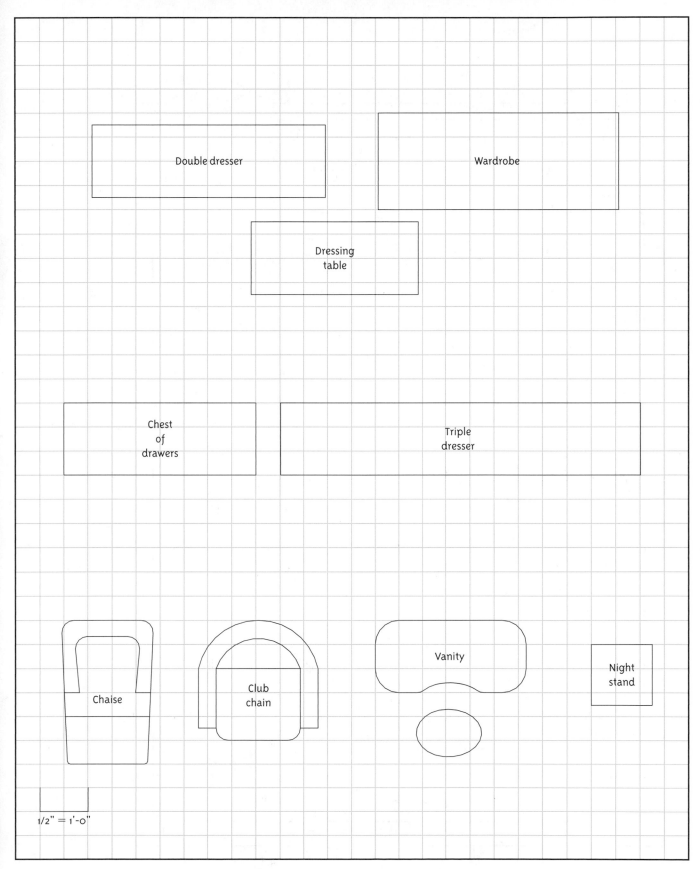

Double dresser

Wardrobe

Dressing table

Chest of drawers

Triple dresser

Chaise

Club chain

Vanity

Night stand

1/2" = 1'-0"

Figure 6-92 (*continued*)

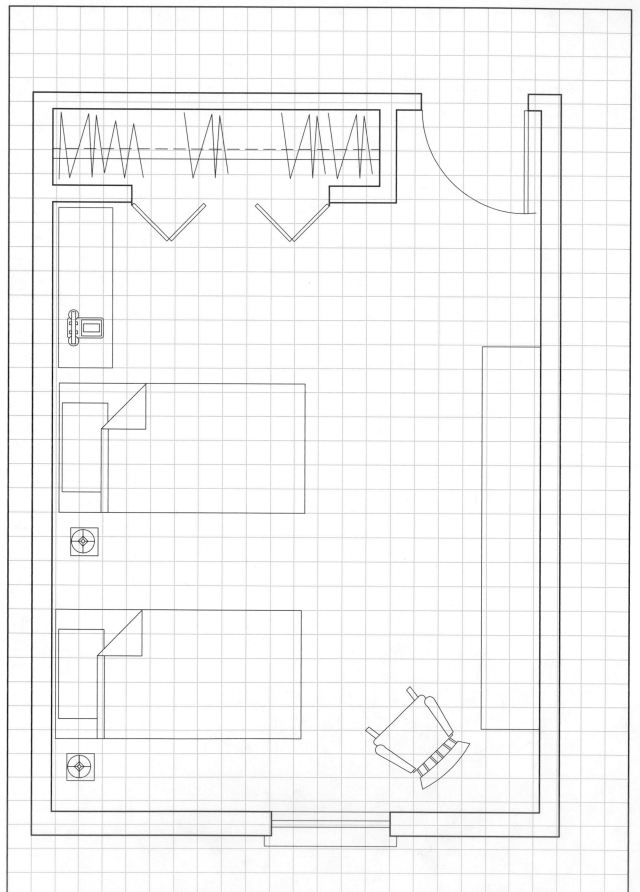

Figure 6-93
Bedroom clearances on a ½"=1'-0" grid.

Closets

There are many different kinds of closets: wardrobe (clothing); pantry (food); broom (cleaning supplies); linen (bedding and towels); and utility (furnace, water heater, washer and dryer) (Figure 6-94). A closet's dimensions need to accommodate what will go in it. Closets should be at least 24" deep and 36" wide; 30" deep and 48" wide is preferable. Shelf depth should not more than 16". For wardrobes, a shelf height of 5' is suitable for shirts, pants, and dresses; 6' is better for coats and gowns; and 40" to 48" is needed for accessibility and for children's areas (Figure 6-95). Closet packages with adjustable rods and shelves are available from several companies. Locate closets on interior walls to provide a noise barrier between rooms; place a coat closet near the entry. The space under a staircase can house a small closet or shelves. A 3'-0" wide stairway offers enough space for storage without compromising the stairs' structural support.

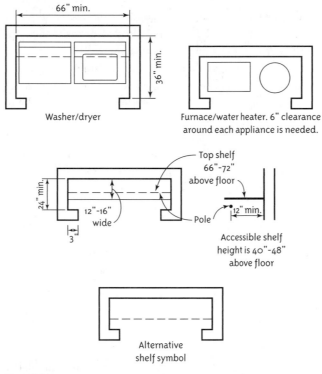

Figure 6-94
Closet clearances.

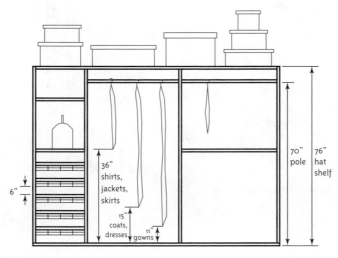

Figure 6-95
Efficient closet configuration.

Laundry Room

Historically, the laundry room (Figure 6-96) was in the basement or on a back porch. This made sense, as those spaces accommodated drips from a wringer-washer and provided a place to hang clothes in bad weather. Today, finding a handy location and creating an efficient layout for this heavily used space is desirable. In addition to the washer and dryer, components and options include counter space, sink, drying bar (normally placed over the sink), storage, shelving, a built-in ironing board, and a laundry chute. Figure 6-97 shows miscellaneous items likely to be found in a laundry or storage room.

Placement near the kitchen is popular, but second-floor laundry rooms are also useful because bedrooms are a large source of dirty clothes. The laundry room should not be located in a traffic area.

Leave 48" in front of each appliance to walk around open doors. If you are designing cabinetry around the appliances, clearance for a top-loading washer lid plus 6" is needed. Do not customize countertops and cabinetry for front-loaders or make them in the exact appliance size, because the current occupants might upgrade or the future occupants may have top-loaders and larger appliances. Coordinate the swing of front-loading appliance doors with the location of their piping connections so that the doors open opposite each other. The dryer vent should be short and straight, as efficiency is lost with each turn and foot of length. Stacked washer/dryer units are available for small spaces. A laundry sink should be deeper than a lavatory; 18" wide and 9" deep with a single-lever, high-neck faucet is appropriate.

Provide a floor drain to accommodate flooding from a broken hose. It can be in a shower base with the washer positioned over it, or in the center of the room with the floor slightly sloped (1/4" per foot) towards it.

Foyers and Mudrooms

The foyer is the home's front entry (Figure 6-97). If it is a separate space, it will serve as a buffer between front-door traffic and the living spaces. As the first area seen when entering, it should be an inviting space, containing a coat closet and room for accessorizing décor such as a bench, rug, artwork, or plants.

Figure 6-96
Spacious laundry room containing cabinets, poles, and counter space. *Courtesy of Bosch Home Appliances.*

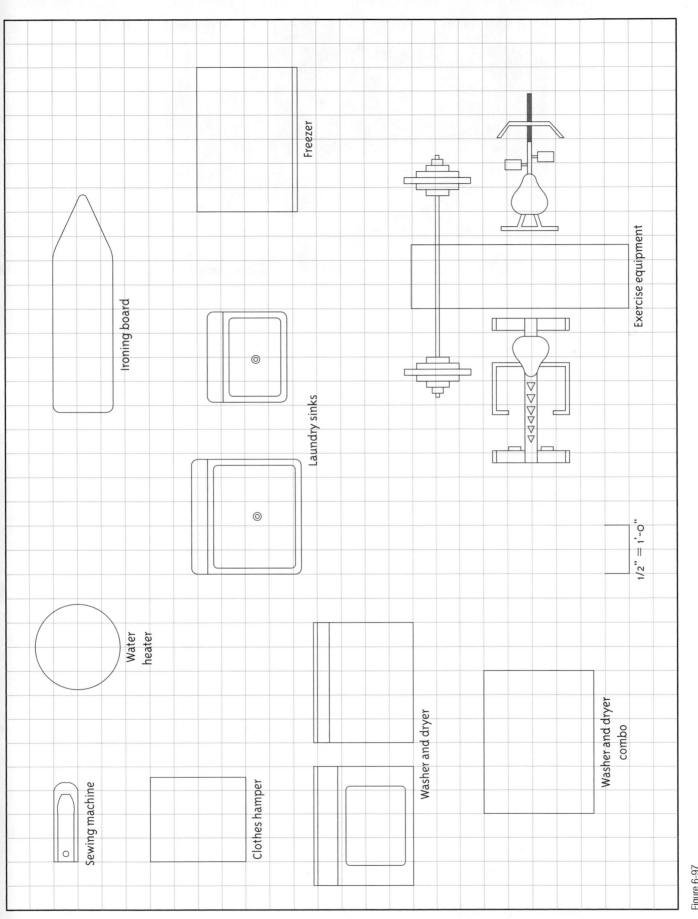

Figure 6-97
Miscellaneous household items on a ½" = 1'-0" grid.

Figure 6-98
An inviting foyer. *Courtesy of Flexsteel.*

Mudrooms are back entrances that offer a place to clean up. An elaborate one might have a shelf to store wet umbrellas, a bench to sit on when removing boots, pegs to hang up keys, a wall-hung sink, shoe racks, cubbies for backpacks and purses, vertical lockers for sports equipment, an extra freezer, a dog kennel, a closet for seasonal clothes, a hamper, or a recessed doormat. A central drain will allow the floor to be hosed down; a garden hose placed outside near the door facilitates this plus allows equipment and pets to be rinsed off. Figure 6-98 gives generic sizes of some common household equipment.

Foyers and mudrooms are high-traffic areas subject to a lot of dirt, hence they shouldn't be combined with other rooms. Doors to the outside should be 36" wide.

Summary

Space planning is an integral part of the design process. It involves synthesizing project requirements and researching sizes and clearances. After all such factors are known, a hard-line floor plan can be drawn.

Suggested Activities

1. Draw a bubble diagram for a building of your choice.
2. Research items that will go inside the building in no. 1, and list their size/clearance requirements. Use the references listed at the end of this chapter.
3. Study floor plans from magazines or online sites for their layout and traffic patterns.
4. Research some ergonomic hand tools and list the ways they differ from non-ergonomic tools.
5. Research specific ways in which proxemic data is used in floor plans.

Questions

1. What is done during the synthesis stage of a floor plan design?
2. What is the kitchen work triangle?
3. Name one way proxemics is used in space planning.
4. What is a desirable accessible base cabinet height?
5. Name three types of bathrooms.

6. What is the NKBA's recommended clearance for the center of the toilet to a side wall?

7. What is a vanity?

8. Name three requirements for a comfortable home office.

9. What is the FHA minimum recommended area for a bedroom?

10. What should a minimum closet depth be?

Further Reading and Internet References

Galvin, P. (1997). *Bathroom Basics: A Training Primer for Bathroom Specialists.* Hackettstown, NJ: NKBA.

Galvin, P. (1998). *Kitchen Basics: A Training Primer for Kitchen Specialists.* Hackettstown, NJ: NKBA.

Guthrie, P. (2004). *Interior Design Portable Handbook : First-Step Rules of Thumb for Interior Architecture* (2nd ed.). New York: McGraw-Hill Professional.

Karlen, M. (2004). *Space Planning* (2nd ed.). Hoboken, NJ: Wiley.

Kumlin, R. (1995). *Architectural Programming: Creative Techniques for Design Professionals.* New York: McGraw-Hill.

A.I.A. (2000) *Architecture Graphic Standards* (10th ed.). Hoboken, NJ: Wiley.

Reznikoff, S. C. (1986). *Interior Graphic and Design Standards.* (need town): Watson-Guptill.

http://www.americanstandard-us.com
American Standard website contains information and downloadable CAD details for American Standard plumbing products.

http://www.bassettfurniture.com/
Bassett Furniture's website has a room planner feature with a space planning tool.

www.boconcept.co.uk/
BoConcept's website contains a downloadable file for drawing floor plans and 3D renderings viewed from different angles.

http://www.boschappliances.com/
Bosch website; contains information on their appliances.

http://www.dreamhomesource.com
Source for floor plans.

http://www.eplans.com
Source for floor plans.

http://www.ikea.com/
Ikea Furniture's website has a downloadable file for creating a kitchen planning tool.

http://www.jordans.com/
Jordan's Furniture website. Room planner feature contains generic furniture templates.

http://www.keidel.com/
Bath, plumbing, kitchen planning guide.

http://www.us.kohler.com
Information on Kohler kitchen and bath fixtures, and downloadable AutoCAD details for Kohler plumbing products.

http://www.letsrenovate.com/
Home remodeling guide website. Contains information on all rooms of a house.

http://www.wbdg.org/
Whole Building Design website; contains discussions of the building design process and different building space types.

CHAPTER 7

Basic Building Construction

OBJECTIVE

- This chapter discusses basic building construction; building components, how they go together, terminology, and the drawings that show them.

KEY TERMS

actual size	frost line	precast
arch	girder	rafter
balloon framing	grade	rebar
basement plans	grade beam	ribbon board
beams	joinery	rigid frame
bond	load-bearing	sheathing
cavity wall	lintel	slab-on-grade
cast concrete	loads	space frames
clear spans	masonry	spandrels
construction drawings	modular	specifications
course	nominal size	spread footing
curtain wall	on center (OC)	steel sections structural
decking	partition	backing
details	piers	stud
dimensional mill lumber	pilasters	substructure
engineered wood products	plate	superstructure
fabricated members	platform framing	truss
footings	post-and-beam framing	veneer
foundation	post-on-pier	wythe
foundation plans	posts	

Knowing a home's general components (Figure 7-1) facilitates competent design. You may not draw the plumbing or mechanical systems, but the ability to interpret them will enable you to check if fixtures and appliances you've specified have been acknowledged by others involved in the project, and if all is in place to receive them. For instance, before hanging a heavy, wall-mounted sink or ceiling-mounted light, there must be blocking to attach to. Identifying structural vs. non-structural walls avoids removal of structural elements during a renovation. Communication with tradespersons will be facilitated if the designer has a working knowledge of construction vocabulary and building parts.

Figure 7-1
Knowing the components of a building and how the pieces go together facilitates competent design.

A building consists of its **substructure** and **superstructure**. The substructure is everything below **grade**, or ground. The superstructure is everything above grade. The superstructure is the building's structural system and everything that is hung on it.

Superstructures are made of wood, concrete, or steel; most buildings contain all three but are primarily constructed of one. Homes, apartment complexes, and light commercial buildings typically have wood or steel frames. Larger buildings are made of concrete or steel, because wood framing is not strong enough for the **loads** and **clear spans** (interior spaces unobstructed by columns or load-bearing walls) required. An additional factor is that wood does not meet the more stringent fire codes.

Structural engineers design the foundation. They also determine the size of and distance between all structural components needed to support the building's stresses and loads, or required by applicable building codes.

The Foundation

The foundation is the base upon which a building is placed. It provides a level surface to build on, forms the basement walls, carries the building's **loads**, or weights, and keeps moisture-sensitive materials off the ground to prevent rot and insect infestation. Foundations are designed to fit the needs of the structure. There are different kinds depending on geographic location, site, soil, building type, and code requirements. **Foundation plans** (Figure 7-2), **basement plans** (Figure 7-3), and their accompanying **details** (Figure 7-4) describe the foundation's construction. Unlike floor plans, these are drawn from top-of-the-wall level. The foundation plan shows walls, footings, grade beams, and pilasters. The basement plan shows these elements plus interior spaces.

Before discussing foundation types, it is helpful to know some related construction terms.

Foundation Terms

Concrete Concrete is a mixture of cement, water, aggregate, and possibly admixtures. Aggregate is a blend of sand, rock, crushed gravel, and cinder ash, all of which provide different strength levels. An admixture is a chemical that makes concrete stronger or more workable.

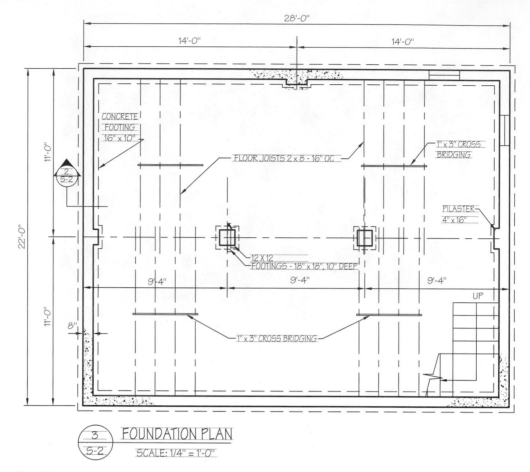

Figure 7-2
Foundation plan. The section symbol links the plan to a detail drawing.

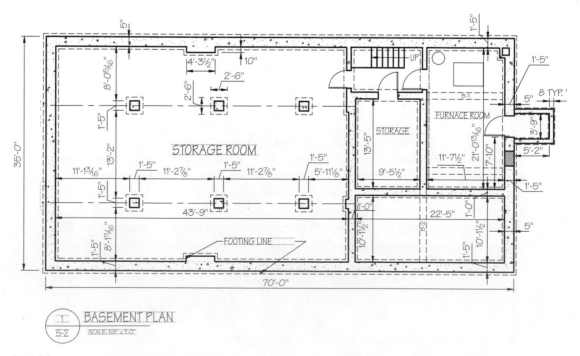

Figure 7-3
Basement plan. It shows the same information as a foundation plan plus interior spaces.

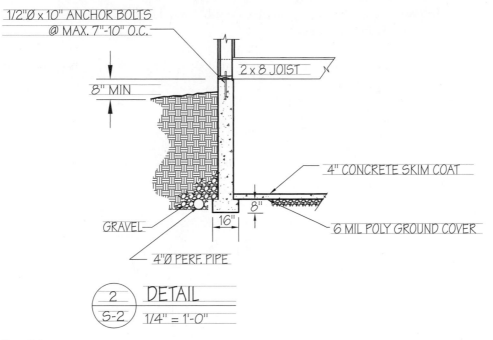

1/2"Ø x 10" ANCHOR BOLTS
@ MAX. 7'-10" O.C.

8" MIN

2 x 8 JOIST

4" CONCRETE SKIM COAT

GRAVEL

16"   8"

6 MIL POLY GROUND COVER

4"Ø PERF. PIPE

2 / S-2   DETAIL
1/4" = 1'-0"

Figure 7-4
Detail drawing referenced in Figure 7-2.

Footing A footing (Figure 7-5) is the widened bottom of a foundation wall, pier, or column. It supports and distributes the weight of the loads resting on it.

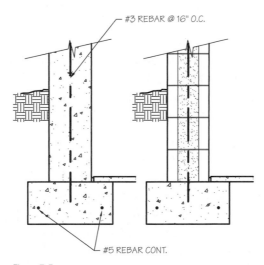

#3 REBAR @ 16" O.C.

#5 REBAR CONT.

Figure 7-5
Footings; one supporting a cast-in-place concrete wall, the other supporting a concrete block wall. A footing base is twice the width of the wall.

Structural Structural means load bearing. A structural member carries the weight of other components. It cannot be easily removed; if it is, the loads it carries must be distributed another way. A building's structural system is the frame that holds it up. A **load-bearing** wall is a wall that carries the weight of an upper story or roof, as opposed to a **partition** wall, which is an interior, non-structural wall that only carries its own weight.

Beam: A **beam** is a horizontal structural member.

Grade beam A **grade beam** is the portion of a slab that is thicker than the rest. Its purpose is to support load-bearing walls above it.

Pier A **pier** (Figure 7-6) is a short post found under buildings, as in crawlspaces or porches.

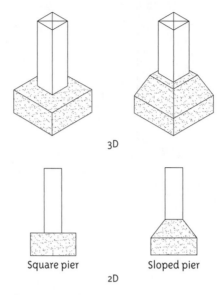

Figure 7-6
Pier types.

Post A **post** is a vertical, structural member outside a wall that supports **beams**. Posts run from floor to floor.

Column A column is a vertical, structural member outside a wall that supports beams. Columns run continuously from ground to roof.

Pilaster A **pilaster** (Figure 7-7) is a post or column attached to a wall. Its purpose is to strengthen the wall where heavy beams will rest.

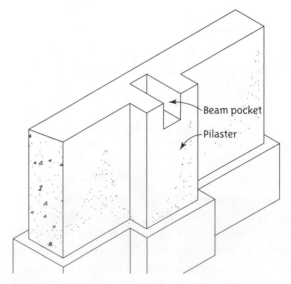

Figure 7-7
Pilaster and pocket to support a heavy beam.

Foundation Types

The most common foundation types are the **slab-on-grade**, T, and **post-on-pier.** Most buildings combine elements of all three (Figure 7-8).

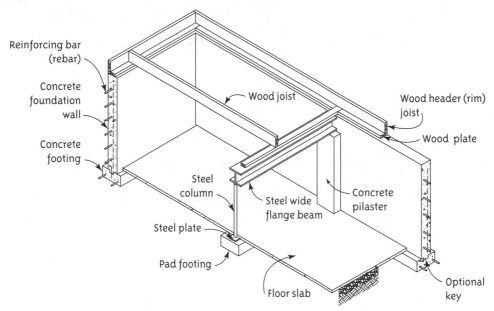

Figure 7-8
Pictorial showing grade beam, post, pier, column, foundation wall, and footing.

Slab on grade   Also called a monolithic foundation, this is a concrete slab on the ground (Figures 7-9, 7-10, 7-11). A shallow excavation may be made, or the slab may be poured into forms above ground. The perimeter is poured deeper than the rest of the slab to support the weight of load-bearing walls, which bear the weight of the upper floors and roof. There may be intermediate grade beams to support load-bearing walls.

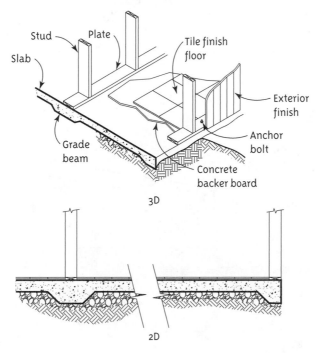

Figure 7-9
3-D and 2-D views of a slab-on-grade foundation with an interior grade beam.

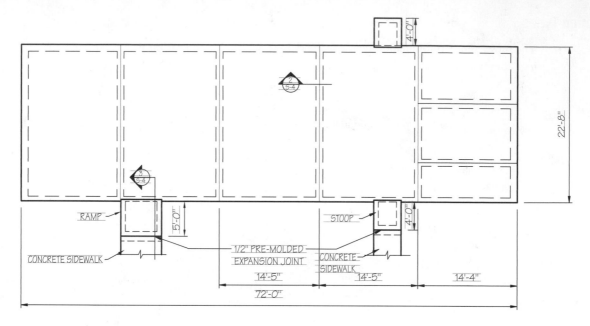

SLAB ON GRADE
SCALE: 1/8" = 1'-0"

1 / S-2

Figure 7-10
Slab-on-grade foundation with section symbols referencing details.

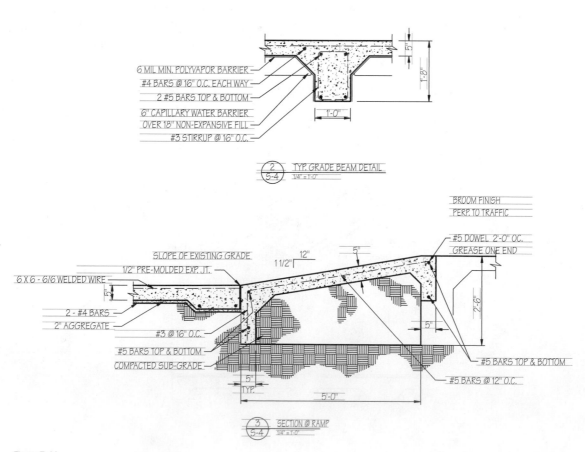

6 MIL. MIN. POLYVAPOR BARRIER
#4 BARS @ 16" O.C. EACH WAY
2 #5 BARS TOP & BOTTOM
6" CAPILLARY WATER BARRIER
OVER 18" NON-EXPANSIVE FILL
#3 STIRRUP @ 16" O.C.

2 / S-4 TYP. GRADE BEAM DETAIL
1/4" = 1'-0"

BROOM FINISH
PERP. TO TRAFFIC

SLOPE OF EXISTING GRADE
1/2" PRE-MOLDED EXP. JT.
6 X 6 - 6/6 WELDED WIRE
2 - #4 BARS
2" AGGREGATE
#3 @ 16" O.C.
#5 BARS TOP & BOTTOM
COMPACTED SUB-GRADE

#5 DOWEL 2'-0" OC.
GREASE ONE END

#5 BARS TOP & BOTTOM
#5 BARS @ 12" O.C.

3 / S-4 SECTION @ RAMP
1/4" = 1'-0"

Figure 7-11
Details for the slab-on-grade foundation in Figure 7-10.

T Also called a **spread footing** or perimeter foundation, this is a wall built on top of a footing, which is a wide base. (Figures 7-12, 7-13, 7-14). In cross-section this system looks like an upside-down T. It enables the construction of basements and crawlspaces. The footings are made of cast concrete and placed below the **frost line,** the level at which the ground no longer freezes (this varies with geographic location). The foundation wall is usually made of block or cast concrete, but in rare cases is made of wood, brick, or stone. Sometimes the foundation wall requires **pilasters.**

Figure 7-12
Forms and rebar for the footings of a T foundation. *Courtesy of strawbale.com.*

Figure 7-13
Concrete is poured and leveled to a smooth surface on which the foundation walls are built.

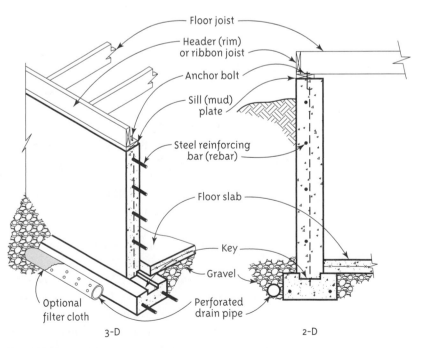

Figure 7-14
3-D and 2-D views of a T foundation.

Pad-and-pier, pad and column, footing and pier Piers and columns are vertical structural members on square footings that support beams (Figure 7-15). Piers are shorter than posts and are located under a house, often in a crawl space or porch. Both are typically used in conjunction with other foundation types. Piers are usually made of wood and posts are usually made of metal. Piers and posts on pads are sometimes called girder jacks or girder posts.

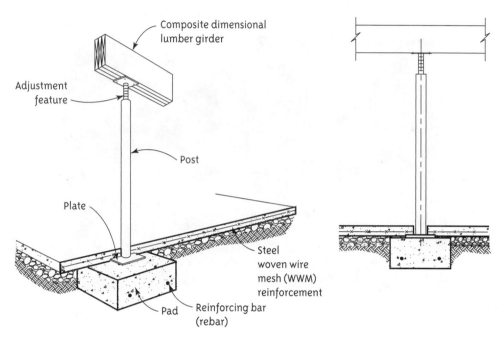

Figure 7-15
Round steel pipe column on a square pad, supporting a beam.

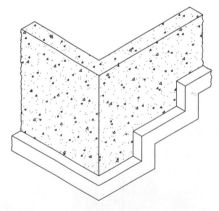

Figure 7-16
Stepped footing, appropriate for building into a hill.

Others There are other foundation types, such as stepped footings (Figure 7-16) for hilly or sloped sites. Piles are piers without footings, used for extremely heavy buildings such as skyscrapers. Preservative-treated wood foundations are occasionally used, because their design prevents the kind of moisture problems that conventional basements may have. Some foundations are made of brick or stone. Wood, brick, and stone foundations are more commonly found in older homes but are rarely used in new construction.

Foundation Materials Cast or poured concrete and concrete block are the materials most commonly used to make foundations. Concrete is a mixture of cement, water, and aggregate, sometimes enhanced with admixtures. Cement is an ingredient of concrete; the two are not synonymous. Concrete block (Figures 7-17, 7-18), also called concrete masonry units (CMU), is a **precast** item, meaning it is poured in a form at a factory and shipped to the site. Concrete block is available in different shapes, sizes and weights, and typically is manufactured in lengths of 16", heights of 8" and widths of 4", 6", 8", 10", and 12". These are **nominal sizes**; the **actual sizes** of concrete blocks are 3/8" smaller to accommodate a mortar joint of that size. Thus, a block may be nominally 8" × 8" × 16" and called out as such, but will actually measure 7⅝" × 7⅝" × 15⅝" and will be drawn that size. Cast concrete is mixed in a drum on site and poured into forms, whose shape it takes. Concrete can also be used for decorative purposes in the superstructure (Figure 7-19). All concrete is reinforced with steel in the form of reinforcing bars (called **rebar**) or wire mesh.

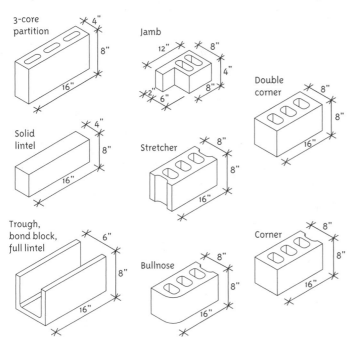

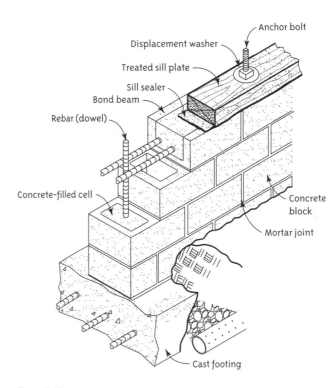

Figure 7-17
Different block types and their modular (nominal) sizes, which vary by manufacturer. A stretcher block is the most common and is laid with its length parallel to the face of the wall. A double corner is used for piers, pilasters, or any use where both block ends are visible. A bullnose serves the same purpose as a corner block but is used where round corners are preferred. A jamb block is combined with stretcher and corner blocks around window openings; the recess allows placement of the casing.

Figure 7-18
A concrete block wall. It is strengthened with reinforcing bar (rebar) and a bond beam, which is a course of bond blocks.

Figure 7-19
Concrete can be used for the superstructure as well as the substructure. Here, thinset concrete is used to create a sculptural wall, and for stairs. *Courtesy of Bomonite Corporation, Madera, CA.*

Wood Construction

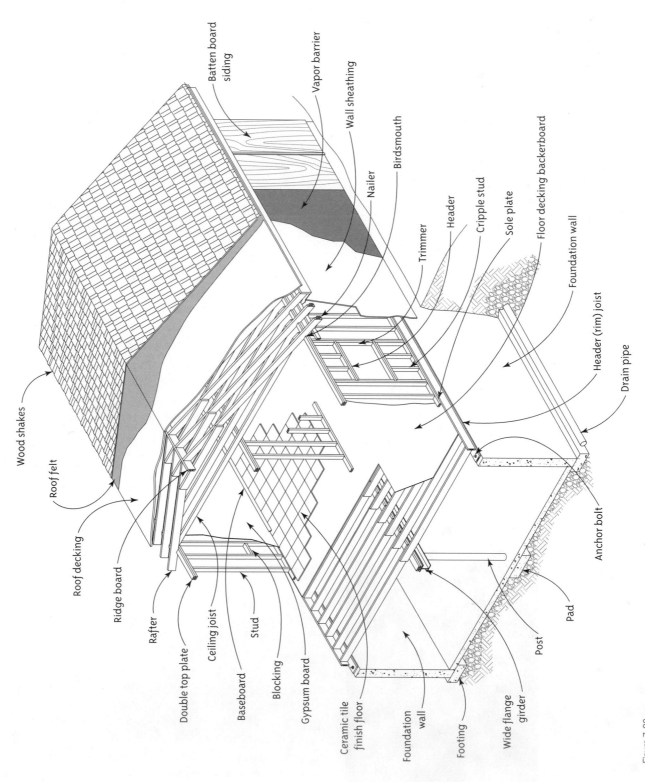

Batten board siding

Vapor barrier

Wall sheathing

Nailer

Birdsmouth

Trimmer

Header

Cripple stud

Sole plate

Floor decking backerboard

Foundation wall

Header (rim) joist

Drain pipe

Anchor bolt

Pad

Post

Wide flange girder

Footing

Foundation wall

Ceramic tile finish floor

Gypsum board

Blocking

Baseboard

Stud

Ceiling joist

Double top plate

Rafter

Ridge board

Roof decking

Roof felt

Wood shakes

Figure 7-20
Components of a wood skeleton–framed house.

Wood Construction Terms

Before a discussion of wood framing can commence, some component definitions are needed. Figure 7-20 shows the major components in a wood-framed house.

Dimensional Lumber Most wood used for framing is **dimensional mill lumber** (Figures 7-21, 7-22, 7-23). This is usually wood from conifer trees such as hemlock, Douglas fir, or Southern yellow pine. It is cut to specific sizes such as 2" × 4" or 2" × 6". Those are nominal or mill sizes; the actual dimensions are smaller. A 2" × 4" is actually 1½" × 3½".

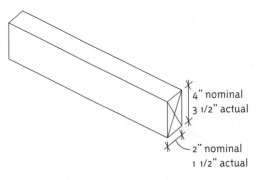

Figure 7-21
A piece of dimensional lumber.

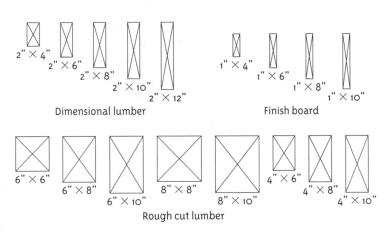

Figure 7-22
Cross-sections of different beam types.

LUMBER SIZES IN INCHES	
NOMINAL	**ACTUAL**
2" × 4"	1½ × 3½
2" × 6"	1½ × 5½
2" × 8"	1½ × 7½
2" × 10"	1½ × 9½
2" × 12"	1½ × 11½
4" × 6"	3⁹⁄₁₆ × 5½
4" × 8"	3⁹⁄₁₆ × 7½
4" × 10"	3⁹⁄₁₆ × 9½
6" × 6"	5½ × 5½
6" × 8"	5½ × 7½
6" × 10"	5½ × 9½
8" × 8"	7½ × 7½

Figure 7-23
Chart of actual and nominal dimensional lumber sizes.

Engineered Wood Products (EWP) **Engineered wood products** are wood veneers and fibers that have been laminated (adhered and pressed together) to produce longer-spanning, greater-load-bearing pieces than dimensional lumber. Their advantage is that they can be made into strong, aesthetically interesting shapes and can have holes cut in them for running ducts and conduit. Examples are glue laminated (Glulam) lumber and laminated veneer lumber (LVL).

Glulam is made of thin, parallel layers or **veneers** of wood bent to the desired shape and then glued and clamped together under pressure (Figures 7-24, 7-25, 7-26). When the glue dries, the piece retains its bent form. Since multiple pieces can be joined together at their ends, almost any length and depth can be produced. LVL is made of cellulose fiber strips. Fibers from poor-quality or small trees, rice, rye, wheat, or straw are crushed into long strands. After being combined with formaldehyde and adhesives, the strands are woven, compressed, and heat treated. The finished shapes are strong, straight, and can span long distances. They are used mostly for columns, beams, and large headers (Figure 7-27).

Figure 7-24
Glulam arches provide the support for this hockey rink. *Courtesy of the APA-Engineered Wood Association, Tacoma, WA.*

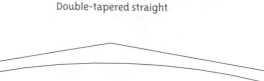

Double-tapered straight

Double-tapered curved

Figure 7-25
Two of the many Glulam beam shapes available.

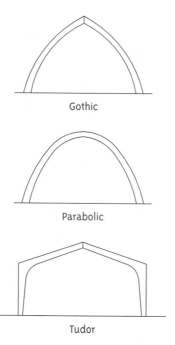

Gothic

Parabolic

Tudor

Figure 7-26
Three of the many Glulam arch shapes available.

Figure 7-27
Engineered wood products, such as the LVL components here, enable heights, open plans, and strengths unattainable with dimensional lumber. *Courtesy of i-Level by Weyerhaeuser, Federal Way, Washington.*

Built-up or composite items are engineered wood products that consist of separate pieces glued or bolted together; wood I-joists and trussed rafters are examples (Figure 7-28).

Oriented Strand Board This is an engineered wood product made of layers of wood and glue pressed together to create 4' × 8' panels. It is often used for sheathing, sub-flooring, and siding.

Beam This is an umbrella term for a horizontal load-bearing member (Figure 7-29). It can be wood, steel, concrete, or a composite of those materials. A beam has different names depending on where it is in the building; a **girder** is a large beam that supports smaller beams; a **rafter** is an inclined beam at the roof; **a joist** is a horizontal beam in ceilings and floors; and a **lintel** (Figure 7-30) is a beam over a door or window. Solid wood beams made of dimensional

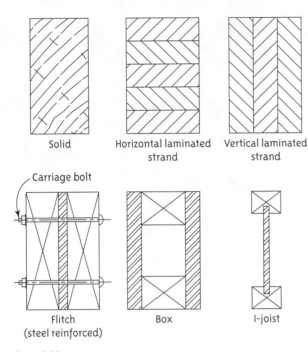

Solid

Horizontal laminated strand

Vertical laminated strand

Carriage bolt

Flitch (steel reinforced)

Box

I-joist

Figure 7-28
Cross-sections of different beam types.

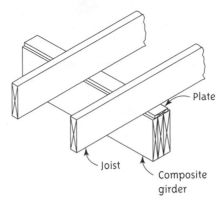

Plate

Joist

Composite girder

Figure 7-29
Beams. The bottom one is a girder, because it supports the smaller joists.

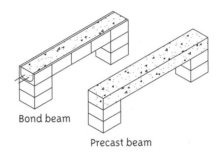

Bond beam

Precast beam

Figure 7-30
Concrete lintels (a lintel is a beam over a door or window opening).

lumber are available in thicknesses ranging from 3" to 12", but sizes over 8" tend to warp. Built-up beams, EWPs and **fabricated members** are used instead. A built-up beam is made of multiple pieces, resulting in a stronger, more stable beam than one made of a solid piece of wood (Figures 7-31, 7-32).

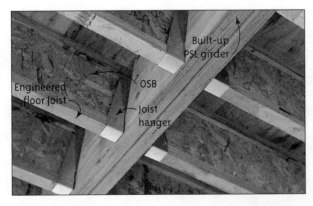

Built-up PSL girder

Engineered floor joist

OSB

Joist hanger

Figure 7-31
I-joists supported by an PSL girder. *Courtesy of the APA- Engineered Wood Association, Tacoma, WA.*

Figure 7-32
LVL header. *Courtesy of i-Level by Weyerhaeuser, Federal Way, Washington.*

Trussed rafter Also called a **truss**, this is a fabricated member placed at the roof (Figure 7-33). It consists of an upper chord, a lower chord, and a web. Gusset plates hold the pieces together.

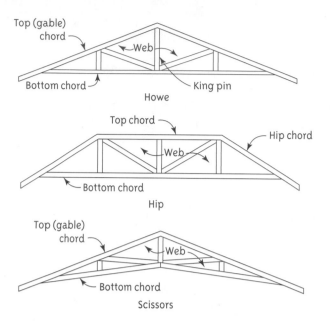

Howe

Hip

Scissors

Figure 7-33
Examples of the many truss types available.

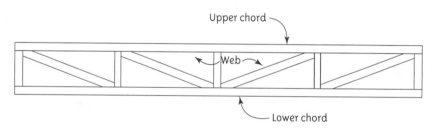

Figure 7-34
Scissors trusses being installed. *Courtesy of i-Level by Weyerhaeuser, Federal Way, Washington.*

Trussed joist Also called a flat truss, this is a fabricated member similar to a trussed rafter except that the upper and lower chords are parallel (Figures 7-35, 7-36). It's often used as a header. Residential designers used them as floor joists.

Figure 7-35
Trussed joist.

Figure 7-36
Components in a wood framed wall. Note the trussed joist used as a header for a window opening. Fabricated beams such as this allow longer clear spans due to their greater load-bearing capability.

Stud A **stud** is a vertical load-bearing member inside a wall (Figures 7-37, 7-38). Cripples are short studs placed above or below a wall opening.

Plate A **plate** is a horizontal board. Bottom plates evenly distribute loads placed on them; top plates tie studs together. Plates can be found throughout a building and are named according to their location; for instance, mudsill or sole plates are at ground level, sill plates are at the bottom of windows, and double top plates are at ceilings and above doors.

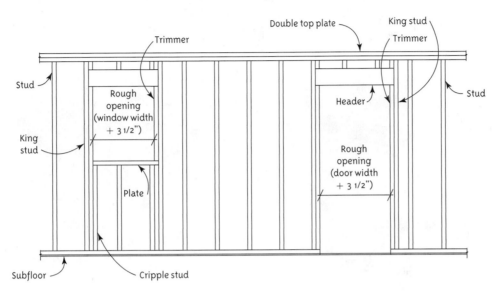

Figure 7-37
Wood framed wall.

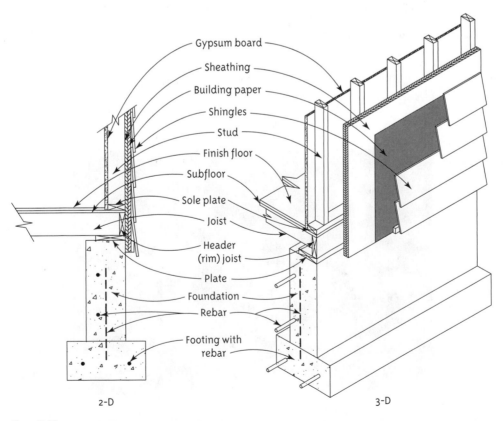

Figure 7-38
2-D and 3-D views of a wood framed wall.

Sheathing, Decking, Sub-floor, Underlayment **Sheathing** is a vertical covering of boards on exterior walls that goes under the final finish. **Decking** is a horizontal covering of boards on the roof or floor (Figures 7-39, 7-40, 7-41, 7-42). Common decking and sheathing materials are OSB, plywood, concrete slab, and corrugated steel sheets. Floor decking is sometimes referred to as the sub-floor, since it is a layer of board over the floor joists. An underlayment, which is a separate layer of plywood or particle board, and the finish floor are applied over the sub-floor. Often the sub-floor and underlayment are combined and are made of one layer of material. Underlayment is used under hardwood or to level uneven surfaces. Concrete backer board is used under ceramic tiles.

Figure 7-39
Floor decking over engineered wood joists. *Courtesy of i-Level by Weyerhaeuser, Federal Way, Washington.*

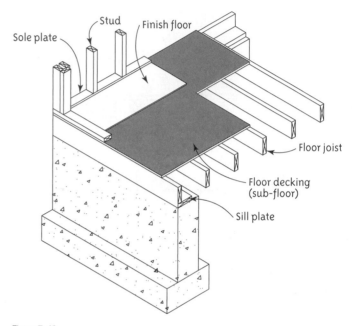

Figure 7-40
Pictorial of floor construction.

Figure 7-41
Tile finish floor adhered to a concrete backer board sub-floor. *Courtesy of United States Gypsum Company, Chicago, IL.*

Figure 7-42
Roof decking.

Wood Frame Types

Wood framed buildings predate stone ones. Archeological evidence of this exists in ancient Egypt, and there is also evidence that Greek stone temples imitated earlier wood ones. While most frame structures were destroyed over the millennia through the ravages of fire and war, this construction method can be long lasting, as the millions of medieval wood-framed buildings still in use around the world as churches and homes attest to (Figure 7-43).

There are two categories of wood framing: **post-and-beam** (Figure 7-44) and skeleton. Post-and-beam, also called timber framing, is the oldest framing method and was the method of wood building used throughout the world for 2,000 years. Trees were cut, hewn into square and rectangular shapes, and connected with mortise-and-tenon or dovetail notches to create big and small buildings. The art of **joinery**, or connection technique (Figures 7-45, 7-46), became a skilled craft. Over the centuries, each country developed its own style of timber framing. In fact, axe and adze marks, types of joints, and shapes of hewn logs have been used to identify the builder's home country and year built of American log cabins.

Figure 7-43
The Paul Revere house, Boston, MA. It was built in 1680, and is of post-and-beam construction. *Courtesy of the Paul Revere Memorial Association.*

Figure 7-44
Components in a post-and-beam framed house. *Courtesy of Timbersmith, Bloomington, IN.*

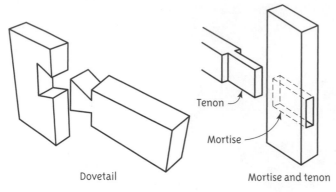

Figure 7-45
Traditional wood joints. A mortise-and-tenon joint consists of a 'tongue', (the tenon) in one board that fits into a hole (the mortise) in another. A dovetail joint consists of a trapezoidal-shaped pin that interlocks with a trapezoidal-shaped tail cut into another board.

Figure 7-46
Modern post-and-beam joinery. *Courtesy of the APA- Engineered Wood Association, Tacoma, WA.*

Post-and-beam framing is still used today for specialty applications and is characterized by large wood (e.g., 6" × 6") rough-cut structural elements spaced far apart, such as on 4-foot centers (Figures 7-47, 7-48). Most of the building's weight is carried by the posts. This makes the walls **curtain walls,** which are non-load bearing walls whose purpose is to shield and enclose the building. Post-and-beam construction enables larger windows and clear spans than are possible with other framing types.

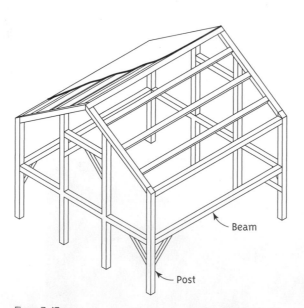

Figure 7-47
Post-and-beam framing.

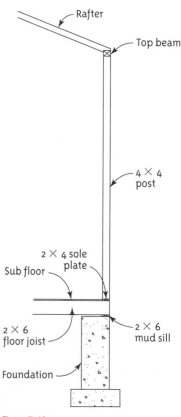

Figure 7-48
Post and beam wall section.

The 19th century brought forth the Industrial Revolution, and machines and factories were built. Sawmills produced smaller, lightweight, **modular**-sized pieces of lumber, which now could be produced in standardized sizes and dimensions. This lumber could be handled more easily and assembled more quickly and cheaply. Factories began mass-producing nails.

The result was skeleton or stick framing, a method of wood construction that suited the needs of the young, westward-growing United States better than post-and-beam technology, whose massive timbers were hard to transport and required more skill to erect.

Modular construction seeks to use as many standardized and mass-produced components as possible, which helps eliminate cutting and waste. Floor, ceiling, and wall materials are manufactured in 4' widths. This dimension should be incorporated into the design and cost estimating processes.

There are two categories of skeleton framing: balloon and platform.

Balloon framing originated in Chicago in the 1830s. It is characterized by long studs that run continuously 20' high or more (Figures 7-49, 7-50). Second floors rest on a **ribbon board** attached to the studs. While the balloon method is still used in occasional stucco or masonry veneer buildings, the long studs required are expensive and harder to find, so starting in the 1930s it was largely replaced with **platform framing** (Figures 7-51, 7-52). This is the standard for most modern homes. Walls are built on top of the floors, not attached to them, with the floor joists providing a platform on which to work. The studs are not continuous from the floor of the first level to the ceiling of the second.

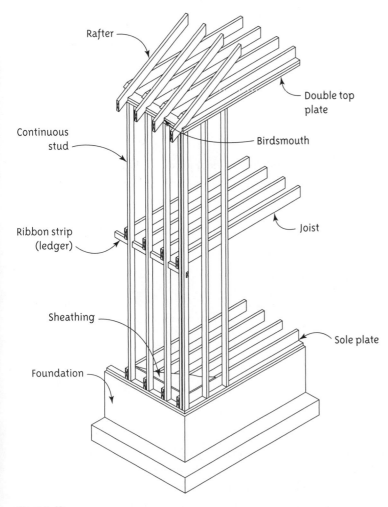

Figure 7-49
Pictorial of balloon framing.

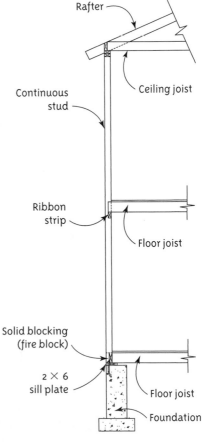

Figure 7-50
Balloon framed wall section.

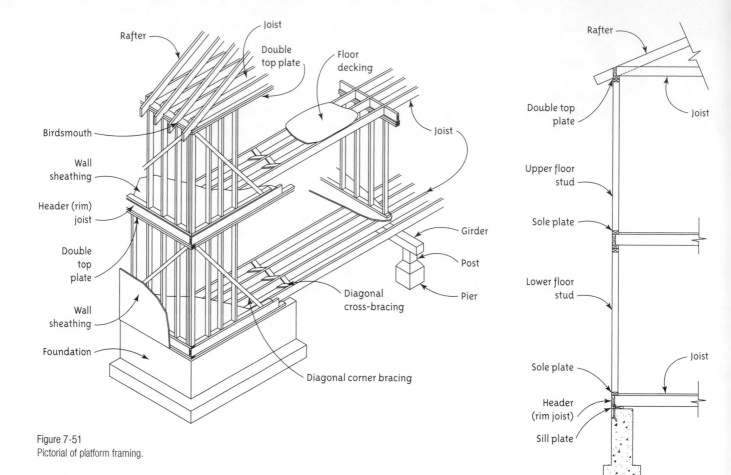

Figure 7-51
Pictorial of platform framing.

Figure 7-52
Platform framed wall section.

In both balloon and platform framing, wall studs, ceiling rafters, and floor joists are placed every 16" or 24" **on center,** abbreviated **O.C.** (Figure 7-53). This is the distance from the center of each face of a structural member. Such spacing enables the use of modular materials.

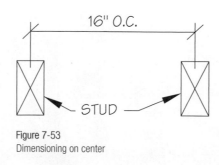

Figure 7-53
Dimensioning on center

Older homes usually have 2" × 4' studs spaced 16" O.C.; newer homes have 2" × 6" studs spaced 16" or 24" O.C., depending on the requirements of building codes or the structural design (Figure 7-54). Within the frame are elements including cross-bracing, which strengthens the walls; cross-bridging, which strengthens the floors; and blocking, which provides a strong surface from which to hang heavy wall- and ceiling-mounted objects.

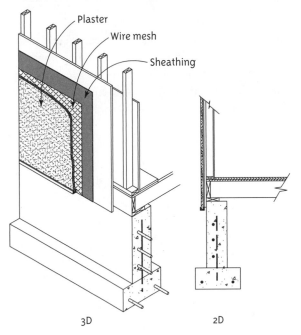

Figure 7-54
3-D and 2-D sections of a wood frame exterior wall with a stucco finish.
It is applied over wire mesh.

Masonry

Masonry is an umbrella term for stone, brick, structural clay tile, glass block and terracotta building materials (Figures 7-55, 7-56). Like wood, the use of masonry materials for construction is as old as civilization. The Egyptians mixed mud and straw to make bricks in 3000 BC. They also used mortar to build the pyramids. The Chinese used cement-like materials to hold bamboo boats together and to build the Great Wall. Sophisticated stone construction technologies evolved in the last few centuries B.C. The ancient Romans used concrete; pozzolana cement was mined in Pozzuoli, Italy to build the Appian Way, the Coliseum and Pantheon, and Roman baths. Today, masonry units are available in dozens of textures, colors, and finishes, both hand-made and machine-generated, and offer many design options for walls and paving.

Figure 7-55
Courses of textured concrete block and brick combined with precast concrete sills and arches give this wall interest. *Courtesy of CertainTeed, Valley Forge, PA.*

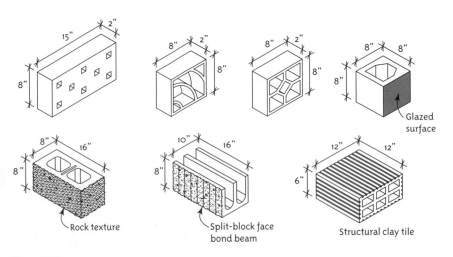

Figure 7-56
Examples of decorative concrete block and structural clay tile. Modular (nominal) sizes are given.

Masonry Terms

Bond A **bond** is an arrangement of brick or blocks in a wall (Figures 7-57, 7-58, 7-59). There are many types, all with different textural interest. The running bond is the most popular for modern construction, because its staggered joints provide good structural stability; it makes maximum use of whole bricks, minimizing the need to cut bricks on site; and it is quick to lay, hence economical. Sometimes patterns are added via contrasting brick colors.

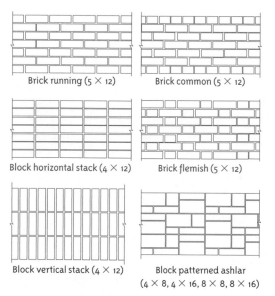

Brick running (5 × 12)   Brick common (5 × 12)

Block horizontal stack (4 × 12)   Brick flemish (5 × 12)

Block vertical stack (4 × 12)   Block patterned ashlar
(4 × 8, 4 × 16, 8 × 8, 8 × 16)

Figure 7-57
Examples of brick and concrete block bonds.

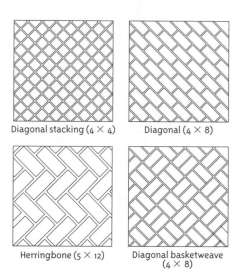

Diagonal stacking (4 × 4)   Diagonal (4 × 8)

Herringbone (5 × 12)   Diagonal basketweave
(4 × 8)

Figure 7-58
Examples of brick paver bonds.

Figure 7-59
Diagonal basketweave pattern stamped in concrete. *Courtesy of Bomonite Corporation, Madera, Ca.*

Mortar Mortar is a mixture of cement, sand, and water that hardens and is used as a masonry binding agent. A mortar joint is the bonding of mortar between masonry units that holds them together (Figure 7-60). Joints are 1/4", 3/8", or 1/2" thick and can be tooled for different appearances. The concave joint is the most common. It is a weatherproof joint, as it directs water away from the building. So do the V and weather joints. The others are mainly aesthetic and are suited for interiors. They are not weatherproof; for instance, a rake joint creates a ledge for water collection. Flush joints, common on split-ribbed concrete block, make cracks between the joint and masonry obvious.

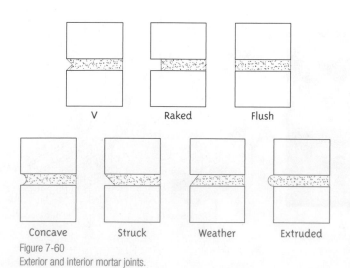

V   Raked   Flush

Concave   Struck   Weather   Extruded

Figure 7-60
Exterior and interior mortar joints.

Wythe (*pronounced* **why'*th***) A **wythe** is a continuous vertical section of a masonry wall, one unit in thickness (Figure 7-61). A single-wythe wall is composed of a single unit, or thickness, of masonry, such as a one-brick or one-block–thick wall. A double-wythe wall is composed of two units, or thicknesses, of masonry. Walls in which more than one thickness of masonry are bonded together are considered one *wythe*.

Veneer A **veneer** is a non-load bearing, aesthetic masonry facing attached to, and supported by, a **structural backing**. The foundation and concrete block backing are the load-bearing portions of this wall type.

Brick A brick is a rectangular mass of clay hardened by heat (Figure 7-62). It is the oldest, most basic unit of masonry construction. Early bricks were adobe, a sun-hardened mixture of straw and clay. The brick's size has historically been small enough to be held in the hand, a standard that remains the same today.

Individual bricks are loosely categorized as face and common (not to be confused with a common bond). Face bricks are evenly sized and colored and used where they will be exposed to view; common bricks have inconsistent color and sizing and are used where they will be hidden from view. Common and face bricks are sized differently. Common bricks come in standard, oversized, and modular sizes. Face bricks come in standard, Norman, and Roman sizes, with different core patterns in each type. As with lumber and concrete block, bricks have nominal and actual dimensions. Nominal dimensions equal the unit's actual size plus the mortar joint thickness. Nominal brick dimensions are based on multiples of 4" (or, in metric, multiples of 100 mm). When designing with this medium, wall lengths and heights should be based on modular sizes.

Brick dimensions are given as width × height × length. Height and length are sometimes called face dimensions, as these are the dimensions showing when the brick is laid as a stretcher.

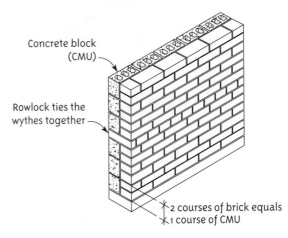

Figure 7-61
A one-wythe masonry wall. Brick is bonded to concrete via header courses.

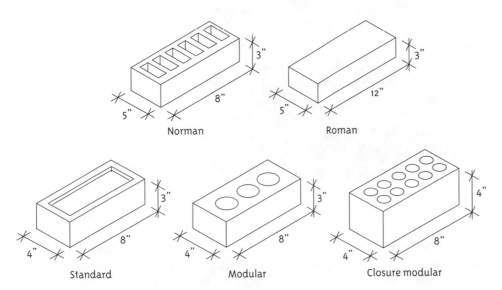

Figure 7-62
Examples of face brick types and modular sizes. There is no true standard for face brick sizes. Identically named bricks may vary in size among different manufacturers. Specify face bricks by size first and then by name. (Nomenclature may also vary among manufacturers.)

Bricks are laid in **courses**, which are horizontal rows. They can be positioned six ways and are identified by the positions in which they are placed (Figures 7-63, 7-64).

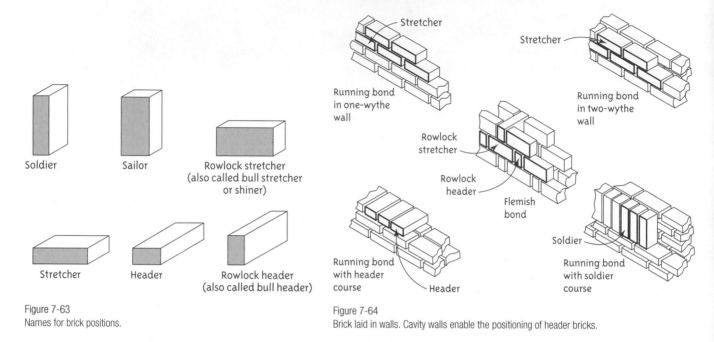

Figure 7-63
Names for brick positions.

Figure 7-64
Brick laid in walls. Cavity walls enable the positioning of header bricks.

Stone

Stone is rock or hard mineral matter. The stone used in building construction is either natural or manufactured. Today, stone is primarily used for veneers and decorative items. Stone is laid as ashlar or random rubble, which does not form a bond; mortar holds the pieces in place (Figure 7-65). Limestone, sandstone, dolomite, marble, granite, and slate are natural stones typically used in construction.

Figure 7-65
Ashlar stone in a fireplace. *Courtesy of Timbersmith, Bloomington, IN.*

Other Masonry Materials

Other masonry materials include glass block and acrylic block (often referred to as glazed masonry units or GMU), structural clay tile, and terra-cotta. Glass block comprises small, hollow, translucent cubes of glass, available in various sizes, thicknesses and colors (Figure 7-66). It is used for windows, walls, and specialty floor surfaces. Some glass block is even load bearing. Structural clay tile is a fired clay block that resembles brick, is lightweight, and can be used in both load and non-load bearing walls. It has properties that give it fire and humidity control

Figure 7-66
Glass or acrylic block can be used as a decorative and functional wall. *Courtesy of Hy-Lite, New York, NY.*

Figure 7-67
Terra cotta sculptural block on a wall outside the Johnson County Community College theatre.

applications in specialized buildings such as warehouses and laboratories. Terra-cotta is a reddish, waterproof ceramic that can be molded in intricate detail, making it popular for sculptural wall decoration (Figure 7-67). It is also used for pipes, bricks, and roof shingles.

Types of Masonry Walls

There are four masonry wall construction types: solid, cavity, faced, and veneer (Figure 7-68). Solid masonry walls are usually made of concrete block, although sometimes brick and block are combined as one load-bearing wythe. **Cavity walls** consist of two separate, parallel walls built several inches apart. Their advantage is that they weigh less and provide better temperature and humidity insulation. Faced walls consist of two wythes of different materials, such as block and brick or brick and structural clay tile. Metal masonry ties of various shapes hold them together to form one structural unit. Veneer walls also consist of two separate and parallel walls, but they are not tied together structurally. A veneer wall is non-load bearing. It can be made of brick, stone, or clay tile.

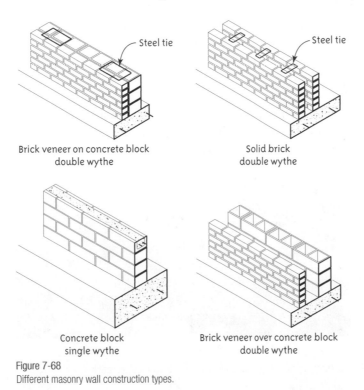

Figure 7-68
Different masonry wall construction types.

Steel Frame

Steel is a mixture of iron and carbon. Iron and a weak form of steel have been around since 250 B.C., but commercial-quality steel became possible in 1855 with the invention of the Bessemer converter, an early blast furnace.

Steel is strong, ductile, relatively cheap, and changed the look of commercial architecture. It makes large (up to 150') clear spans possible, which are spaces unobstructed by intermediate load-bearing posts or walls (Figure 7-69). It also enables great heights. With masonry, buildings are height limited, as each level up requires a thicker foundation. In contrast, steel's ability to support heavier loads without a proportionately thick foundation enabled buildings to ascend. In 1885 the first skyscraper, the Home Insurance Building in Chicago, went up. Since then, buildings have gone ever higher; records are broken every decade. Theoretically, if not yet logistically, a mile-high building made of steel is possible.

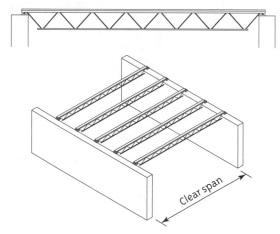

Figure 7-69
Open-web steel joists provide a greater clear span than wood beams.

Steel Terms

Arch This is a bent truss. **Arch** designs are a series of ribs that form the roof and sides of the building. They are used for warehouses, sports stadiums, large gyms, garages, and sheds (Figure 7-70). Their ease of construction makes residential applications popular with do-it-yourselfers. This type of building is semi-customizable (as modular construction generally is). Its construction allows for doors and windows in the end walls, not the sides, and the overhead clearance drops with the distance away from the center of the building.

Figure 7-70
Steel arches provide the large clear span needed for this sports arena.

Rigid Frame A **rigid frame** (also called a **bent**) consists of two columns and a beam or truss combined with steel skeleton framing and flat steel panels for the roof and walls (Figures 7-71, 7-72). Doors and windows can be placed in any wall.

Figure 7-71
Steel rigid-frame construction provides a large clear span in this prefabricated warehouse.

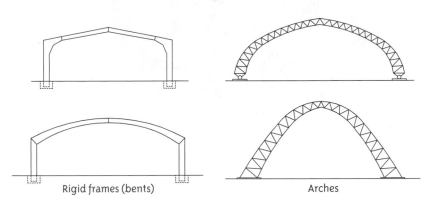

Rigid frames (bents)   Arches

Figure 7-72
Steel arches and bents.

Prefab Short for *prefabricated*, this refers to buildings whose components are made into units and assembled almost completely at a factory. The units are shipped to the construction site for assembly.

Types of Steel Construction

There are two types of steel construction: steel skeleton (also called steel cage), and large-span construction. Steel skeleton framing is similar to wood skeleton framing; trusses, columns, girders, and beams are defined the same way with steel as they are with wood (Figure 7-73). Wide-flange, standard-flange, channels, and angles are the most common steel sections or cross-sectional shapes (Figures 7-74, 7-75). These are single shapes made by extrusion through a form and cutting to the length needed (up to 40' is possible). Composite shapes are made by bolting or welding individual pieces of steel together (Figure 7-76). Steel joists, studs, and rafters have pre-punched holes for attaching exterior and interior finishes and running pipes and wires through (Figure 7-76).

Figure 7-73
Steel skeleton framing for an apartment building.

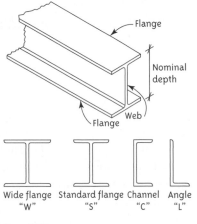

Flange

Nominal depth

Web

Flange

Wide flange   Standard flange   Channel   Angle
"W"   "S"   "C"   "L"

Figure 7-74
Common steel sections. They are manufactured in hundreds of weights and sizes.

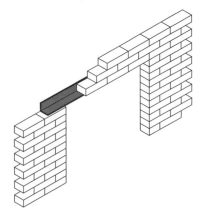

Figure 7-75
A steel angle used as a lintel.

Figure 7-76
These steel open-web joists are composite shapes. They span greater distances and carry heavier loads than single steel sections.

Figure 7-77
Holes in these steel structural members allow conduit and rods to be run through.

Large-span construction is used for commercial buildings such as sports arenas, gymnasiums, warehouse stores, and convention centers. Trusses, arches, and **space frames** (Figure 7-78) are used, as these components provide the clear spans such buildings require. A space frame is a three-dimensional truss.

Figure 7-78
Interior view of the space frame ceiling (looking up at it) at the Jacob K. Javits Center in NYC.
Courtesy of NYCC.

With both steel-frame and large-span construction, all building components are made at the factory and erected on site. Any interior work is done after the construction is complete. However, some steel buildings are **modular**, also called prefab. These are almost completely fabricated in modular units and assembled at the factory, where interior and exterior walls, roofing, doors, wiring, and carpeting are installed; then they are moved to the site and erected there. Modular buildings are smaller and have a more traditional, finished appearance. They use traditional construction materials for finishes, such as wood, stucco, brick and gypsum board, and are often used for medical labs, classrooms, offices, and retail stores. Since they are designed to be transported on trucks, modular buildings usually have an interior ceiling height limit of 8', as opposed to typical interior steel building heights of 30' and up, and each section of a modular building is limited to between 10' to 18' wide and 36' to 76' long. Some modular units are designed to be assembled on site into much larger buildings, up to tens of thousands of square feet. Modular buildings can also be stacked on top of each other to create buildings up to three or four stories high.

Steel construction design drawings have their own drawing conventions, symbols, and dimensioning standards that describe their specialized fastening and intersection methods. They tend to be more symbolic than architectural design drawings; in the plan view, steel beams are shown as heavy, solid lines that are identified by their shape, size, and weight for example, a wide flange beam that is 8" deep and weighs 67 pounds per lineal foot will be notated W 8 × 67.

Figure 7-79 is a 3-D drawing of a steel modular structure and its framing plan. Note the modular bays, the main divisions of the structure. "Bay" refers to the space between two bents or between four posts. Note the spandrels, steel girders that span from one perimeter column to another. Also note the roof decking; it is made of corrugated steel (Figure 7-80).

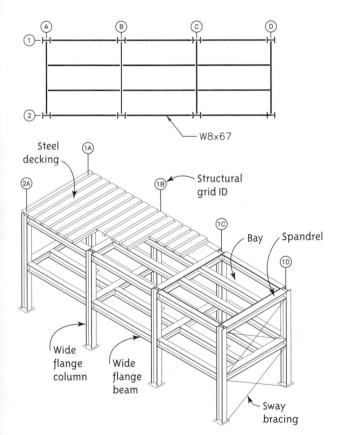

Figure 7-79
Framing plan of a multi-bay steel framed building.

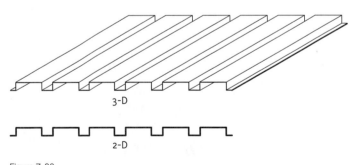

Figure 7-80
Pictorial and cross-sectional views of steel roof decking. Different cross-sections exist to accommodate piping, wiring, and floor outlets.

Finish Materials The structural system is covered with layers of other materials that comprise the finished floors, walls, and ceilings. Finish floor materials include wood, ceramic tile, vinyl, concrete, or carpet. Furring strips are thin (2" × 2" or 1" × 3") strips of dimension lumber attached to a masonry wall interior, to which panels of finished wall materials can be attached (Figure 7-81). Finish roof materials include clay tile, copper, wood shingles, or rolls of asphalt

Figure 7-81
If an interior finish is applied to a masonry wall, treated furring strips are needed or a polymer vapor barrier must be installed between the wall and the strips. An alternative to furring strips is to offset a framed wall from the masonry wall. Furring strips need to be whatever thickness is needed to clear any piping or ductwork running along the wall. This photo shows batting insulation placed between the strips. *Courtesy of CertainTeed, Valley Forge, Pa.*

Figure 7-82
Painted gypsum board is a common interior finish. *Courtesy of United States Gypsum Company, Chicago, IL.*

shingles. Exterior wall finishes include wood or vinyl lap siding, board-and-batten siding, stucco board, stucco over wire mesh, and wall assemblies that combine an exterior finish with insulation. The most common interior wall finish is painted **gypsum board** (Figure 7-82).

New materials are constantly being developed and it is the designer's responsibility to keep abreast of new technologies, processes, materials, and products.

Gypsum board, also called drywall, gypboard, or plasterboard (Sheetrock® and Gyproc® are popular brand names) is the most common interior wall surface. It consists of gypsum powder pressed between two heavy sheets of building paper. Wallboard is a sheet of compressed wood fibers, used for floor and wall sheathing. Plywood, particleboard, oriented strand board, and medium density fiberboard (MDF) are types of wallboard. Plywood is made from multiple layers of thin veneers glued together and is often stained and used for finish work. Particleboard is made of small wood particles and chips mixed and glued together. It is mostly used as an underlay. MDF is made of wood fibers glued together and is used as a laminate. All are engineered wood products.

Construction Drawings

Construction drawings are the drafted documents that describe building components and systems (Figures 7-83, 7-84, 7-85). There are many such drawings, all describing different parts. Producing a set of construction drawings is a team effort to which many people contribute. The architect is the chief designer and has overall project responsibility. Large portions of the project are given to consultants, such as civil, mechanical and electrical engineers, landscape architects, and interior designers. Depending on the scope of the project, consultants may also include lighting designers, building code specialists, artists, and behaviorists, all of whom use floor plans and drawings provided by the architect. All have different opinions and points of view on the design process, so knowing how to work together, integrate everyone's ideas, and interpret their submissions, is critical.

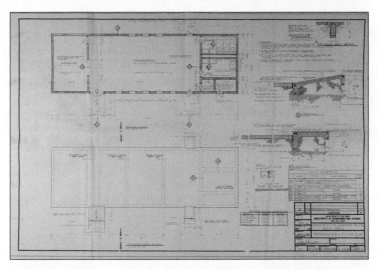

Figure 7-83
Foundation plan, floor plan, and foundation details. *Courtesy of Department of Public Works, Fort Sill, OK.*

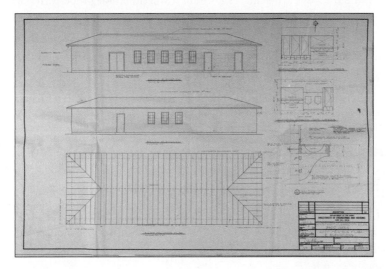

Figure 7-84
Exterior elevations, roof framing plan, interior elevations, detail. *Courtesy of Department of Public Works, Fort Sill, OK.*

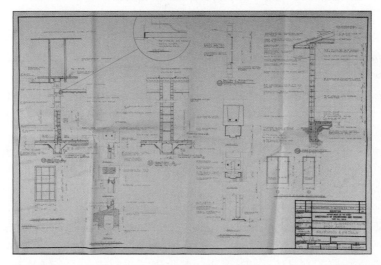

Figure 7-85
Details, wall section. *Courtesy of Department of Public Works, Fort Sill, OK.*

Specifications accompany each discipline's drawings. Specifications are text documents that include written descriptions of the quality and type of materials shown in the drawings. In a large project they are bound in a separate book. In a small project, they may be placed directly on a drawing sheet.

Here is an overview of several of the different types of construction drawings needed to describe a building.

Title page *or* cover sheet This contains the index, project name and address, designer and consultant name, and sometimes a location map and/or rendering (artist's drawing) of the project.

Survey This is a type of plan that shows the property's boundaries and includes the legal description.

Structural discipline drawings These are done by a civil engineer. They include the site and foundation plans and framing plans.

- *Site plan* Also called the plot or plat plan, this is a birds-eye view of the existing or proposed footprint (outline) of the building as it sits, or it will sit, on the property. Topography such as paving, parking, access roads, utility lines, trees, sewers, drains, and contour lines are shown. Site plans that show how the ground is to be excavated are called grading plans.

- *Foundation plan* This is a horizontally-sliced picture of the supporting structure underneath the building that distributes the building's weight onto the ground. Its outline looks similar to that of the floor plan.

- *Roof framing* This is a horizontal view of the roof's structural components.

Architectural Discipline Drawings These drawings are the basis for, and footprint of, the structural, mechanical, electrical, and civil discipline drawings. They include the floor plan, roof plan, other plans, elevations, sections, details, wall framing plans, and schedules.

- *Floor Plan* This is an architectural discipline drawing. It is a horizontal-slice picture that shows the building's layout. Walls, windows, doors, built-ins, fixtures, appliances, stairs, symbols, and more are on it. It is the heart of this set of drawings because most other drawings refer to it.

- *Roof plan* This is a horizontal view of the roof's surface. It shows valleys, hips, slope, skylights, materials, and anything that is mounted on the roof.

- *Other Plans* There are separate plans for lighting, electrical and mechanical fixtures, items to be demolished, plumbing, the reflected ceiling, and furniture arrangements.

- *Wall framing plans/stud layouts* These show how the wall is framed and the arrangement of structural elements in it.

- *Exterior elevations* These are height (vertical) drawings of the exterior walls. Windows, doors, finishes, shutters, and anything else affixed to the walls are shown, as well as the grade line.

- *Interior elevations* These are height (vertical) drawings of the interior walls as seen from one interior corner to the other. Windows, doors, finishes, cabinetry, attached woodwork (chair rails, crown molding, baseboard), and anything else that will hang on the walls are shown.

- *Sections* These are vertical slices through the building from foundation to roof. They can be made through the whole building or through a portion of it, such as a wall.

- *Details* These are vertical or horizontal views made into small parts of a building, such as windows, doors, or stairs. Their purpose is to show how the pieces go together.

- *Wall framing plans/stud layouts* These show how the wall is framed and the arrangement of structural elements in it.

- *Legends (keys) and Schedules* These are tables of information, used because the information contained would otherwise clutter up the drawings. They are needed for doors, windows, finishes, columns, pipes, and other repeating features. Legends and schedules are referenced from the floor plan via callouts. For example, a door callout on the floor plan tells the reader to look at the door schedule and find the number lettered on the callout.

Heating, Ventilation, Air conditioning, Ventilation (HVAC) and Plumbing Discipline Drawings. These are done by a mechanical engineer. They include HVAC and plumbing plans and riser diagrams.

- *HVAC plan* This is a horizontal drawing that shows furnaces, air conditioners, boilers, radiant heat systems, and the ducts, vents, and piping that carry treated air throughout the building.
- *Plumbing plan* This is a horizontal drawing that shows the water heater and plumbing fixtures and the pipes that service those fixtures.
- *Riser diagram* Also called a *schematic*, this is a vertical drawing that shows plumbing fixtures and the configuration of pipes servicing them.

Summary

Buildings are made of wood, steel, and concrete. Their components are made on site or fabricated in standardized sizes in factories. Techniques and systems have evolved throughout the years in step with technology advancements. Understanding buildings' basic construction is necessary to competently draft drawings that describe them.

Suggested Activities

1. Draw a foundation plan for a floor plan you've drawn or seen in a magazine.
2. Visit a prefabricated steel building and identify its structural system and parts.
3. Look for brick paving on courtyards and sidewalks and sketch the patterns.
4. Look at brick walls and identify the bonds and exposed portions of the brick.
5. Visit a building supply store and examine different components used in construction. Note standardized sizes.

Questions

1. What are the three materials chiefly used to construct buildings?
2. Describe the difference between post-and-beam, balloon, and platform framing.
3. Describe the difference between a T foundation and a slab-on-grade one.
4. Name three brick bonds.
5. What do nominal size and actual size mean?
6. What is a truss?
7. What is a joist?

Further Reading and Internet Resources

Ballast, D. K. (2007). *Interior Construction and Detailing for Designers and Architects.* Belmont, CA: Professional Publications.

Ching, F. & Adams, C. (2000). *Building Construction Illustrated* (3rd ed.). Hoboken, NJ: Wiley.

Huth, M. (2004). *Understanding Construction Drawings* (4th ed.). Clifton Park, NY: Thomson Delmar.

Muller, E. J. (2003). *Reading Architectural Working Drawings* (6th ed.). Englewood Cliffs, NJ: Prentice Hall.

http://www.apawood.org/
Web site of the Engineered Wood Association (APA), PO Box 11700 Tacoma, WA, 98411. Contains lots of information on engineered wood products.

http://www.askthebuilder.com/
Searchable website where for information on all aspects of building construction.

http://www.certainteed.com/
Certainteed company website. Contains information on their building material products, design and interior space planning, with relevant codes and standards.

http://www.hy-lite.com/
Hy-lite company website. Contains information on acrylic block.

http://www.kohler.com/
Kohler company website. Contains information on their kitchen and bath products.

http://www.paulreverehouse.org/
Website of the Paul Revere Memorial Association. Contains information on the Revere house.

http://strawbale.com/
A.C. Construction website.

http://www.timbersmith.com/
Timbersmith company website. Contains information on their post-and-beam framed homes.

http://www.trusjoist.com/
Website for i-Level by Weyerhauser. Contains information about their i-joist engineered wood product.

http://www.usg.com/
U.S. Gypsum company website. Contains information on their building material products.

http://www.wbdg.org/
Site provided by the National Institute of Building Design. Has lots of information about building.

CHAPTER 8

Drafting and Dimensioning the Architectural Floor Plan

OBJECTIVE

- This chapter discusses how to draft and dimension a floor plan using straight-edged tools. The single-family residence is the primary focus of this chapter.

KEY TERMS

aligned	dimension line	load-bearing
annotation	dimension notes	non–load-bearing
chase	dimensions	partial
chase walls	extension lines	partitions
codes	fenestration	poché
construction plan	floor plan	presentation plan
curtain	hard-lining	stringers
dimensioning	kiosks	toekicks

What is a Floor Plan?

The **floor plan** is an orthographic, two-dimensional drawing made by inserting a horizontal cutting plane 4'-0" above the ground (Figure 8-1). A floor plan shows lengths and widths, but not heights. It is the centerpiece of a set of construction drawings, to which most other drawings refer. The roof and walls above the cutting plane are "lifted up" and the observer looks straight down into the rooms. Sometimes the cutting plane is offset, meaning that it is staggered at different levels to show features that will not display if just one level of cutting plane is used. This is done on irregular floor plans, such as those of split-level houses.

What Goes on an Architectural Floor Plan?

An architectural floor plan shows finished and unfinished space. Balconies, attached garages, decks, patios, and pools are included; sometimes detached garages and pools are, too. Walls, wall openings, windows, doors, door swings, skylights, exposed beams, columns, soffits, cabinetry, built-ins, appliances, plumbing fixtures, stairs, and fireplaces are all shown on an architectural floor plan. Room names and symbols that reference the plan to schedules, sections, and details are shown. Electrical and mechanical symbols may be shown, but usually have their own plans so as to not clutter the architectural floor plan. Each level of the building has its own plan.

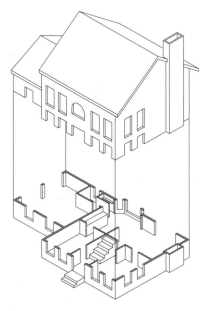

Figure 8-1
A floor plan is made with a horizontal cutting plane 4'-0" above the ground.

Figure 8-2
Presentation drawing, ground floor. *Courtesy of Jay Colestock, Colestock & Muir Architects, Boca Raton, FL.*

Figure 8-3
Presentation drawing, first floor. *Courtesy of Jay Colestock, Colestock & Muir Architects, Boca Raton, FL.*

Figure 8-4
Presentation drawing, second floor. *Courtesy of Jay Colestock, Colestock & Muir Architects, Boca Raton, FL.*

Depending on whether the plan is for presentation or construction, different information and drawing techniques may apply. A **presentation plan's** purpose is to be a selling tool, so it will display color, furniture, and décor to provide eye appeal (Figures 8-2, 8-3, 8-4). These elements would not appear on a **construction plan**, as its purpose is to show building assembly. A presentation drawing might be dimensioned with a simple **annotation,** or note, that gives a room's usable length and width, whereas the dimensioning on a construction drawing will be far more detailed (Figures 8-5, 8-6).

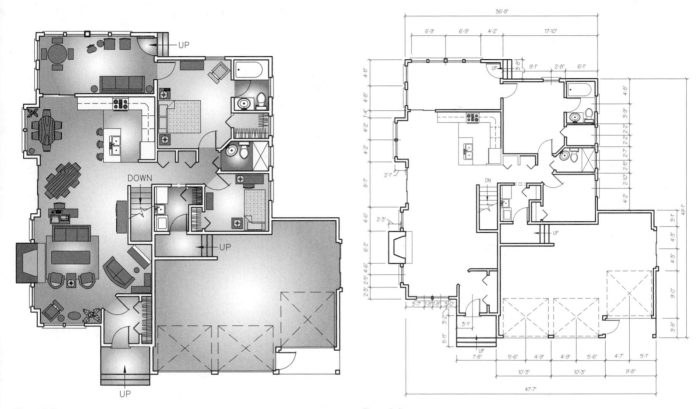

Figure 8-5
Presentation drawing of a house floor plan.

Figure 8-6
Construction drawing of the same house floor plan.

Symbols

To draw a floor plan, knowledge of symbols, **poché**, and wall thickness is needed. Items must be pochéd with line symbols that represent the material they are made of. Walls must be drawn with a thickness that represents their construction. Doors, windows, cabinets, drains, draperies, and other common items are drawn to represent their size and type. Figures 8-7 through 8-15 show some common wall thicknesses and symbols found on floor plans. (Door and window symbols are discussed in detail in Chapter 9.)

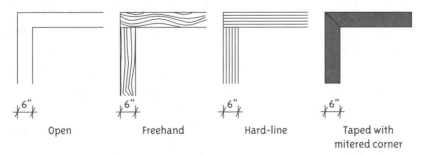

Open   Freehand   Hard-line   Taped with
                                                    mitered corner

Figure 8-7
Different ways to poche a 6" exterior wood-frame wall.

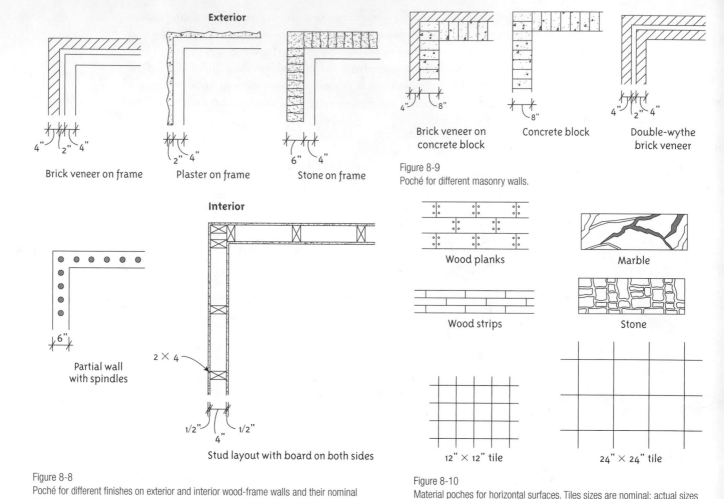

Exterior

Brick veneer on frame

Plaster on frame

Stone on frame

Interior

Partial wall with spindles

2×4

Stud layout with board on both sides

Figure 8-8
Poché for different finishes on exterior and interior wood-frame walls and their nominal thicknesses.

Brick veneer on concrete block

Concrete block

Double-wythe brick veneer

Figure 8-9
Poché for different masonry walls.

Wood planks

Marble

Wood strips

Stone

12" × 12" tile

24" × 24" tile

Figure 8-10
Material poches for horizontal surfaces. Tiles sizes are nominal; actual sizes are 1/4" smaller on each side to account for mortar joints.

Drain

2-way drapery

1-way drapery

Vertical blinds

Closet

Figure 8-11
Item Symbols. Note the floor drain on top. It is used in any location where water accumulates, such as the shower, bathtub, laundry room or garage.

BATT

LOOSE FILL INSULATION

RIGID INSULATION

METAL LATH AND PLASTER

PLASTER

TERRAZZO

EARTH/ COMPACT FILL

GRAVEL

CARPET AND PAD

ACOUSTICAL TILE

MARBLE

EXISTING WALL TO REMOVE

OR

EXISTING WALL TO REMAIN

EXISTING OPENING TO ENCLOSE

WOOD STUD

METAL STUD

Figure 8-12
Material poché for section views.

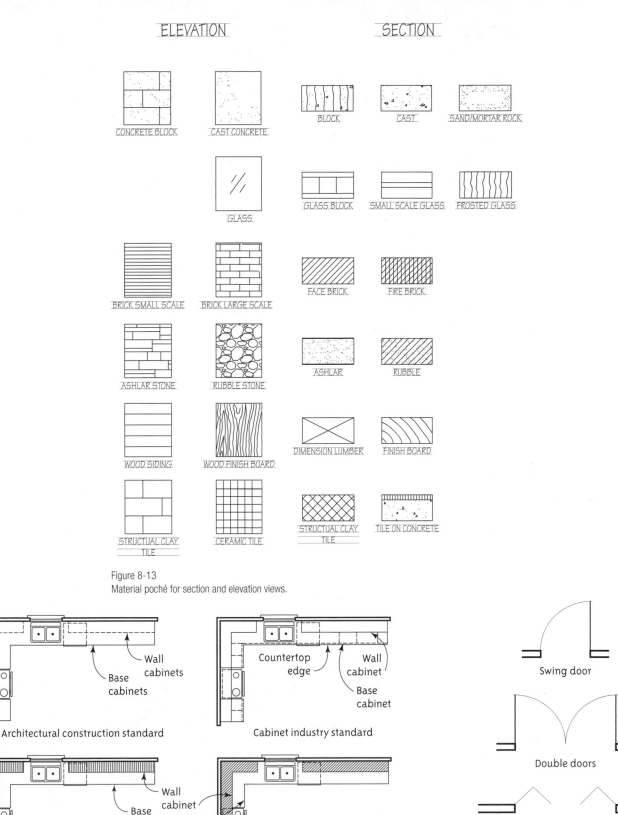

ELEVATION   SECTION

CONCRETE BLOCK   CAST CONCRETE   BLOCK   CAST   SAND/MORTAR ROCK

                    GLASS   GLASS BLOCK   SMALL SCALE GLASS   FROSTED GLASS

BRICK SMALL SCALE   BRICK LARGE SCALE   FACE BRICK   FIRE BRICK

ASHLAR STONE   RUBBLE STONE   ASHLAR   RUBBLE

WOOD SIDING   WOOD FINISH BOARD   DIMENSION LUMBER   FINISH BOARD

STRUCTUAL CLAY   CERAMIC TILE   STRUCTUAL CLAY   TILE ON CONCRETE
TILE   TILE

Figure 8-13
Material poché for section and elevation views.

Architectural construction standard   Cabinet industry standard

Wall cabinets
Base cabinets

Countertop edge
Wall cabinet
Base cabinet

Presentation

Wall cabinet
Base cabinet

Wall cabinet

Figure 8-14
Different ways of showing wall and base cabinets in plan.

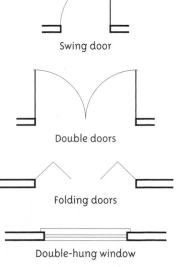

Swing door

Double doors

Folding doors

Double-hung window

Figure 8-15
Door and window symbols.

Drafting a Floor Plan

In earlier chapters the preliminaries to creating a floor plan were covered. Here we will discuss its **hard-lining**, which is using straight-edged tools to draw a presentation or construction-ready product. This is a suggested sequence for drafting the final layout developed in the bubble diagrams in Chapter 6.

1. Decide which scale to use. Most architectural floor plans are drawn at a 1/4" = 1'-0" scale; very large ones are drawn at 1/8" = 1'-0". Floor plans of kitchens and baths drawn to the cabinetry/millwright industry standard are 1/2" = 1'-0". Use a hard lead to construct the entire plan, then trace with a soft lead or ink.

If the floor plan will go on a sheet with other drawings, complete all drawings before inking any of them. Do not ink drawings individually as they are completed, as this will result in a poorly composed sheet. Lay out all the drawings on a mock-up sheet first to ensure there is enough space for them plus their ID labels, dimensions, the title block, and the border.

2. Draw the plan's overall width and length (Figure 8-16). This helps keep all the smaller interior measurements on track.

Figure 8-16
Draw a rectangle of the plan's length and width.

3. Draw the exterior walls (Figure 8-17). Show walls with a double line; the space in between will represent the wall's thickness. This thickness is based on nominal, not actual material sizes. For instance, an exterior wall framed with 2" × 4" studs, with gypsum board on one side and sheathing/siding on the other, is 5½" wide. However, it is difficult to manually draw with such precision at the 1/4" = 1'0" scale used for most plans, so a 6" thickness is substituted for convenience.

   Use the parallel bar to draw a horizontal line representing a wall. Mark the wall's length. Draw a vertical line at the wall's end with a triangle placed squarely on the bar, then measure and mark its length.

4. Draw the interior walls (Figure 8-18) There should be a noticeable difference in thickness between the exterior and interior walls; exterior walls are usually thicker. They are drawn as double lines spaced their material's nominal thickness apart. Again, this may not be the actual size; for example, an interior wood-framed wall with ½" gypsum board on each side is 4½" thick, but 4" or 5" is easier to draw. To avoid having to constantly measure walls, use a pair of dividers that is set to the wall's thickness.

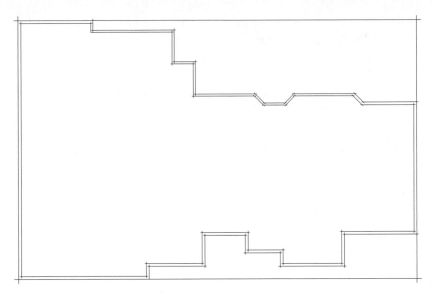

Figure 8-17
Draw the exterior walls.

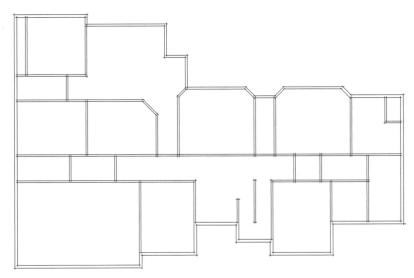

Figure 8-18
Draw the interior walls.

When drawing walls to enclose a specific square footage in each room, it is important to be consistent in how those rooms are measured. For instance, if you draw one room's dimensions from interior wall to interior wall, you must draw all rooms' dimensions that way. Do not measure another room from the interior line on one side to the exterior line on another. Know that while the rooms are drawn to enclose a certain usable square footage, they are not dimensioned (meaning their size is not annotated) this way.

Periodically check corners for squareness by placing a triangle in them (rest the triangle on a parallel bar; (Figure 8-19). If both the horizontal and vertical walls align with the triangle's edges, the corner is square. If the walls do not align with the triangle's edges, the walls are not perpendicular to each other.

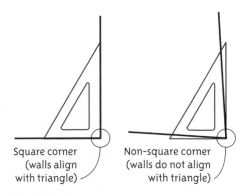

Square corner
(walls align
with triangle)

Non-square corner
(walls do not align
with triangle)

Figure 8-19
Place a triangle on the floor plan's corners to check for wall perpendicularity.

5. Draw door and window openings (Figure 8-20). In wood frame construction, these openings are measured to their centers.

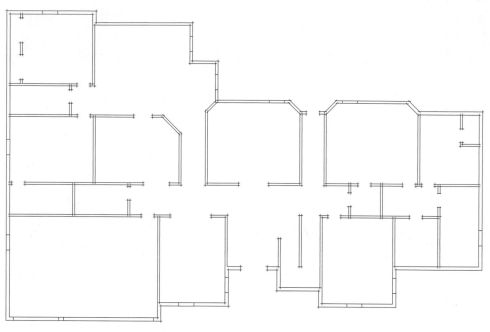

Figure 8-20
Draw door and window openings.

6. Draw the stairs and fireplace (Figure 8-21). The number of stairs and their tread widths must be known. Measure the staircase width, then draw treads with lines equally spaced the tread width apart. The exact size of the fireplace also must be known. Draw half the full stair run and then place an arrow in the middle with the note "UP" at its foot. On the second floor, reverse the arrow and letter "DN" for down. Stairs and fireplaces are discussed in detail in Chapter 11.

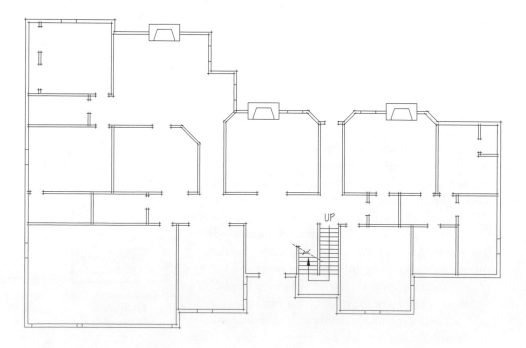

Figure 8-21
Draw the stairs and fireplace.

7. Darken the walls so smaller features may be accurately added (Figure 8-22). Appropriate line weights should be introduced at this point; walls should be thicker than any other lines on the floor plan.

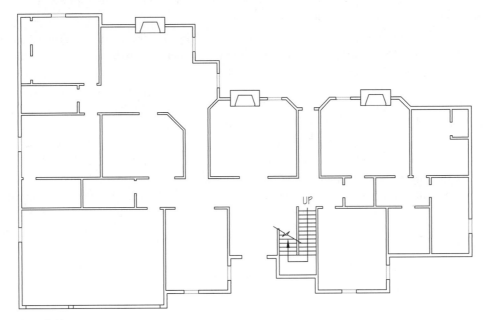

Figure 8-22
Darken the walls.

8. Draw cabinetry, appliances, plumbing fixtures, and door and window symbols, and outline the front and back porches (Figure 8-23). Doors and windows are discussed in detail in Chapter 9. Place floor-mounted plumbing fixtures slightly away from the wall, rather than directly against it.

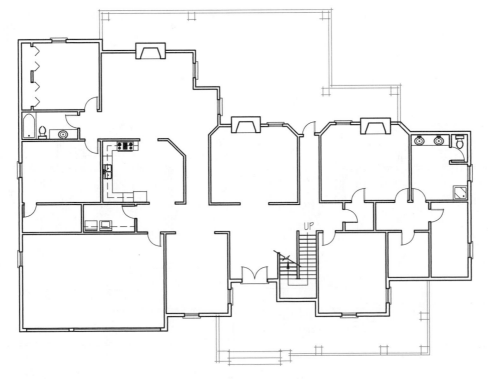

Figure 8-23
Draw cabinetry, appliances, and door and window symbols; lay out the front and back porches.

There are different ways to represent wall cabinets. Architectural construction drawings show base cabinets drawn with a solid line and wall cabinets drawn with a dashed (hidden) line, because they are above the 4'-0"cutting plane. Cabinet industry drawings show base cabinets with dashed lines, their overhanging countertops with solid lines, and the wall cabinets above them with solid lines.

9. Darken the front and back porches. Add the roof overhang. Finally, add details (Figure 8-24). Those should be drawn last because they have the most flexibility to be worked around the other items. If the plan is for construction, add symbols that reference it to details and schedules. Add room names, electrical and mechanical symbols if appropriate, and floor finish changes and annotate ceiling heights. Add poché (also called hatch or section-lining), a symbol that represents the material. Any portion of the building that has been sliced through is pochéd. On presentation drawings, poché is often solid black and applied with tape. Poché flooring as it appears when seen from directly above. The whole floor doesn't need to be pochéd, just scattered patches; the goal is to convey the essence of the material. Any partial walls not tall enough to reach the 4'-0" cutting plane are distinguished from the other walls with a note or by being left open (not pochéd). If they have spindles that go to the ceiling, the spindles are pochéd, because they are sliced by the cutting plane.

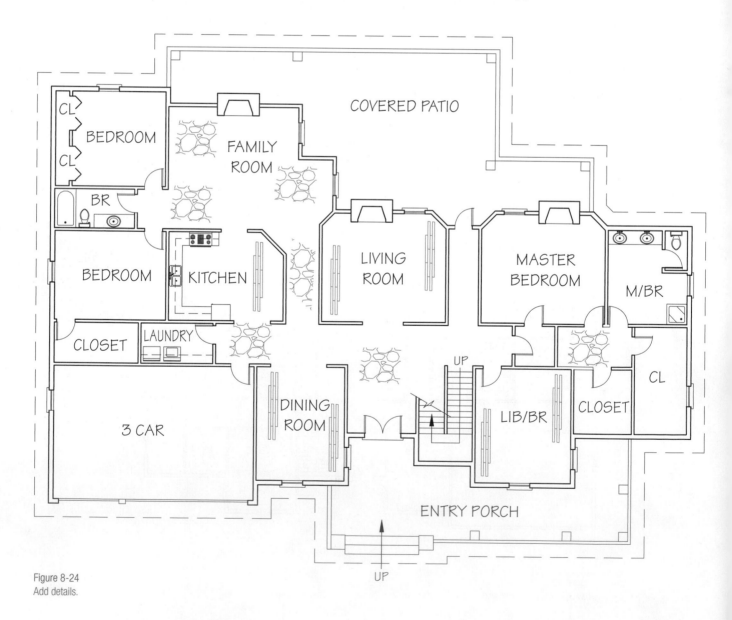

Figure 8-24
Add details.

It is critical that line weights, symbols, poché, notes, and dimensions are drawn, placed, and arranged so that the plan is easy to read. All final lines should be equally dark; their functional differences should be expressed by their width. Walls are drawn thickest; furniture, fixtures, and doors thinner (with equal weight); door swings and window glass thinnest. There should be no construction lines on the final drawing. Lettering guidelines may remain, but they should be light and barely visible when mechanically copied.

10. Draw a second-floor plan by overlaying a piece of tracing paper on the first floor plan. Align stairwells, chimneys, load bearing, chase, and major partition walls. The exterior **fenestration** (arrangement of windows and doors) design may require first- and second-floor windows to align. Overlay the floor plan with all other plans, such as basement and mechanical, to check for alignment and general accuracy.

11. If you need a furniture plan, add furniture with appropriate clearances (Figures 8-25, 8-26). Avoid blocking doorways and circulation paths and utilize focal points and views. Draw each piece's entire footprint. This is not necessarily just the space between the four legs; for example, a couch may have large rolled or flared arms that lengthen its footprint. The online programs maintained by furniture manufacturers referenced in Chapter 6 are sources for plan and elevation symbols of people, pets, and other accessories that you can use to accessorize a presentation plan.

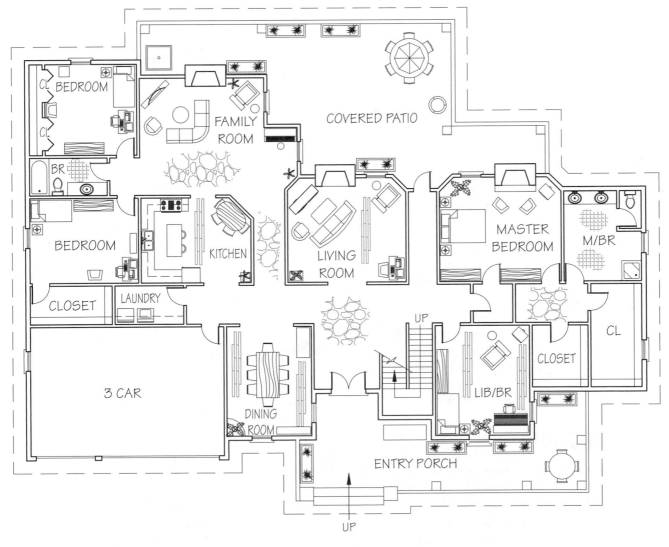

Figure 8-25
If a furniture plan is needed, add furniture.

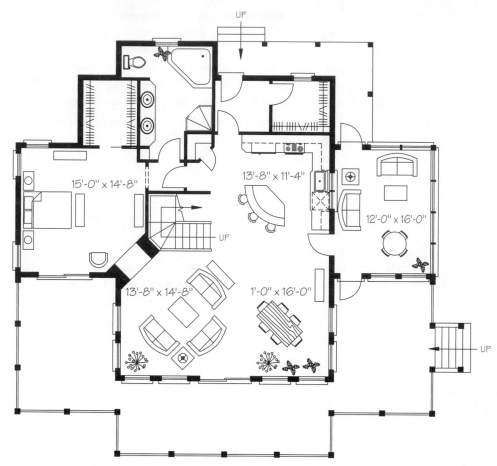

Figure 8-26
Furniture plan.

Draw furniture slightly away from the wall instead of touching it, which acknowledges the design trick of leaving a little space between the piece and the wall. (Empty space behind furniture is perceived as bigger than when furniture is pushed against the wall, so what is lost physically is gained visually). Applying color to the furniture with markers or watercolors and/or inking shades and shadows on part of each piece gives a dramatic effect, but this should only be done once competency with these media is established, as poor color application can ruin an otherwise well-drafted drawing.

This drawing process is actually quite similar to what would be done if you used computer drafting software. Drafting software presents the drafter with advantages such as easy editing as well as the challenge of learning a precise sequence of commands. But the thought process that gets the drafter from start to finish remains the same.

Types of Walls

Walls can be generally categorized as follows.

Load bearing **Load-bearing** walls can be interior, exterior, or both. They bear the weight of upper floors and/or the roof. Floor plans don't annotate which walls are load-bearing; sometimes this can be determined by studying their placement on the drawings. A wall's size and position relative to overhead beams or to grade beams on the foundation plan might indicate its status.

Non-load bearing Also called *non-structural* or *non-seismic*, **non–load-bearing** walls do not bear any of the building's weight; hence they are not factored into structural computations. They can be exterior walls, such as **curtain** walls, which are facades hung on a load-carrying steel frame, but are usually interior walls. Non–load-bearing walls carry only their own weight. Their purpose may be aesthetic or functional. (For example, they can divide a large space.)

Partial **Partial** walls are not full height. Codes require that they be at least 36" above the floor. They are sometimes topped with spindles that go to the ceiling.

Partition **Partitions** are non-load bearing interior walls, available as modular wall systems that can be delivered to a site (Figure 8-27). They are used to create cubicle spaces in offices and factories. They are also used for **kiosks**, which are independent stands from which merchandise is sold. Partition systems come in many heights, materials, and colors. Some are short, others reach to the ceiling. They can be made of wood, metal, or laminate; finished in different colors and fabrics; and have slats to insert shelves and openings for doors and hardware. They can accommodate ductwork inside, be fire and noise rated, be hinged in pairs and manually operated, or be continuously hinged and roll on casters that are electrically operated. Although they are not load-bearing, they may support attached loads such as cabinetry, shelving, or grab bars.

Figure 8-27
Partition walls are common in office buildings. This is a decorative glass one.

Chase walls **Chase walls,** also called stack or wet walls, house plumbing pipes. A **chase** is a passageway between floors or walls. Chase walls are thicker than other interior walls because they contain vertical pipes that drain wastes. A 6" stud wall is the minimum thickness needed; more thickness may be needed depending on fixture number and type. The following are minimum thicknesses for the arrangements shown. Note that the shapes of floor- and wall-mounted fixtures are different, as is their placement. Floor-mounted fixtures are drawn slightly away from the wall; wall-mounted ones are drawn flush to it. Figures 8-28 through 8-31 show chase walls for different materials and fixtures.

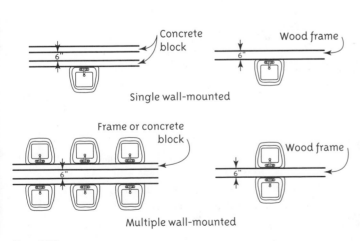

Figure 8-28
Chase wall thickness for wall-mounted lavatories.

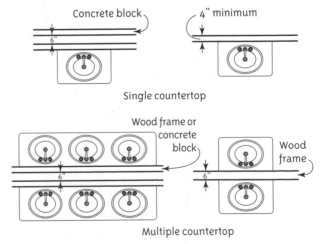

Figure 8-29
Chase wall thickness for countertop-mounted lavatories.

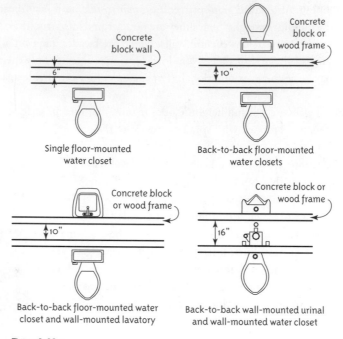

Single floor-mounted
water closet

Back-to-back floor-mounted
water closets

Back-to-back floor-mounted water
closet and wall-mounted lavatory

Back-to-back wall-mounted urinal
and wall-mounted water closet

Figure 8-30
Chase wall thickness for floor and wall-mounted fixtures.

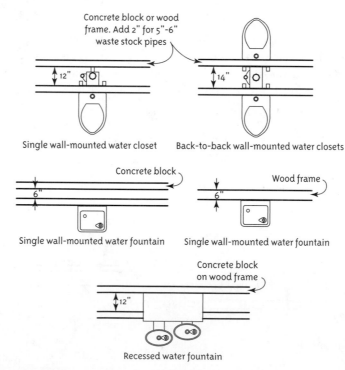

Single wall-mounted water closet

Back-to-back wall-mounted water closets

Single wall-mounted water fountain

Single wall-mounted water fountain

Recessed water fountain

Figure 8-31
Chase wall thickness for floor and wall-mounted fixtures.

Dimensions **Dimensions** are numbers that describe the size of features and their location from other features. **Dimensioning** is the process of indicating size and location in space. Different types of drawings—architectural, cabinetry, mechanical, electrical, cabinetry, modular, and metric—all have their own standards. Practices also vary between offices. We will discuss two standards, one for architectural construction drawings and one for cabinet industry drawings (also called millwright or shop drawings).

Dimensioning Architectural Construction Plans

There are a lot of rules for dimensioning architectural drawings: accuracy, completeness, and a single, legible interpretation are the essential ones. Here are some additional requirements.

- Architectural dimensioning is **aligned.** This means the **dimension notes,** which are numbers that describe size and location, are drawn parallel to the *dimension lines.* These notes are placed above or to the left of the dimension lines and are usually centered. Dimension lines, also called **stringers,** are linear dimensions strung together to form a line, or string. They run perpendicular to **extension lines,** which are lines that emanate from the endpoints of the portion being dimensioned. Dimension lines end with arrowheads or slashed lines that touch the extension lines to indicate an exact measurement. They are thinner than the object lines so as not to be mistaken for them, but are just as dark (Figure 8-32).

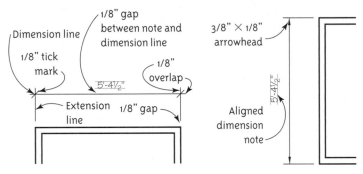

Figure 8-32
Dimension arrowheads are opaque and three times as long as they are wide. Dimension tick (slash) marks are 1/8" long, hard-lined at a 45° angle. The overlap of the extension and dimension lines is also 1/8". Dimension text is aligned or parallel to the dimension lines, centered, and spaced 1/8" above or to the left of them in architectural construction drawings. There is a 1/8" gap between object and extension lines. Dimension and extension lines are thinner than wall lines.

- Imperial (also called English or U.S. Customary) dimension notes are written in feet and inches. If a dimension is smaller than 12", place a 0'- in front (0'-2"). If the dimension is a whole foot number, place a -0" at the end (2'-0"). When space is tight it is acceptable to abbreviate a dimension, such as 6° (pronounced "six –oh") instead of lettering 6'-0" and/or to place the dimension note outside the stringer with a leader line pointing to it. Dimension numbers are the same size as the characters in lettered notes, which are typically 1/8" tall. Neither letters nor tick marks should be made smaller to squeeze into a small space. When there is not enough room to place a dimension note inside the space, place it outside with a leader line. Draw fractions with a diagonal slash. Each number of the fraction should be the entire height of the guideline (Figures 8-33, 8-34)

Figure 8-33
Dimension notes are lettered in guidelines. Proper architectural format requires both feet and inches place markers, even for whole numbers or for numbers less than one foot.

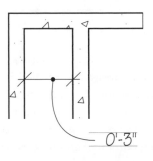

Figure 8-34
Place a dimension note outside a space if there isn't enough room to place it inside.

- Circles and arcs are dimensioned from their center points (Figure 8-35). That center point is located with two extension lines to the walls, and its radius is noted. Endpoints of a curve are referenced to the walls.

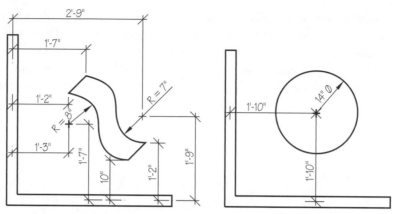

Figure 8-35
Locate a circle's or curve's center from two directions.

- There are typically three stringers (Figure 8-36). The one closest to the object contains the smallest dimensions, the second contains some small and some larger dimensions, and the third contains one overall dimension. Stringer spacing should be uniform, but this is trumped by completeness, so if the room available doesn't allow for uniformity, spacing may vary. The distance between the object and the first stringer should be between 1/4" and 1". A common protocol is to draw the first stringer 3/4" from the drawing and subsequent stringers 1/2" apart. Spacing on all sides should be the same.

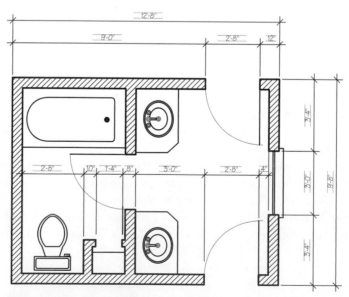

Figure 8-36
Stringer placement on a masonry plan.

- Place exterior dimensions outside the plan to avoid overcrowding. They should clear roof overhangs, steps, porches, and notes. At the same time, dimensions shouldn't be too far from the features they describe. Place interior dimensions inside the plan. Extension and dimension lines should be placed so they do not cross. When this is not possible, "loop" the crossing portion of one line (Figure 8-37).

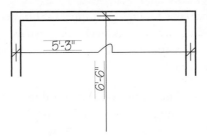

Figure 8-37
Loop the crossing portion of an extension or
dimension line.

- The placement of dimension and extension lines varies depending on the dimensioning standard used and whether the building is wood frame or masonry. Generally, fixtures, appliances, beams, pilasters, and columns are dimensioned to their centers. Place a centerline running in both directions through a column (Figure 8-38).

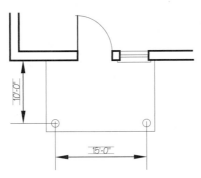

Figure 8-38
Columns are dimensioned to their centers.

- Dimension comprehensively; workers should not have to guess where a feature is nor have to hold a scale to the drawing. Analyze whether enough information has been given to allow workers to build from the drawing. You should not repeat the same dimension on a drawing, but you may have to show that dimension on several different drawings. On floor plans, major dimensions such as length and width of the overall building, individual rooms, closets, halls, and wall thicknesses are given. Dimensions of small portions are described either by a note on the floor plan or on separate, enlarged drawings. Some dimensions may be provided in local note form; for example, instead of dimensioning the walls around a shower, a drafter may simply letter "36" × 36" inside the shower. Assumptions may sometimes be made, such as when a door is centered between two walls or when it's located against a wall (the dimension would be the minimum distance possible from the wall). Depending on the audience and/or the purpose, some standard-size interior features such as cabinet depths may not need dimensions on one drawing, but may be required to have dimensions on another.

- Dimension notes describe actual sizes, which may contradict the drawing. For instance, a note saying "2 × 4 stud" may be on a wall with an actual measurement of 1½" × 3½".

- Doors and windows must be located on the drawings but are rarely given size dimensions. Instead, they are referenced to in schedules.

Architectural drawings are dimensioned differently for wood frame and masonry (concrete or solid brick) construction. These differences are as follows.

Wood Frame Wood framed buildings are dimensioned from the outside face of the stud on exterior walls. Interior walls, doors, and windows are dimensioned to their centers. This is because carpenters build the frame first and then apply the finishes. However, when manually drafting a wood-frame building with wood exterior siding, extension lines are placed flush with the exterior corners for ease of drafting. The dimension note will be the distance from exterior face of stud to exterior face of stud.

Masonry Buildings made of masonry are dimensioned to their edges. Extension lines run from corner to corner of the building, windows, and doors. This is because masonry units are modularly sized and such dimensioning is easier for the workers to follow.

Masonry Veneer on Wood This is dimensioned the same way as wood frame, because the wood frame is the structural system and the masonry is applied like any other finish. When manually drafting a wood-frame building with a masonry veneer, the extension lines are placed level with the exterior studs and additional extension lines are needed for the thickness of the masonry veneer (Figures 8-39, 8-40, 8-41).

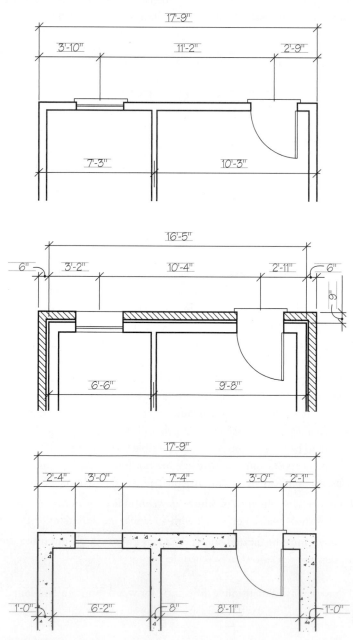

Figure 8-39
Dimensioning practices for wood, masonry, and veneer construction.

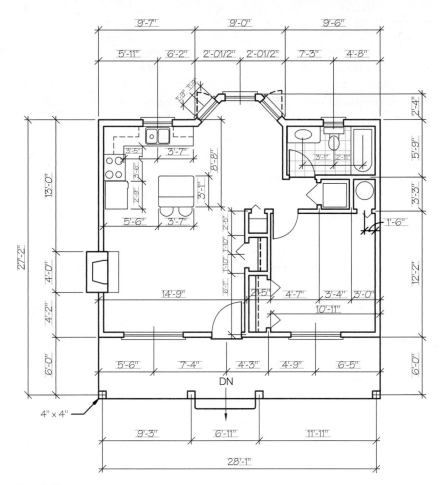

Figure 8-40
Wood-frame dimensioning. Dimensions are taken from opening centers.

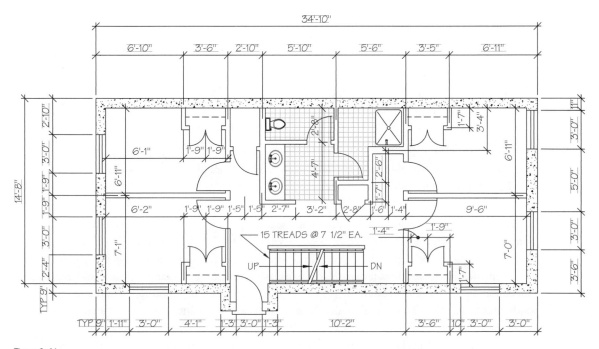

Figure 8-41
Masonry and wood combination. Each material is dimensioned for its type.

Dimensioning Cabinet Industry Plans

The cabinet industry dimensions and notates its drawings differently from architectural construction drawings (Figures 8-42, 8-43). Cabinet installers typically deal with interior finished spaces and appliance fit, so they need to receive information in a different format than that used by framing carpenters and masons. A millwright-joinery standard is used. Here are its rules.

- Drawings are done to a 1/2" = 1'-0" scale.

- Wall (upper) cabinets are drawn with a solid line. Base (lower) cabinets are drawn below the counter top with a hidden line. The countertop overhang is drawn with a solid line. Lines for **toekicks**, the indented spaces at the bottom of cabinets, are not drawn.

- Three stringers are typically used to dimension a room. The first shows all opening sizes and locates partition walls, columns, chimneys, and other fixed items. The second locates appliance centerlines. The third shows the room's overall length. Stringers are spaced an equal distance apart from each other and from the wall. That distance may vary, with 1/2" being a common one.

- Appliances (cooktops, refrigerators, sinks, wall ovens, microwave ovens, fans, lights, heating and air conditioning ducts, and radiators) are dimensioned to their centers. A centerline is projected to the dimension line, with the symbol "CL."

- Cabinets are not dimensioned. Instead, they are annotated with **codes,** which are number and letter combinations that contain size and detail information. An example of a wall cabinet code is "W2136-L." The *W* means wall cabinet; it is 21" wide, 36" tall, and has one door hinged on the left side (as seen when facing it). An example of a base cabinet code is "B30-R." The *B* means base cabinet, it is 30" wide and its one door is hinged on the right side. Height and depth are not given in the base cabinet code; instead, they are given in specifications or on section views. A base cabinet is almost always 34½" (excluding countertop) high and 24" deep. The countertop is almost always 1½" tall, bringing the total height to 36".

- Islands and peninsula cabinets are dimensioned from the edges of their countertops. Their exact location is identified by dimension lines running in two directions to adjacent walls or to the face of cabinets and appliances opposite them.

- Special features such as pull-out shelves and lazy Susans are not drawn. Instead, a circled number is placed inside that cabinet on the drawing that keys it to a schedule.

- All numbers are given in inches or inches and fractions of inches. (In metric, numbers are given in centimeters and decimals of centimeters.) Dimension notes are placed inside a broken dimension line and are unidirectional, reading straight up. They are not aligned over a continuous stringer like architectural construction drawings are.

- Height from finished floor to ceiling is noted on the plan.

- Openings to receive cabinets or appliances at a later date are dimensioned with the appliance's exact height, width, and depth.

There are many specialty cabinets, such as corner base, corner wall, utility, laundry, and hamper. Many cabinets contain specialized features, such as revolving shelves, roll-out trays, built-in wine racks, partitions for lid and tray storage, and trash can concealment. Other feaures include integrated breadboxes, lazy Susans, tableware storage, stemmed glass holders, wall china displays, wall microwave shelves, and spice racks. There are also many types of case goods (items that store household goods). Non-cabinet case goods include mirrors, medicine chests, entryway ensembles, and entertainment centers. All of these are coded. There are hundreds of codes reflecting the different types; here is a list for common millwork items. Codes may differ between manufacturers.

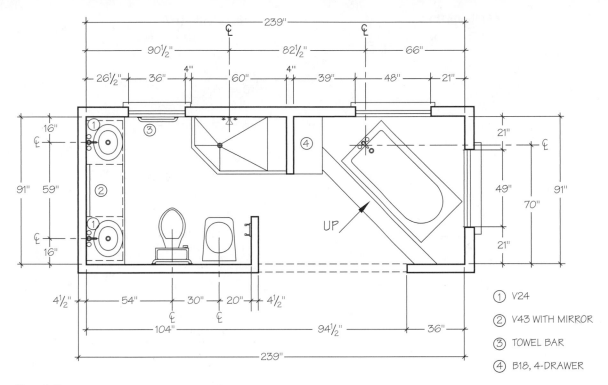

Figure 8-42
Bathroom dimensioned to cabinet industry standards. Note that fixtures are dimensioned to their centers. Circled numbers reference millwork to a schedule.

① V24
② V43 WITH MIRROR
③ TOWEL BAR
④ B18, 4-DRAWER

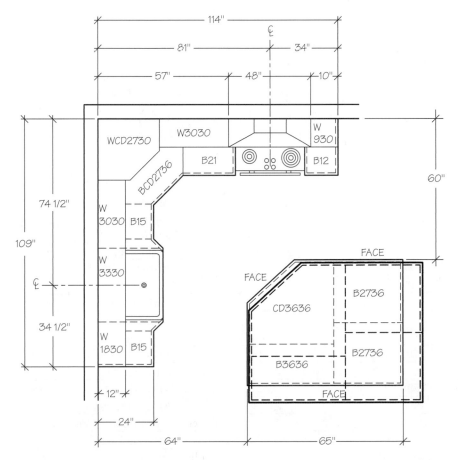

Figure 8-43
Kitchen dimensioned to cabinet industry standards. Note the codes that describe the cabinet sizes.

ABBREVIATION	MILLWORK ITEM
ADD	Adjustable drawer dividers
BBC	Base blind corner
LS	Lazy Susan
BC	Base column
BDW	Base dishwasher cabinet
BDA	Base drawer angle
BEC	Base end corner
BFF	Base fluted filler
BMC	Base microwave cabinet
BOBC	Base open basket cabinet
BPP	Base pantry pullout
BPPS	Base pot and pan storage
BSDC	Base spice drawer cabinet
BWR	Base wine rack
BRM	Braid molding
BB	Bread box
CR	Chair rail
CD	Corner diagonal
CBT	Chopping block table
DROT	Deep roll-out trays
DE	Dentil edging
DDR	Desk drawer cabinet
DS	Drawer storage cabinet
DSA	Drawer storage angle
EDE	Egg and dart edge
EC	Entertainment cabinet
FM	Framed mirror
FBD	Furniture base drawer
FBS	Furniture bookshelf
GD	Glass door inserts
GC	Grape corbel
GRPO	Grape onlay
MC	Mirror cabinet
MSP	Multi-storage pantry
OGM	Ogee molding
OVR	Oval rosette
OC	Oven cabinet
PFIB	Peninsula floating island base
RTKR	Recessed toekick
ROT	Roll-out tray
U	Utility roll-out tray
V	Valance
VB	Vanity base
VBBC	Vanity base blind corner
VC	Vanity cabinet
VSB	Vanity sink base
VSDC	Vertical spice drawer
WBC	Wall blind corner cabinet
WCD	Wall china display
WC	Wall column
WMS	Wall microwave shelf

"As a cabinet builder and installer, here are some observations I've made. Many cabinet installers prefer tolerances; that is, clearance between the appliance and the opening it fits into. If the exact appliance to be installed is known, I'll go to the factory website and download installation instructions. Not all appliances of a specified width are the same. Many appliances are smaller than their specified width; for instance, many 30" stoves are actually 29⅞", as the designation 24", 30", etc., is not the actual appliance width, but the recommended opening dimension. Or a 36" wide refrigerator might require a 36½" opening, a 30" range might require a 30¼" opening, a 24" dishwasher might require a 24¼" opening. Drop-ins, slide-ins, pop-up vents, double ovens, freestanding, overlaid, inset, French sinks, microwave trim kits, different handle options for a refrigerator all require a different trim panel configuration.

Refrigerator openings often have baseboard at the bottom and edge banding on the countertop, so a larger opening may be needed to slide the fridge in and open its doors all the way. The latter is often an issue; many refrigerators don't have zero-clearance doors so a spec for the specific appliance must be used or the designer must provide standard openings and appliances that fit will be picked. There are few standard widths for appliances. Cabinets cannot be competently built when guessing at appliance dimensions."

Summary

Floor plans are hard-lined once the layout is final. Fixed architectural features are shown in symbol form; other symbols reference the plan to other drawings. Dimension notes show the size and location of the features. Dimensioning is done differently depending on the target audience for the drawing.

Suggested Activities

Sketch, measure and hard-line the floor plan of the room you are in, and then lay out extension and dimension lines for it.

1. Measure the actual width of some appliances and compare with their specifications sheet sizes. Measure the width of the openings they are encased in and the tolerance clearance.

2. Obtain some floor plans (from a magazine or other source) and sketch extension and dimension lines on them.

Questions

1. What is the difference between a load-bearing wall and a non–load-bearing wall?

2. What is the purpose of a chase wall?

3. What is the actual thickness of an interior 2" × 4" wood frame wall with 1/2" gypsum board on both sides?

4. How does dimensioning for wood-frame construction differ from masonry construction?

5. Name some differences between architectural construction dimensioning and cabinet industry dimensioning.

6. What is a common protocol for stringer spacing on architectural construction drawings?

7. What does the cabinet code "W1215-L" mean?

CHAPTER 9

Doors and Windows

OBJECTIVE

• This chapter discusses doors and windows. Specifically discussed are types, sizes, plan and elevation symbols, construction details, residential schedules, and code requirements.

KEY TERMS

accordion	fan	lights
active panel	fenestration	locks
awning	fixed	louvered
bay	flush	masonry opening
bolts	folding	Palladian
box bay	French	pane
bow	garden	panel
buck	glazing	passive panel
call-out	glider	picture
casement	handles and knobs	pocket
casing	hardware	rails
clear width	head	right hand
clerestory	header	right hand reverse
closers	hinges	rough opening
combination	holders	R-value
double	hollow core	sash
double action	hopper	schedule
double hung	jalousie	sill
Dutch	jamb	single-hung
egress	left hand	stiles.
exit devices	left hand reverse	

There are many types of doors and windows. A designer selects them based on appearance, operation, purpose, architectural style, environment, energy performance, and building codes. Material choices include wood, fiberglass, aluminum, steel, bronze, vinyl, or cladding. (Metal or vinyl cladding is glued or snapped onto wood). Some are insulated; all have varying **R-values** (energy conservation property) and fire-resistant properties to comply with building requirements.

Interior designers frequently select doors and windows, and in that capacity may need to draw them. Knowing how these building components are made, framed, and operated is key to effectively drawing them.

Doors

A door's purpose is to provide access, privacy, security, fire protection, and ventilation. A designer selects a door to meet appearance, functionality, security, energy efficiency, and building code criteria.

Types

Doors are sold as whole, assembled units. "Entry systems" consisting of a door pre-hung in its frame, complete with threshold, hardware, weather-stripping, and sidelights, are also available (Figure 9-1). Doors are made of wood, wood composites, steel, fiberglass composite, or mixtures of these. Some steel doors have wood veneers. Other choices include hollow core, solid core, flush, paneled (Figure 9-2), louvered, and fire-rated doors. Any door can have a foam insulation core for energy efficiency. Some doors contain windows, called **lights** (also spelled *lites*). Doors may be active, passive or stationary. The **active panel** is the hinged, or operating, one. The **passive panel** may only be opened once the active panel is opened. A **stationary panel** is fixed; it doesn't move at all. However, it is easiest to classify doors according to hinge type.

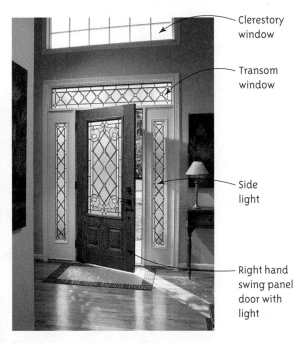

Clerestory window

Transom window

Side light

Right hand swing panel door with light

Figure 9-1
In residential construction doors open into the house and into the rooms.
Courtesy of Pella Windows and Doors, Pella, IA.

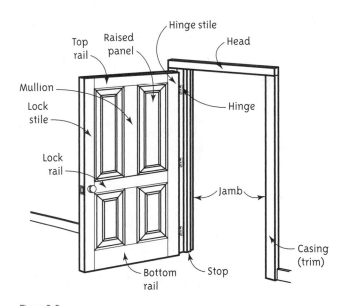

Top rail

Raised panel

Hinge stile

Head

Mullion

Lock stile

Hinge

Lock rail

Jamb

Bottom rail

Stop

Casing (trim)

Figure 9-2
Parts of a panel door

A fire door assembly is a commercial construction door complete with hardware and frame that protects against the passage of fire. It is placed in areas where fire walls are required, such as areas of refuge and egress (see Chapter 5) It must have a latch and a self-closing device. Fire doors are assigned a fire rating, which represents the amount of time it can withstand an intense fire without failing or permitting the fire to penetrate.

Swing (hinged) A **swing** door is hinged on one side and swings in or out of the room (Figures 9-3, 9-4). Most swing doors open 90° to the wall; some open 180° to lie against the wall. The latter is typically installed to satisfy commercial building code egress requirements.

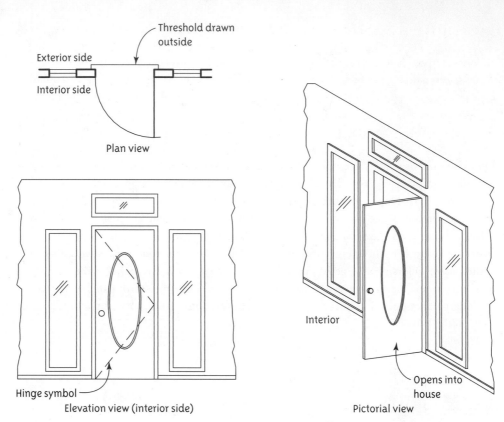

Threshold drawn outside

Exterior side

Interior side

Plan view

Hinge symbol

Elevation view (interior side)

Interior

Opens into house

Pictorial view

Figure 9-3
Exterior swing door with sidelights and transom window.

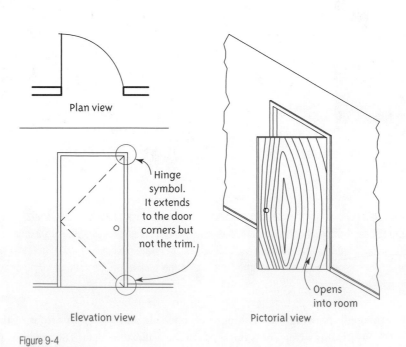

Plan view

Hinge symbol. It extends to the door corners but not the trim.

Elevation view

Opens into room

Pictorial view

Figure 9-4
Interior swing door.

Swing doors may be **hollow core, solid core, flush, panel,** or **louvered.** Solid-core doors are made of solid wood or solid wood blocks and covered with a veneer. They are normally for exterior entrances. Hollow-core doors have interiors of honeycombed wood strips and are also veneered. They are used for interiors. A flush door has a smooth, even face. A panel door has rectangular raised or recessed areas framed by rails and stiles (vertical and horizontal framing elements). Although panel construction originated with wood doors and its original purpose was to minimize cracking and warping by giving the panels enough room to shift with temperature changes, panel patterns are also embossed in metal doors. Some of the panels may be lights (windows in a door, also spelled lites). Louvered doors are constructed like panel doors but have louvers instead of panels, which are useful in areas such as closets because they provide air circulation. Door frames may be wood or metal; a metal frame is called a **buck**.

Swing doors are also classified by hand (Figure 9-5). The hand of a swing door is determined by which side it is hinged on and which way it swings. You need to know the hand when buying pre-hung doors, hinges, or locksets, because privacy and keyed locks have different inside and outside handles. To determine the hand, assume you are looking at the door from the outside (standing in a corridor or outside the building). When the knob is on your left and the door swings away from you, it is a **left-hand** (in-swing) door. If the knob is on your left and the door swings toward you, it is a **left-hand reverse** (out-swing) door. When the knob is on your right and the door swings away from you, it is a *right hand* (out-swing) door. When the knob is on your right and the door swings toward you, it is a **right-hand reverse** (in-swing) door.

Figure 9-6 shows plan-view symbols of a swing door. Preferences for which one to use vary among designers, but the office standard should be used throughout a drawing set. Figure 9-7 shows how to draw the one showing the door perpendicular to the wall. Start by drawing a line the nominal door size. Start the line from the hinge corner and side jamb that the door swings from. The opening and door must be the same size. Use a compass, door symbol template, or

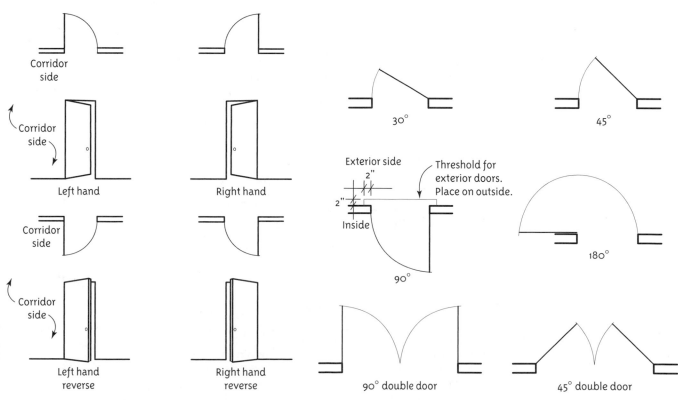

Figure 9-5
Door hands.

Figure 9-6
Swing door symbols in plan. Exterior doors have a threshold.

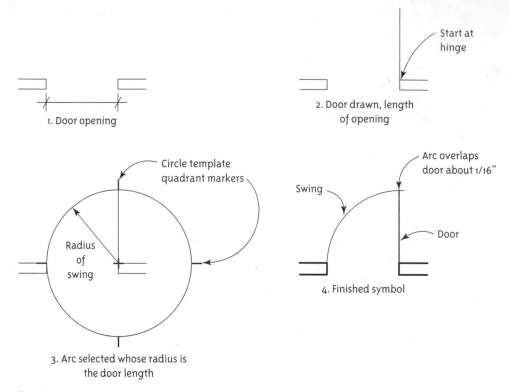

1. Door opening

2. Door drawn, length of opening

Start at hinge

Circle template quadrant markers

Radius of swing

3. Arc selected whose radius is the door length

Arc overlaps door about 1/16"

Swing

Door

4. Finished symbol

Figure 9-7
Steps for drawing a plan symbol of a swing door using a circle template. Note that in the finished symbol all elements (door, arc, and wall) touch each other. There are no gaps between them.

circle template to draw the arc. A template may be easiest to use for inking because a compass ink attachment is not required. When using a circle template, a circle with a radius the same length as the door opening must be chosen for the swing; a random circle cannot be chosen.

The symbol for an exterior swing door includes a threshold; interior doors do not. In the floor plan, place swing doors in corners, opening against the adjacent wall (Figure 9-8). All doors need a minimum of 3" of wall for the casing; 4" is preferable, and 5" or 6" can be useful if the wall is to be papered. This is because construction irregularities are more obvious under wallpaper. Having some extra wall space will help minimize that appearance.

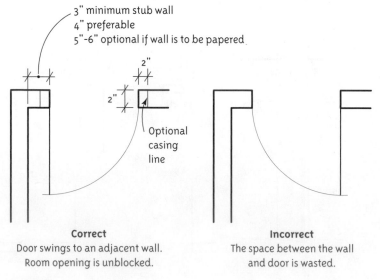

3" minimum stub wall
4" preferable
5"-6" optional if wall is to be papered

2"

2"

Optional casing line

Correct
Door swings to an adjacent wall.
Room opening is unblocked.

Incorrect
The space between the wall and door is wasted.

Figure 9-8
Proper placement of swing doors in room corners.

If a door template is used, make sure that the template's door and swing touch their respective openings; do not allow gaps. (Figure 9-9).

Figure 9-9
Using a door template. Note that the jambs touch the arc and the door.

Swing variations A **double door** (Figure 9-10) is two swing doors hung from opposite jambs in one opening. A **French** (Figure 9-11) door is a double door with lights. Double and French doors are used to obtain a larger, more attractive entrance. The **Dutch** door (Figure 9-12) is a swing door that is cut into top and bottom halves which operate independently. It is common in venues such as daycare centers, because the top can be opened for light and interaction while the bottom stays closed for security. A **double-action** (Figure 9-13) door swings both ways and is common in high-traffic areas such as between a restaurant kitchen and dining room.

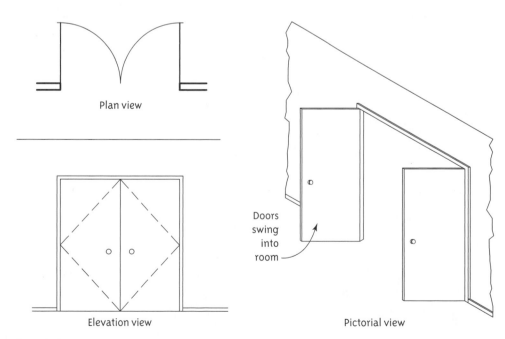

Plan view

Elevation view

Doors swing into room

Pictorial view

Figure 9-10
Interior double door.

Figure 9-11
French doors (double doors with lights). *Courtesy of Pella Windows and Doors, Pella, IA.*

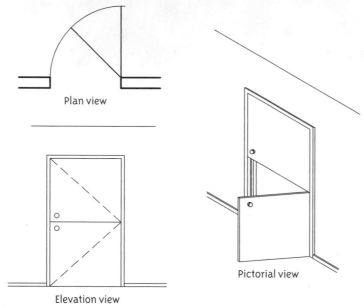

Plan view

Elevation view

Pictorial view

Figure 9-12
Interior Dutch door.

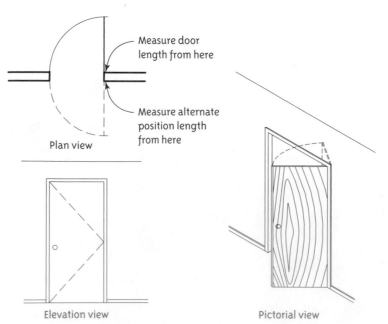

Measure door length from here

Measure alternate position length from here

Plan view

Elevation view

Pictorial view

Figure 9-13
Interior double-action door.

Sliding **Sliding** doors (Figures 9-14, 9-15) are placed where there isn't enough floor space for swing doors, such as in small rooms or tight corners. They are hung from a metal track in the frame head. Double sliding doors usually have one movable door that slides in front of one fixed door. A glass sliding door works the same way as a double sliding door and serves as both a door and window. A **pocket** door (Figure 9-16) is a single sliding door that retracts into a wall pocket.

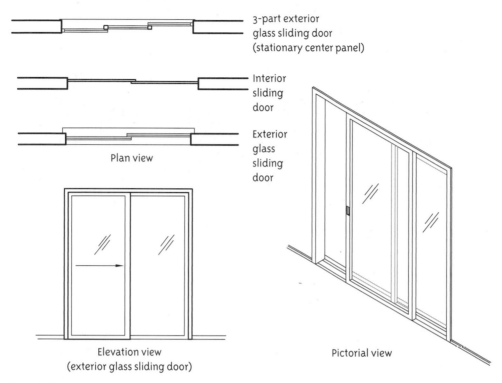

3-part exterior glass sliding door (stationary center panel)

Interior sliding door

Exterior glass sliding door

Plan view

Pictorial view

Elevation view
(exterior glass sliding door)

Figure 9-14
Sliding doors.

Figure 9-15 ·
Sliding glass door with transom window and muntins.
Courtesy of Pella Windows and Doors, Pella, IA.

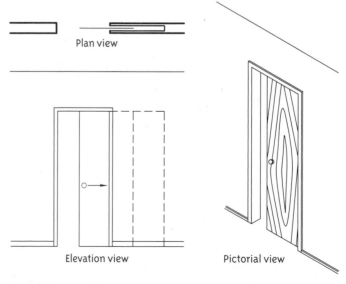

Plan view

Elevation view

Pictorial view

Figure 9-16
Pocket door.

Folding. A **folding** door is part hinged and part sliding (Figure 9-17). Two leaves (sections of a door) are hinged together, and one is also hinged to the jamb. Both leaves slide in a track. Although it requires more floor space than the folding door, it completely opens up the doorway. A single folding door has two leaves and a double folding door (also called bi-fold) has four.

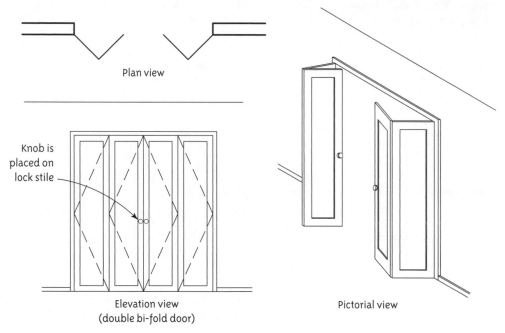

Plan view

Knob is placed on lock stile

Elevation view
(double bi-fold door)

Pictorial view

Figure 9-17
Double folding (bi-fold) door.

Accordion An **accordion** door operates like a folding door but has multiple leaves (Figure 9-18). They are made of hinged wood or flexible plastic. An accordion door provides a movable wall between rooms. When retracted, it can be, but is not always, hidden in a wall pocket.

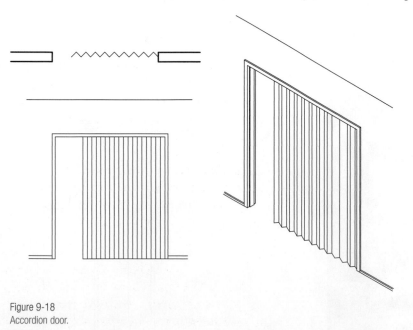

Figure 9-18
Accordion door.

Overhead Sectional This is a large door made of wood or steel that is assembled in sections and moves vertically along a track (Figure 9-19). Its hardware is an assembly of hangers, cables, pulleys, tracks, torsion springs, and an opener. Some have lights (windows) and are decorative. Overhead sectional doors are primarily found on garages and warehouses.

Revolving   This door type is used in commercial and industrial buildings (Figure 9-20). Its advantage is that it accommodates heavy traffic while minimizing drafts and energy loss.

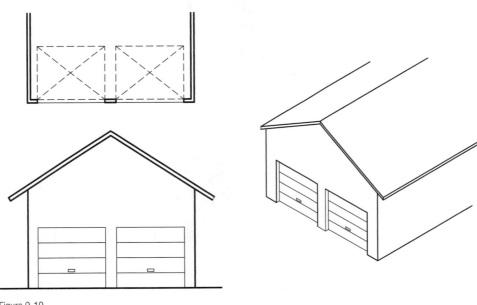

Figure 9-19
Overhead sectional door.

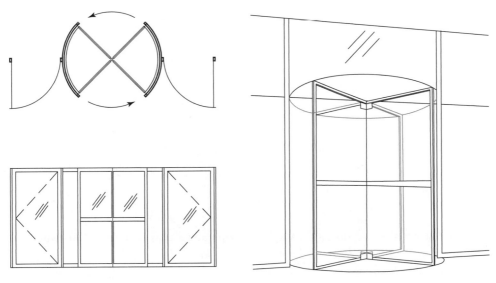

Figure 9-20
Revolving door.

Cased Opening   This is an opening with no door (Figure 9-21) Cased openings may be door height or open to the ceiling. Orthographically, the floor plan symbol looks the same whether the opening is straight or curved; the word "arch" represents intent. When drawing a door-height cased opening, be sure that the dashed lines, which represent the wall overhead, touch the solid lines, which represent the jambs.

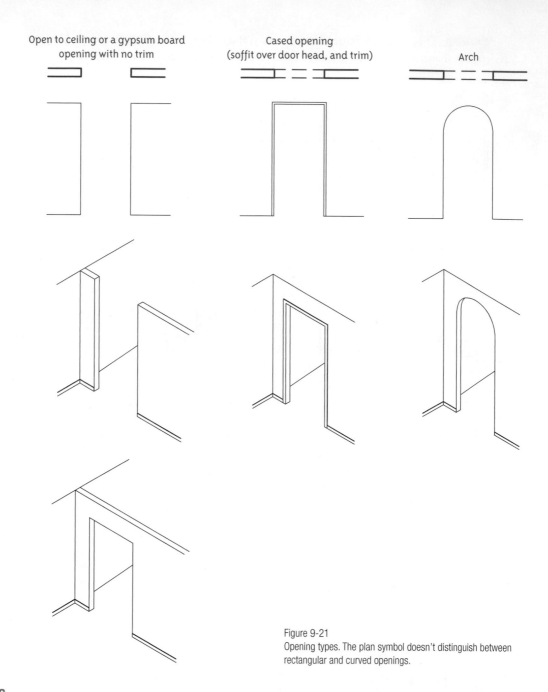

Open to ceiling or a gypsum board
opening with no trim

Cased opening
(soffit over door head, and trim)

Arch

Figure 9-21
Opening types. The plan symbol doesn't distinguish between
rectangular and curved openings.

Door Sizes

Exterior residential swing door sizes vary, but a common size is 6'-8" tall, 3'-0" wide and
1¾" thick. In wheelchair-accessible homes, the entry should have a 34" clearance around it.
The typical interior swing door is 6'-8" tall and 1 3/8" thick. Standard widths range from
2'-0" to 3'-0" in increments of 2". The minimum width for a bedroom door width should be
2'-6", but, where possible, a wider door, such as 2'-8" or 2'-10", is more desirable so furniture
can be easily moved in and out. Wheelchair-accessible doors must be at least 2'-8" wide; 3'-0"
is preferable. The **rough opening**, or hole in the wall, for an interior door is usually framed
3" higher than the door height and 2 ½" more than the door width to provide space for the
frame. The term *rough opening* applies to wood-framed buildings; **masonry opening** applies to
concrete or brick buildings.

On panel doors, narrow door stiles are 2" wide, medium door stiles are 4" wide, and wide
door stiles are 5" wide. A narrow-stile door usually has a 4" bottom rail and a medium- or
wide-stile door has a 6" bottom rail. High bottom rails are 10".

Sliding doors are generally 6'-8" tall with a total unit width of 6'-0" or 8'-0". Larger unit widths of 9'-0" or 12'-0" are available, which generally include a 4'-0" stationary glass panel between the doors.

Code Considerations

Egress doors Egress refers to a continuous, unobstructed path of travel to move occupants through a building. Egress doors are placed specifically to provide this clear travel path. Building codes, ICC (International Codes Council), and ANSI (American National Standards Institute) standards and ADAAG (Americans with Disabilities Act Accessibility Guidelines) usually require egress doors to be swing doors at least 6'-8" high and, when open, provide 32" of **clear width**, which is the distance between the jamb protrusions. In commercial buildings egress doors generally swing in the direction of exit travel to avoid the possibility of people piling up against a door that opens inward. An egress door cannot reduce any required stair landing dimensions or corridor width by more than 7" when fully open, nor can it impinge on more than half of the required corridor width at any open position. Depending on whether the approach to the door is from the push, pull, hinge, or latch side, the wall in which that the door closes should meet minimum width requirements as shown in Figures 9-22 and 9-23.

A door that is at least 36" wide is needed to comply with the 34" clear width requirement. The codes do not allow any leaf on an egress door to be more than 4' wide. Therefore, if 60" of exit width are required, more than one door is needed. Two 36" doors provide 68" of clear width. This exceeds the requirement, but is the closest increment. Remember, a 36" wide door only provides 34" of clear width. If 40" of exit width is needed, one 48" door works, but one 36" door does not, and there may not be room for two 36" doors. To obtain the required door widths in smaller rooms, doors can swing into the room. If for design reasons the door must swing out toward the travel path, options include using a 180° door to allow it to fully open against the wall, recessing the door into the room via an alcove (Figure 9-24), enlarging the landing at the door, or widening the corridor.

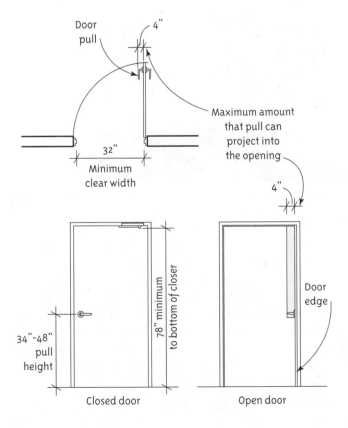

Figure 9-22
Hardware clearance requirements for a means of egress door.

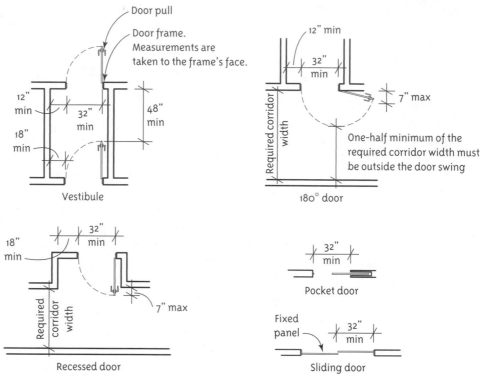

Figure 9-23
Opening clearance requirements for a means of egress door.

Figure 9-24
Door recessed in an alcove so it will not block traffic flow in a busy corridor.

Revolving and power-operated doors may sometimes be used with manual swing doors to satisfy egress requirements. However, the codes impose additional requirements; revolving doors must collapse to provide the minimum clear width and power-operated doors must be able to be manually opened. Still, allowable use of these doors for emergency egress is limited, and even when allowable, can only occur in institutional and mercantile occupancies. Some codes credit revolving doors with meeting 50 percent of the exit requirements; others do not credit them at all.

Sectional overhead door height is 6'-0" to 8'-0" in 3" Most residential models are 7'-0" (four sections high) or 8'-0" (five sections high). Standard widths are 8'-0", 9'-0", 10'-0", 12'-0", 14'-0", 15'-0", 15'-6", 16'-0", and 18'-0". The rough opening is the same size as the door, because the door fits against the opening from the inside.

Door Hardware

Door **hardware** is everything that operates and holds a door in place (Figure 9-25). It includes **hinges, locks, handles, knobs, exit devices, bolts, closers, holders,** and **stops**. It also includes accessories such as kick plates and pull/push plates. The door's function, fire rating, how many people will use it daily, and building codes are considered in hardware selection.

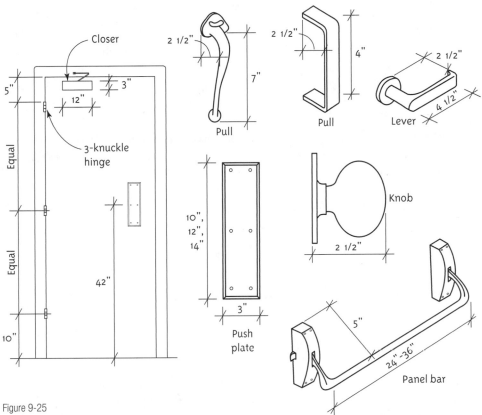

Figure 9-25
Commercial and residential door hardware.

Hardware, especially on fire doors and building exits, is regulated by local and national codes. The Americans with Disabilities Act and the National Fire Protection Association's Life Safety Code are the model codes most referenced. While requirements vary, all codes' basic premise is to provide free egress at all times. In a commercial building, occupants must be able to exit a door in one operation with one simple maneuver. For example, an occupant can operate a lever by simply pushing it down. If a deadbolt is added, requiring an extra step, the door will not meet code.

Hinges These hold the door to the wall. There are full mortise, half mortise, full surface, half-surface, spring, swing-clear, and the conventional "three knuckle" hinges. Spring hinges contain springs used as automatic closers and are best for doors that do not receive heavy traffic yet need to be closed each time they are opened. Swing-clear hinges are bent so that a door can open fully, allowing a clear opening.

Handle and Knob The handle and knob operate a latch that opens the door. Lever handles are used in commercial buildings; knobs are used exclusively in residential ones.

Locks These provide security to a door. They have three basic components: a latch bolt, a lock strike, and a cylinder. A bolt is the metal bar that protrudes from the lock and holds it to the strike, which is the metal compartment in the door that receives the bolt. Spring-loaded latch bolts, which retract when the knob is turned, are used in almost all doors. The cylinder contains the springs and tumblers that allow the keys to operate.

There are four types of locks, named for how they are installed in the door: bored, preassembled, mortised, and integral. Each includes some or all the following: bolts, latch bolts, dead latches, deadbolts, lock strikes, and cylinders. Mortise locks are stronger and more versatile than other types of locks. Magnetic locks utilize an electromagnet.

Bolts These provide supplemental security to a lock or hold the inactive half of a double door in place.

Exit Devices These are latches released by pressing a panic bar (also called a cross bar). They give a crowd quick, easy egress in an emergency. They consist of the latch, a panic bar, and trim. Codes strictly regulate their placement. Some states require them to be placed between 30" and 44" above the finished floor. Others require them to be placed at least 36" above the finished floor.

Closers hese automatically close a door after it has been manually opened. They are selected based on the door's traffic level. A closer has a spring, a checker that controls closing speed, and an arm that connects the door to the frame.

Holders and Stops These protect walls and equipment from the door as it swings open. They are typically metal with rubber bumpers on the top or front. Holders, attached to the door and frame, keep a door in an open or quasi-open position.

Some door hardware provides access control, which means it is connected to a security system that records and monitors door use. Locks support the monitoring system, and hinges have interior spaces inside where wires can be run wires. Electrified hardware allows remote access.

Windows

A window's purpose is to provide light, air, privacy, security, a view, and sometimes emergency egress. Proper window design and placement is needed to fulfill illumination and emergency egress requirements. At the same time, **fenestration**, which is the arrangement, proportioning, and design of windows and doors in a building, is a major design consideration. So although windows are selected to fulfill interior needs, their selection must also consider overall building appearance.

A **sash** (Figure 9-26) comprises the window glass plus the frame that holds it. This frame consists of two **rails** and two **stiles**. A rail is the horizontal structural component and a stile is the vertical structural component of a window or door's top and bottom edge. A separate frame, called **trim** or **casing**, is installed around the sash. The sash may contain muntins, thin bars that divide a large piece of glazing (glass) into smaller pieces. Purely decorative muntins are added directly to the glass to obtain a divided-light appearance (Figures 9-27, 9-28) and sometimes are removable as one large grille to facilitate cleaning.

Mullions are larger horizontal and vertical bars between window units.

Glazing The words *glass, lights, glazing,* and *pane* are often used synonymously; technically, **glazing** is glass and a **pane** is a framed sheet of glass within a window. If there are no muntins, the pane is the whole piece of glass, stile to stile, rail to rail. Glass choices include obscure, single, double- and triple-pane, laminated, tempered, insulating, low-emissive, gas-filled, impact, and art. All provide different levels of energy efficiency, sun protection, and privacy. Most windows have double-pane glass. This is a type of glazing where the panes are hermetically sealed, separated with a spacer, and sometimes filled with krypton or argon gas, which provides additional insulation. Low-emissive, or "low-e" glazing is double-pane glass in which the outside pane has a thin, clear, heat-reflective coating. There may be a heat mirror between the panes, which is a low-e film that either reflects heat away from the house or retains it.

Figure 9-26
Installation of a double-hung window. *Courtesy of Pella Windows and Doors, Pella, IA.*

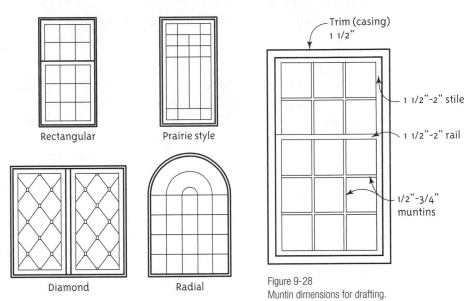

Rectangular

Prairie style

Diamond

Radial

Figure 9-27
Muntin patterns. Exterior view of a double-hung window.

Trim (casing) 1 1/2"

1 1/2"-2" stile

1 1/2"-2" rail

1/2"-3/4" muntins

Figure 9-28
Muntin dimensions for drafting.

Types

Like doors, windows arrive from manufacturers as completely assembled units. They share some of the same construction terms.

There are 11 window types: **double hung**, **fixed**, **slider**, **casement**, **awning**, **hopper**, **jalousie**, **pivot**, **bay**, **bow**, and **skylight**. There are also combination windows, which consist of two or more put together.

Double hung Also called vertical slide windows, these windows have two sashes that slide vertically in grooves inside a frame (Figures 9-29, 9-30). Friction devices hold the sashes where desired.

The plan symbol represents a horizontal slice through the window four feet up from the floor. The centered double lines are the glazing, and the lines on either side of the glazing are the wall sill below. Note the line weights. Sill and glazing lines are drawn lighter than the walls.

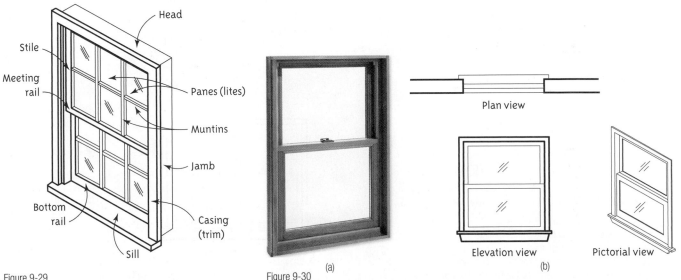

Head

Stile

Meeting rail

Panes (lites)

Muntins

Jamb

Bottom rail

Casing (trim)

Sill

Figure 9-29
Double-hung window as seen from the outside.

(a)

Figure 9-30
Double-hung window as seen from the interior. *Courtesy of Loewen, Manitoba, Canada.*

Plan view

Elevation view

Pictorial view

(b)

Note how all lines are parallel and intersecting lines touch; there are no gaps between them. If the window is drawn at a large scale (greater than 1/4" = 1'-0"), the symbol may include muntins and mullions.

Single hung A single-hung window has one sliding sash and one fixed sash. The complete unit—sashes, frame, weatherstripping, and hardware—is placed in the wall's rough opening and then nailed to the studs on the sides and to the header on top. After the interior walls are finished, trim is nailed around the window to conceal the joints where the window meets the wall. A sloping sill is installed on the sill plate of the rough opening.

Fixed This type of window doesn't open (Figure 9-31). It consists simply of a sash and the frame around it. They come in rectangular, diamond, radial, prairie style and other patterns. They may be embellished with muntin bars, nonstructural pieces that divide a window into multiple panes.

A fixed window can be bought as a preassembled unit or the glass can be custom cut on site and directly attached to the wall frame. Since a fixed window does not open, screens, weatherstripping and hardware are not needed. Fixed windows come in square, rectangular, trapezoidal, round, and half-circle shapes. A large fixed window is often called a picture window. Picture windows usually don't have muntins, as their purpose is to provide an unobstructed view.

A fixed window's symbol looks the same whether the window is square, rectangular, circular, trapezoidal, or triangular.

Slider Also called a glider, a slider is a two-sash window that slides horizontally (Figure 9-32). Both sashes may slide or one may be fixed. Some gliders have three sashes, with the middle one being fixed and the other two being operable.

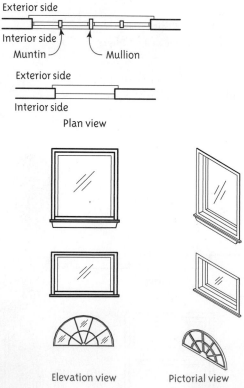

Figure 9-31
Fixed window. The floor symbol is the same no matter what the window's shape. Two different windows are shown in the plan view. The first contains muntins and mullions, and its parallel center lines represent double-paned glazing. The second shows one line, which represents the single sash.

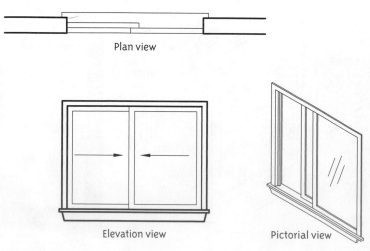

Figure 9-32
Slider window.

Casement A casement window is hinged on the side and swings out (Figures 9-33, 9-34). A casement is usually operated by a hand crank. A casement window may have one sash or it may have multiple sashes separated by vertical mullions. A pair of casement sashes may swing in the same direction, swing together to close on themselves, or swing together to close into a vertical mullion. Some casement windows are projected casements; these have a linking mechanism other than hinges (Figure 9-35). Awning and hopper windows can also have this linkage.

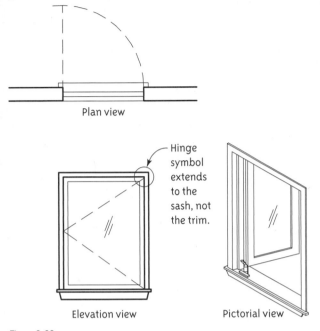

Figure 9-33
Casement window.

Figure 9-34
Two-sash casement window. *Courtesy of Loewen, Manitoba, Canada.*

Awning An awning window is hinged at the top and swings out, resembling an awning when open (Figures 9-36, 9-37). A window unit may have one sash or multiple sashes separated by horizontal mullions. Multiple sashes may swing in the same direction, swing together to close on themselves, or swing together to close into a horizontal rail.

Figure 9-35
Projected 2-sash awning window.
Courtesy of Loewen, Manitoba, Canada.

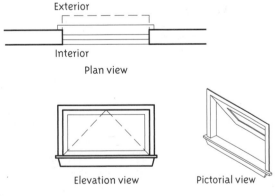

Figure 9-36
Awning window.

Figure 9-37
Awning windows. *Courtesy of Pella Windows and Doors, Pella, IA.*

Hopper A hopper window is hinged at the bottom and swings into the house so rain won't enter (Figure 9-38). Instead of a hand crank, there is usually a handle at the sash top. Its purpose is to provide ventilation because it directs air upward. Hopper windows are usually installed in basements. Both awning and hopper windows open in the direction they do to keep rainwater out.

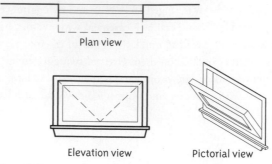

Plan view

Elevation view   Pictorial view

Figure 9-38
Hopper window.

Jalousie Also called a louvered window, a jalousie window has rows of narrow, horizontal glass slats fastened with clips to a frame (Figure 9-39). The slats can be transparent or opaque, and open and close in unison like a mini-blind. Even when closed they are not completely weather tight, so they are not a good choice for harsh climates; an additional consideration is that the arrangement of panes obscures the view. Their advantage is their ability to direct ventilation.

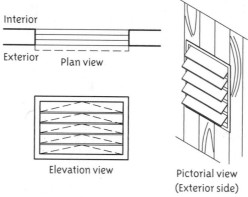

Interior

Exterior   Plan view

Elevation view   Pictorial view
                        (Exterior side)

Figure 9-39
Jalousie window.

Pivot A pivot window has pivots instead of hinges (Figure 9-40). It may pivot horizontally or vertically. Its advantage is that it is easy to clean.

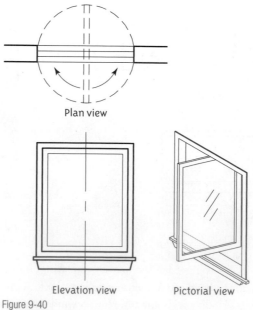

Plan view

Elevation view   Pictorial view

Figure 9-40
Pivot window.

Combination This is a pairing of two or more different window types. A picture window is often flanked by two movable windows. A fixed fan (half-circle) window may be paired with a fixed square. A **Palladian, also called a Venetian,** window is a combination, or three-part window composed of a large, arched central section flanked by two narrower, shorter sections with square tops (Figure 9-41).

Figure 9-41
Palladian window. *Courtesy of VELUX America.*

A window drawn on a floor plan that is intended for placement above the 4' cutting plane is drawn with a hidden line. If multiple windows are stacked vertically with some at and some above the cutting plane, the window symbol is drawn solid, because it represents the window at the cutting plane level. The composition is clarified via a **schedule call-out** and in the elevation. A common example of an above-cutting-plane–level window is a **transom,** which is a fixed window above a door. Drafters are sometimes presented with unusual cases such as the giant half-circle pictured in Figure 9-42. It has a different width at different places along its height. The drafter drew two solid lines representing the width at the place the cutting plane touches were drawn, then added dashed lines on either side of them to indicate the total width (Figure 9-43).

Figure 9-42
Custom fixed window with mullions. *Courtesy of Pella Windows and Doors, Pella, IA.*

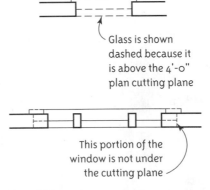

Glass is shown dashed because it is above the 4'-0" plan cutting plane

This portion of the window is not under the cutting plane

Figure 9-43
Plan symbol for the window in Figure 9-42. Portions that are not cut by the standard 4'-0" cutting plane are shown with dashed lines.

Bay A bay window projects 18" to 24" from the wall (Figure 9-44). The most common type goes from floor to ceiling and consists of a fixed center and two double-hung or casements windows. The side windows are set at 30° or 45° angles to the wall. Bay windows are available as units that consist of three or more windows (Figure 9-45) or they may be assembled into a combination window; that is, individual windows may be placed as one composition in a bay-framed wall (Figure 9-46). When the side windows are 90° to the wall the window is called a **box bay** or **garden window** (Figure 9-47); such a window projects 12" to 18" from the wall. Figure 9-48 shows how to draw a bay window.

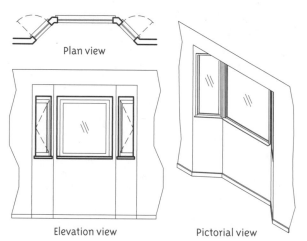

Figure 9-44
Bay window.

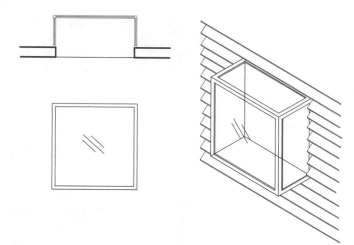

Figure 9-47
Box bay (garden) window.

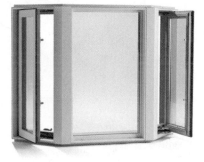

Figure 9-45
Bay window as one complete unit. *Courtesy of Loewen, Manitoba, Canada.*

Figure 9-46
Floor to ceiling separately-framed bay window. *Photo courtesy of Pella Windows and Doors, Pella, IA.*

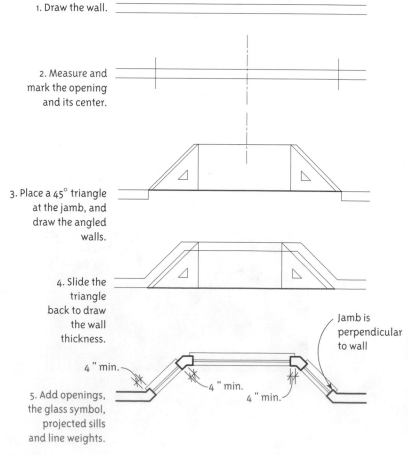

1. Draw the wall.

2. Measure and mark the opening and its center.

3. Place a 45° triangle at the jamb, and draw the angled walls.

4. Slide the triangle back to draw the wall thickness.

5. Add openings, the glass symbol, projected sills and line weights.

Jamb is perpendicular to wall

4" min.

4" min.

4" min.

Figure 9-48
Steps for drawing a three-sash bay window.

Bow Like a bay window, a bow window projects from the wall (Figures 9-49, 9-50). Its panes generally are narrower and more numerous than those of a bay window, and are installed to give the illusion of a curve. Bow windows can be fixed or operable; casements are often chosen. Figure 9-51 shows how to draw a bow window. Figure 9-52 shows some common widths and arrangements for bow and bay windows.

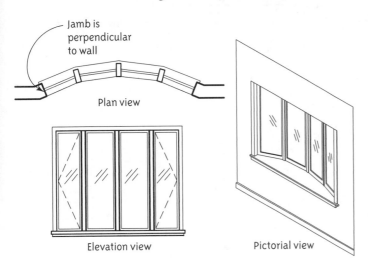

Jamb is perpendicular to wall

Plan view

Elevation view

Pictorial view

Figure 9-49
Bow window.

Figure 9-50
Bow window. *Courtesy of Pella Windows and Doors, Pella, IA.*

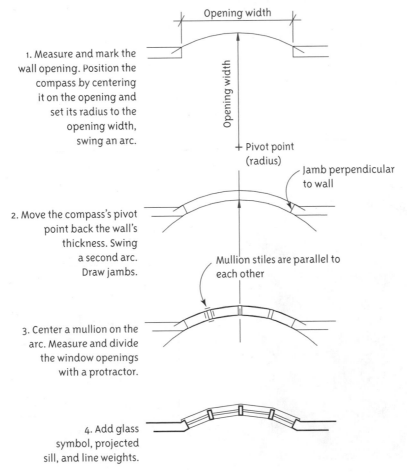

1. Measure and mark the wall opening. Position the compass by centering it on the opening and set its radius to the opening width, swing an arc.

Opening width

Opening width

+ Pivot point (radius)

Jamb perpendicular to wall

2. Move the compass's pivot point back the wall's thickness. Swing a second arc. Draw jambs.

Mullion stiles are parallel to each other

3. Center a mullion on the arc. Measure and divide the window openings with a protractor.

4. Add glass symbol, projected sill, and line weights.

Figure 9-51
Steps for drawing a four-sash bow window.

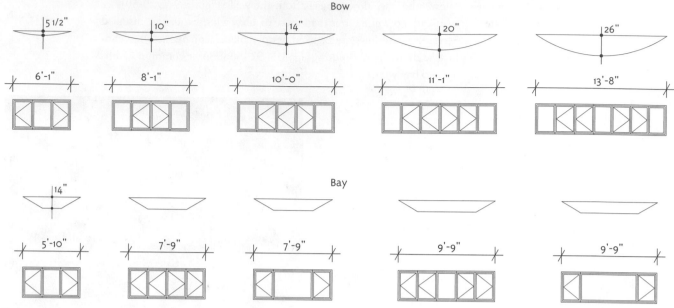

Bow

Bay

Figure 9-52
Bay and bow window sizes and arrangements.

Skylights and Clerestories Skylights are windows in the roof (Figure 9-53). They are used to obtain more natural light than what the windows in the wall can provide. They are usually fixed, rectangular, and sized to fit between the roof framing, but operable models and non-rectangular shapes are available. Skylights can be installed as single units or as multiple units in a grid. They may be drawn on the floor plan with hidden lines (Figure 9-54) or on the reflected ceiling plan with solid lines.

Clerestory windows are not skylights, but are windows set high in the wall, usually directly under the roof (Figure 9-55). Their purpose is to illuminate a difficult space while maintaining privacy. They are usually fixed but can be operable. Since they are above the 4' horizontal cutting plane, it is technically correct to show them dashed in the floor plan.

Figure 9-53
Skylights. *Courtesy of VELUX America Inc.*

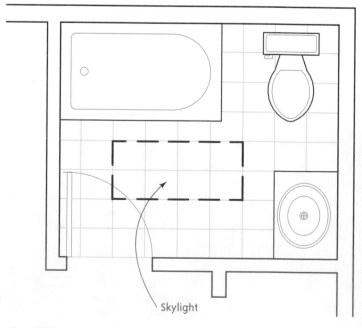

Skylight

Figure 9-54
Skylight floor plan symbol.

Figure 9-55
Clerestory window. *Courtesy of VELUX America.*

Sizes

There are four sizes associated with a window: rough opening, unit, sash and glass. It's necessary to know exactly what is being referenced when you are given window and door sizes. Window and door sizes are given width first and height second.

Rough opening size describes the hole in a wood-framed wall. It is approximately 1/4" larger on each side than the unit size, and is the measurement installers need to level and plumb. Unit size is the overall dimensions of the whole window and is used when drafting. This size is measured from different places depending on the window. For instance, wood double-hung windows sometimes come with attached trim, so the unit size includes this trim. Clad casement windows often come with nailing flanges that are included in the unit size. Sash size measures the glass plus the frame that grips it. Some windows, such as casements, have one sash and others, such as sliders, have two. Sash size refers to each piece; two sashes together are not measured as one sash. Glass size is the measurement of the glass alone; it does not include the immediate frame.

Most windows range from 2'-0" to 12'-0" widths, in 6" increments. Exact size varies with the material and manufacturer; obtain product catalogs and technical guides for size charts and details. Figure 9-56 shows some common, generic window sizes. Figures 9-57–9-63 shows sizes specific to Loewen windows (a company based in Manitoba, Canada).

COMMON WINDOW UNIT SIZES

TYPE	WIDTH	HEIGHT	NOTES
DOUBLE HUNG	1'-10", 2'-0", 2'-4", 2'-6", 2'-8", 2'-10", 3'-0", 3'-4", 3'-6", 3'-8", 3'-10", 4'-0", 4'-6"	2'-6", 3'-0", 3'-2", 3'-6", 4'-0", 4'-6", 4'-10", 5'-0", 5'-6", 5'-10", 6'-0"	
HORIZONTAL GLIDER	3'-8", 4'-8", 5'-8", 6'-0", 6'-6"	2'-10", 3'-6", 4'-2", 4'-10", 5'-6", 6'-2"	For multiple units, overall scale is the sum of each unit plus 3/4" for each mullion used.
CASEMENT	1'-6", 2'-0", 3'-0", 3'-4", 4'-0", 6'-0", 8'-0", 10'-0", 12'-0"	1'-0", 2'-0", 3'-0", 3'-4", 3'-6", 4'-0", 4'-6", 5'-0", 5'-4", 5'-6", 6'-0"	For multiple units, overall size is the sum of each unit plus 1/8" for each mullion used.
FIXED	1'-0", 1'-6", 2'-0", 2'-6", 3'-0", 4'-0", 5'-0", 5'-10", 6'-0"	4'-6", 4'-10", 5'-6", 6'-6", 7'-0", 7'-6", 8'-0"	
HOPPER	2'-0", 2'-8", 3'-6", 4'-0"	1'-4", 1'-6", 1'-8", 2'-0", 4'-0", 6'-0"	
JALOUSIE	1'-8", 2'-0", 2'-6", 3'-0"	3'-0", 4'-0", 5'-0", 6'-0"	
BAY/BOW	4'-0", 6'-0", 8'-0", 10'-0", 12'-0"	3'-0", 4'-0", 5'-0", 6'-0"	
AWNING	2'-0", 2'-6", 2'-8", 3'-0", 3'-4", 4'-0", 5'-4", 6'-0", 6'-5", 8'-0", 10'-0", 12'-0"	1'-6", 2'-0", 3'-0", 3'-6", 4'-0", 5'-0", 6'-0"	For multiple units, overall size is the sum of each unit plus 1/8" for each stack used.
FAN	2'-0", 2'-4", 2'-6", 2'-10", 3'-0", 3'-6", 4'-0", 4'-8", 5'-0", 5'-4", 6'-0"	1'-6", 2'-0", 2'-8", 2'-10", 3'-0", 4'-0"	Height is chord height.

Figure 9-56
Chart of common window unit sizes.

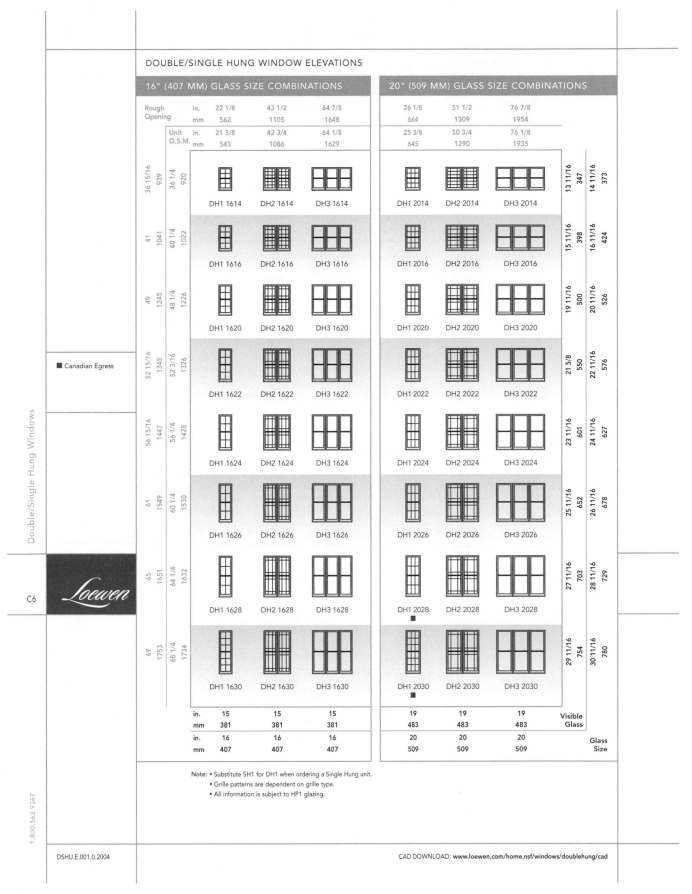

Figure 9-57
Double- and single-hung window sizes.

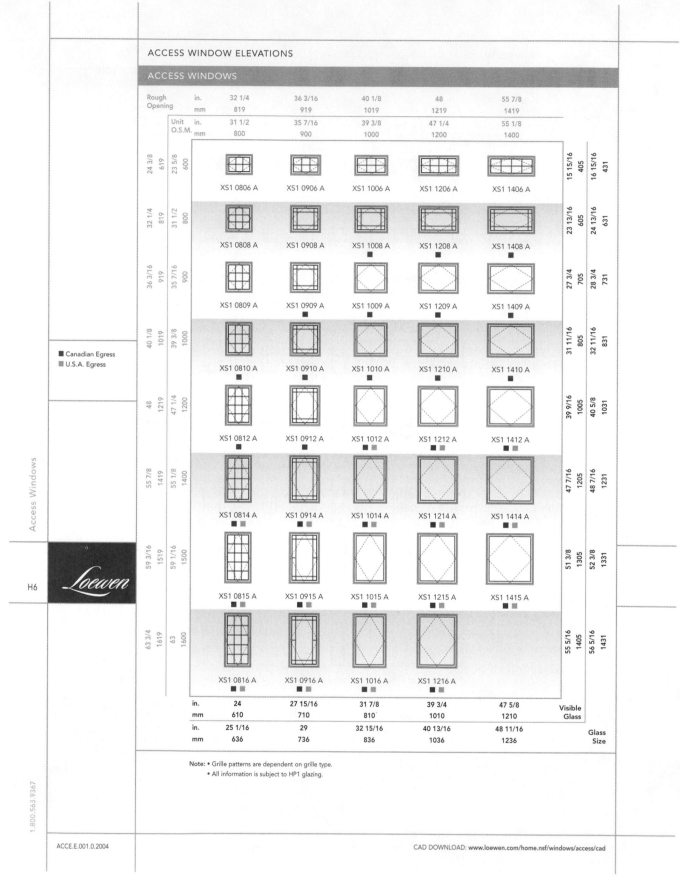

ACCESS WINDOW ELEVATIONS

ACCESS WINDOWS

Note: • Grille patterns are dependent on grille type.
• All information is subject to HP1 glazing.

■ Canadian Egress
■ U.S.A. Egress

Access Windows

H6

Loewen

Figure 9-58
Access (pivot) window sizes.

TRADITIONAL AWNING WINDOW ELEVATIONS

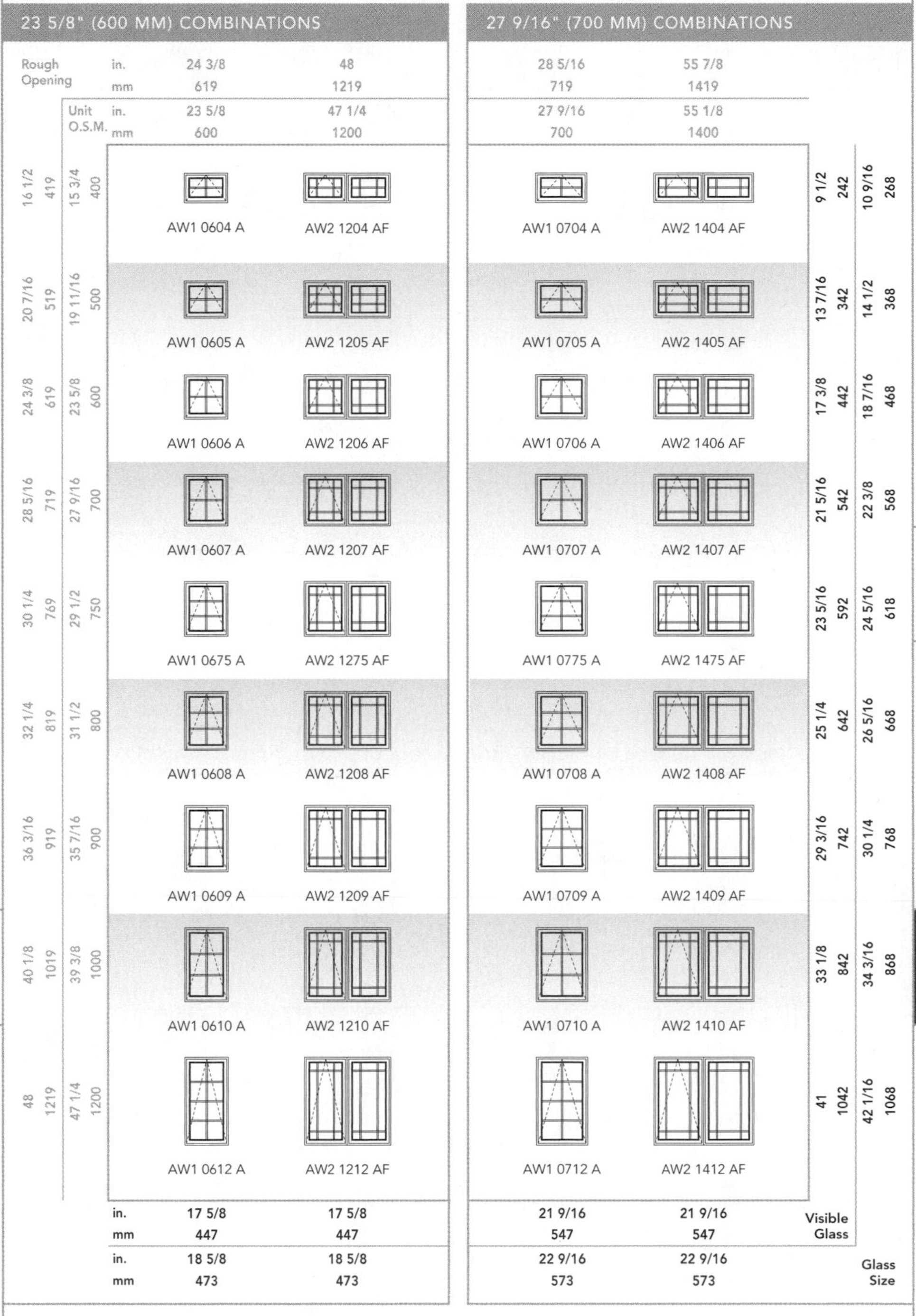

23 5/8" (600 MM) COMBINATIONS | 27 9/16" (700 MM) COMBINATIONS

Rough Opening	in.	24 3/8	48		28 5/16	55 7/8		
	mm	619	1219		719	1419		
Unit O.S.M.	in.	23 5/8	47 1/4		27 9/16	55 1/8		
	mm	600	1200		700	1400		

			23 5/8" COMB		27 9/16" COMB			
16 1/2 / 419	15 3/4 / 400	AW1 0604 A	AW2 1204 AF	AW1 0704 A	AW2 1404 AF	9 1/2 / 242	10 9/16 / 268
20 7/16 / 519	19 11/16 / 500	AW1 0605 A	AW2 1205 AF	AW1 0705 A	AW2 1405 AF	13 7/16 / 342	14 1/2 / 368
24 3/8 / 619	23 5/8 / 600	AW1 0606 A	AW2 1206 AF	AW1 0706 A	AW2 1406 AF	17 3/8 / 442	18 7/16 / 468
28 5/16 / 719	27 9/16 / 700	AW1 0607 A	AW2 1207 AF	AW1 0707 A	AW2 1407 AF	21 5/16 / 542	22 3/8 / 568
30 1/4 / 769	29 1/2 / 750	AW1 0675 A	AW2 1275 AF	AW1 0775 A	AW2 1475 AF	23 5/16 / 592	24 5/16 / 618
32 1/4 / 819	31 1/2 / 800	AW1 0608 A	AW2 1208 AF	AW1 0708 A	AW2 1408 AF	25 1/4 / 642	26 5/16 / 668
36 3/16 / 919	35 7/16 / 900	AW1 0609 A	AW2 1209 AF	AW1 0709 A	AW2 1409 AF	29 3/16 / 742	30 1/4 / 768
40 1/8 / 1019	39 3/8 / 1000	AW1 0610 A	AW2 1210 AF	AW1 0710 A	AW2 1410 AF	33 1/8 / 842	34 3/16 / 868
48 / 1219	47 1/4 / 1200	AW1 0612 A	AW2 1212 AF	AW1 0712 A	AW2 1412 AF	41 / 1042	42 1/16 / 1068

	in.	17 5/8	17 5/8	21 9/16	21 9/16	Visible Glass
	mm	447	447	547	547	
	in.	18 5/8	18 5/8	22 9/16	22 9/16	Glass Size
	mm	473	473	573	573	

Note:
- For a unit with a Mission sash subtract 3/4" (20 mm) from the vertical glass measurement.
- Grille patterns are dependent on grille type.
- Available with Mission sash.
- All information is subject to HP1 glazing.

CAD DOWNLOAD: www.loewen.com/home.nsf/windows/awning/cad

AWNI.E.001.0.2004

Awning Windows

Loewen F7

www.loewen.com

Figure 9-59
Awning window sizes.

TRADITIONAL CASEMENT WINDOW ELEVATIONS

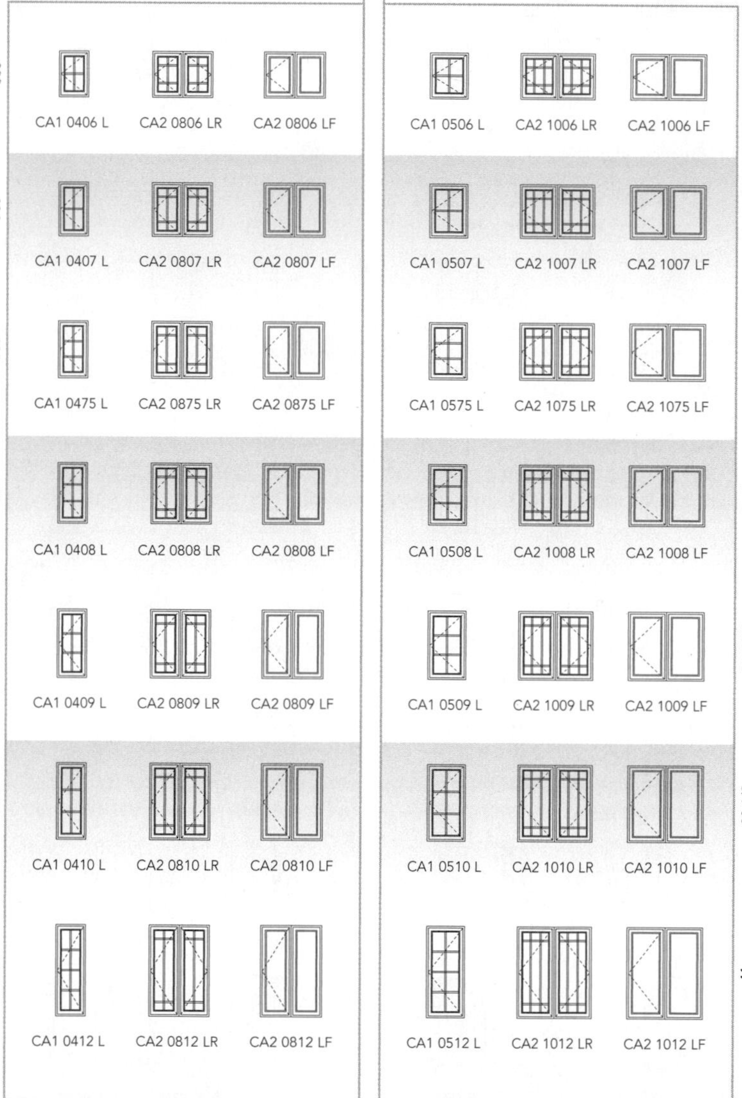

		15 3/4" (400 MM) COMBINATIONS			19 11/16" (500 MM) COMBINATIONS				
Rough Opening	in. mm	16 1/2 / 419	32 1/4 / 819	32 1/4 / 819	20 7/16 / 519	40 1/8 / 1019	40 1/8 / 1019		
Unit O.S.M.	in. mm	15 3/4 / 400	31 1/2 / 800	31 1/2 / 800	19 11/16 / 500	39 3/8 / 1000	39 3/8 / 1000		

Left column row measurements (Rough Opening / Unit O.S.M.):

24 3/8 / 619	23 5/8 / 600	CA1 0406 L	CA2 0806 LR	CA2 0806 LF	CA1 0506 L	CA2 1006 LR	CA2 1006 LF	17 3/8 / 442	18 7/16 / 468
28 5/16 / 719	27 9/16 / 700	CA1 0407 L	CA2 0807 LR	CA2 0807 LF	CA1 0507 L	CA2 1007 LR	CA2 1007 LF	21 5/16 / 542	22 3/8 / 568
30 1/4 / 769	29 1/2 / 750	CA1 0475 L	CA2 0875 LR	CA2 0875 LF	CA1 0575 L	CA2 1075 LR	CA2 1075 LF	23 5/16 / 592	24 5/16 / 618
32 1/4 / 819	31 1/2 / 800	CA1 0408 L	CA2 0808 LR	CA2 0808 LF	CA1 0508 L	CA2 1008 LR	CA2 1008 LF	25 1/4 / 642	26 5/16 / 668
36 3/16 / 919	35 7/16 / 900	CA1 0409 L	CA2 0809 LR	CA2 0809 LF	CA1 0509 L	CA2 1009 LR	CA2 1009 LF	29 3/16 / 742	30 1/4 / 768
40 1/8 / 1019	39 3/8 / 1000	CA1 0410 L	CA2 0810 LR	CA2 0810 LF	CA1 0510 L	CA2 1010 LR	CA2 1010 LF	33 1/8 / 842	34 3/16 / 868
48 / 1219	47 1/4 / 1200	CA1 0412 L	CA2 0812 LR	CA2 0812 LF	CA1 0512 L	CA2 1012 LR	CA2 1012 LF	41 / 1042	42 1/16 / 1068

| | in. mm | 9 3/4 / 247 | 9 3/4 / 247 | 9 3/4 / 247 | 13 11/16 / 347 | 13 11/16 / 347 | 13 11/16 / 347 | Visible Glass |
| | in. mm | 10 3/4 / 273 | 10 3/4 / 273 | 10 3/4 / 273 | 14 11/16 / 373 | 14 11/16 / 373 | 14 11/16 / 373 | Glass Size |

Note: • For a unit with a Mission sash, subtract 3/4" (20 mm) from the vertical glass measurement.
• Grille patterns are dependent on grille type.
• Available with Push Out hardware.
• Available with Mission sash.
• All information is subject to HP1 glazing.

CAD DOWNLOAD: www.loewen.com/home.nsf/windows/casement/cad

Loewen

Casement Windows

B8

1.800.563.9367

Figure 9-60
Casement window sizes.

PICTURE/DIRECT SET WINDOW ELEVATIONS

1 3/8" (35 MM) FRAME PICTURE/DIRECT SET WINDOWS

Rough Opening	in.	16 1/2	20 7/16	24 3/8	28 5/16	30 1/4	32 1/4	36 3/16		
	mm	419	519	619	719	769	819	919		
Unit O.S.M.	in.	15 3/4	19 11/16	23 5/8	27 9/16	29 1/2	31 1/2	35 7/16	Visible Glass	Glass Size
	mm	400	500	600	700	750	800	900		

RO (in./mm)	OSM (in./mm)									Visible Glass (in./mm)	Glass Size (in./mm)
16 1/2 / 419	15 3/4 / 400	PS1 0404	PS1 0504	PS1 0604	PS1 0704	PS1 7504	PS1 0804	PS1 0904		11 5/8 / 296	12 3/4 / 324
20 7/16 / 519	19 11/16 / 500	PS1 0405	PS1 0505	PS1 0605	PS1 0705	PS1 7505	PS1 0805	PS1 0905		15 9/16 / 396	16 11/16 / 424
24 3/8 / 619	23 5/8 / 600	PS1 0406	PS1 0506	PS1 0606	PS1 0706	PS1 7506	PS1 0806	PS1 0906		19 1/2 / 496	20 5/8 / 524
28 5/16 / 719	27 9/16 / 700	PS1 0407	PS1 0507	PS1 0607	PS1 0707	PS1 7507	PS1 0807	PS1 0907		23 7/16 / 596	24 9/16 / 624
30 1/4 / 769	29 1/2 / 750	PS1 0475	PS1 0575	PS1 0675	PS1 0775	PS1 7575	PS1 0875	PS1 0975		25 7/16 / 646	26 9/16 / 674
32 1/4 / 819	31 1/2 / 800	PS1 0408	PS1 0508	PS1 0608	PS1 0708	PS1 7508	PS1 0808	PS1 0908		27 3/8 / 696	28 1/2 / 724
36 3/16 / 919	35 7/16 / 900	PS1 0409	PS1 0509	PS1 0609	PS1 0709	PS1 7509	PS1 0809	PS1 0909		31 5/16 / 796	32 7/16 / 824
40 1/8 / 1019	39 3/8 / 1000	PS1 0410	PS1 0510	PS1 0610	PS1 0710	PS1 7510	PS1 0810	PS1 0910		35 1/4 / 896	36 3/8 / 924

Visible Glass	in.	11 5/8	15 9/16	19 1/2	23 7/16	25 7/16	27 3/8	31 5/16
	mm	296	396	496	596	646	696	796
Glass Size	in.	12 3/4	16 11/16	20 5/8	24 9/16	26 1/2	28 1/2	32 7/16
	mm	324	424	524	624	674	724	824

Note:
- All units subject to HP1 glazing.
- Standard grilles in Picture units will line up with a Casement, and are not equal lite.
- Grille pattern dependent on grille type.

Loewen E13

PICD.E.007.0.2004

www.loewen.com

Figure 9-61
Picture window sizes.

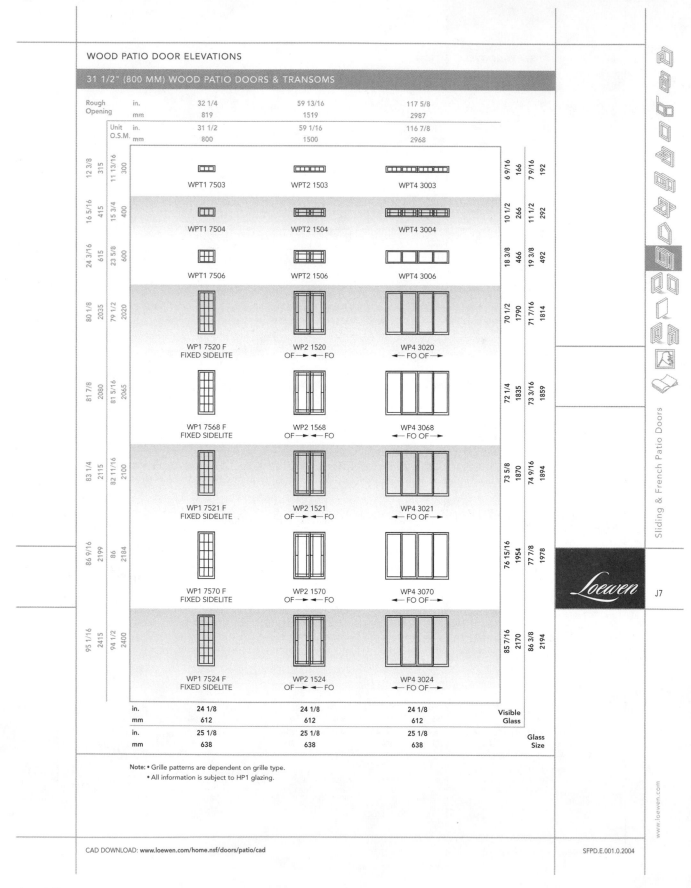

31 1/2" (800 MM) WOOD PATIO DOORS & TRANSOMS

Rough Opening	in.	32 1/4	59 13/16	117 5/8	
	mm	819	1519	2987	
Unit O.S.M.	in.	31 1/2	59 1/16	116 7/8	
	mm	800	1500	2968	

WPT1 7503   WPT2 1503   WPT4 3003

WPT1 7504   WPT2 1504   WPT4 3004

WPT1 7506   WPT2 1506   WPT4 3006

WP1 7520 F FIXED SIDELITE   WP2 1520 OF → ← FO   WP4 3020 ← FO OF →

WP1 7568 F FIXED SIDELITE   WP2 1568 OF → ← FO   WP4 3068 ← FO OF →

WP1 7521 F FIXED SIDELITE   WP2 1521 OF → ← FO   WP4 3021 ← FO OF →

WP1 7570 F FIXED SIDELITE   WP2 1570 OF → ← FO   WP4 3070 ← FO OF →

WP1 7524 F FIXED SIDELITE   WP2 1524 OF → ← FO   WP4 3024 ← FO OF →

	in.	24 1/8	24 1/8	24 1/8	Visible Glass
	mm	612	612	612	
	in.	25 1/8	25 1/8	25 1/8	Glass Size
	mm	638	638	638	

Note: • Grille patterns are dependent on grille type.
    • All information is subject to HP1 glazing.

Sliding & French Patio Doors

Loewen   J7

CAD DOWNLOAD: www.loewen.com/home.nsf/doors/patio/cad   SFPD.E.001.0.2004

www.loewen.com

Figure 9-62
Sliding door sizes.

SPECIALTY WINDOW ELEVATIONS

HALF ROUND AND EXTENDED HALF ROUND WINDOWS

HALF ROUND WINDOWS

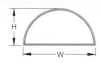

HALF ROUND STANDARD OPTIONAL GRILLE

Unit Sizes				
	Imperial		Metric	
Model Number	Unit O.S.M.	Rough Opening	Unit O.S.M.	Rough Opening
RT 0603	23 5/8 X 11 13/16	24 3/8 X 12 9/16	600 X 300	619 X 319
RT 0735	27 9/16 X 13 3/4	28 5/16 X 14 1/2	700 X 350	719 X 369
RT 0804	31 1/2 X 15 3/4	32 1/4 X 16 1/2	800 X 400	819 X 419
RT 0945	35 7/16 X 17 11/16	36 3/16 X 18 7/16	900 X 450	919 X 469
RT 1005	39 3/8 X 19 11/16	40 1/8 X 20 7/16	1000 X 500	1019 X 519
RT 1206	47 1/4 X 23 5/8	48 X 24 3/8	1200 X 600	1219 X 619
RT 1407	55 1/8 X 27 9/16	55 7/8 X 28 5/16	1400 X 700	1419 X 719
RT 1608	63 X 31 1/2	63 3/4 X 32 1/4	1600 X 800	1619 X 819
RT 1809	70 7/8 X 35 7/16	71 5/8 X 36 3/16	1800 X 900	1819 X 919
RT 2010	78 3/4 X 39 3/8	79 1/2 X 40 1/8	2000 X 1000	2019 X 1019
RT 2412	94 1/2 X 47 1/4	95 1/4 X 48	2400 X 1200	2419 X 1219
RT 2613	102 3/8 X 51 3/16	103 1/8 X 51 15/16	2600 X 1300	2619 X 1319
RT 2814	110 1/4 X 55 1/8	111 X 55 7/8	2800 X 1400	2819 X 1419
RT 3015	118 1/8 X 59 1/16	118 7/8 X 59 13/16	3000 X 1500	3019 X 1519

EXTENDED HALF ROUND WINDOWS

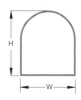

EXTENDED HALF ROUND STANDARD OPTIONAL GRILLE

Unit Sizes				
	Imperial		Metric	
Model Number	Unit O.S.M.	Rough Opening	Unit O.S.M.	Rough Opening
ER 0810	31 1/2 X 39 3/8	32 1/4 X 40 1/8	800 X 1000	819 X 1019
ER 0812	31 1/2 X 47 1/4	32 1/4 X 48	800 X 1200	819 X 1219
ER 0814	31 1/2 X 55 1/8	32 1/4 X 55 7/8	800 X 1400	819 X 1419
ER 1010	39 3/8 X 39 3/8	40 1/8 X 40 1/8	1000 X 1000	1019 X 1019
ER 1012	39 3/8 X 47 1/4	40 1/8 X 48	1000 X 1200	1019 X 1219
ER 1014	39 3/8 X 55 1/8	40 1/8 X 55 7/8	1000 X 1400	1019 X 1419
ER 1016	39 3/8 X 63	40 1/8 X 63 3/4	1000 X 1600	1019 X 1619
ER 1018	39 3/8 X 70 7/8	40 1/8 X 71 5/8	1000 X 1800	1019 X 1819
ER 1212	47 1/4 X 47 1/4	48 X 48	1200 X 1200	1219 X 1219
ER 1214	47 1/4 X 55 1/8	48 X 55 7/8	1200 X 1400	1219 X 1419
ER 1216	47 1/4 X 63	48 X 63 3/4	1200 X 1600	1219 X 1619
ER 1218	47 1/4 X 70 7/8	48 X 71 5/8	1200 X 1800	1219 X 1819
ER 1414	55 1/8 X 55 1/8	55 7/8 X 55 7/8	1400 X 1400	1419 X 1419
ER 1416	55 1/8 X 63	55 7/8 X 63 3/4	1400 X 1600	1419 X 1619
ER 1418	55 1/8 X 70 7/8	55 7/8 X 71 5/8	1400 X 1800	1419 X 1819
ER 1616	63 X 63	63 3/4 X 63 3/4	1600 X 1600	1619 X 1619
ER 1618	63 X 70 7/8	63 3/4 X 71 5/8	1600 X 1800	1619 X 1819
ER 1818	70 X 70 7/8	71 5/8 X 71 5/8	1800 X 1800	1819 X 1819
ER 2418	94 1/2 X 70 7/8	95 1/4 X 71 5/8	2400 X 1800	2419 X 1819

Note: • Grille patterns are dependent on grille type.
• All information is subject to HP1 glazing.

• Specialty shape units are available in both Casement and Picture/Direct Set configurations. Additional sizes beyond those shown here are available. Contact your local representative or Loewen Architectural Services for assistance.

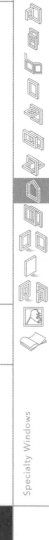

CAD DOWNLOAD: www.loewen.com/home.nsf/windows/custom/cad

SPTY.E.001.0.2004

Figure 9-63
Half-round (fan) window sizes.

Placement in the wall varies with ceiling height and function (Figure 9-64). Sill (bottom) height of windows is based on furniture, room arrangement, and view. Head height may vary depending on ceiling height, but is generally the same 6'-8" height above finished floor as the door heads. When figuring placement of a window in an elevation drawing, it is helpful to sketch a figure of a standing or sitting person to give scale. A mistake beginning drafters often make is placing windows at unrealistic heights because they are not cognizant of occupant eye level.

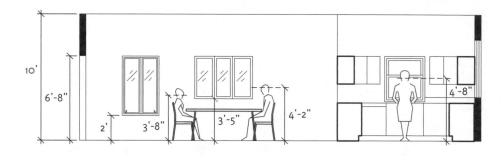

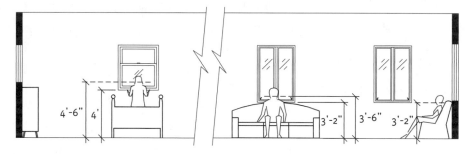

Figure 9-64
Elevation heights of eye levels and window sills. Note how the tops of windows and doors are aligned.

Windows may be used as emergency exits as long as they are operable from the inside without tools. The sill of emergency egress windows must be within 44" of the floor, with a minimum area of 5.7 square feet (grade-floor windows may have a minimum area of 5 square feet). They must have a net clear opening at least 20" wide and 24" high.

Selection and Placement

Here are 10 things to consider when choosing and placing windows.

1. Align the tops of windows on multistory buildings and choose windows that complement the building's architectural style.

2. Glass area should be at least 20 percent of the room's floor area.

3. Ventilation windows should be at least 10 percent of the floor area.

4. Placement on the south walls will bring in more light, but also will bring in more heat. Minimize this placement in warm climates; maximize it in cold.

5. Several small windows on multiple walls illuminate a room more evenly, and with better contrast, than one large opening on one wall.

6. Windows high on a wall provide more light penetration than windows placed low. Window shape also affects light penetration; tall, narrow windows give thin and deep rays of light, while short, wide ones produce broader but shallower ones.

7. Windows placed opposite each other attract cross-breezes and provide good air circulation.

8. Like mirrors, large expanses of glass can make a room look larger.

9. Design window openings to match standard window sizes.

10. Place windows to reveal the best views. Avoid undesirable views, especially for important spaces.

Door and Window Construction Details

Although doors and windows arrive as completely assembled units and details of their construction are available from the manufacturer for tracing, some drafter customization is needed to show how these components are to be installed in the walls. Therefore, it is necessary to know basic rudiments of their construction (Figures 9-65 and 9-66).

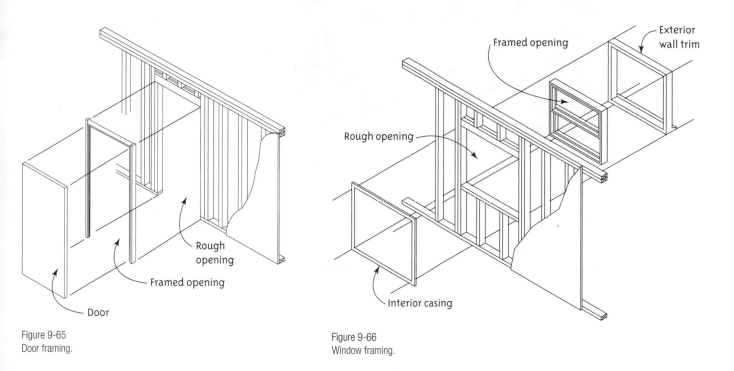

Figure 9-65
Door framing.

Figure 9-66
Window framing.

The rough opening is framed in the wall. With doors, the frame is leveled, plumbed, and attached to the rough opening and the door is attached with a hinge to the frame. With windows, the window is placed in the rough opening, leveled and adjusted, and then nailed to the frame. Trim or casing may be applied to the wall to conceal the joint between the window and the wall.

Construction details are created similarly for doors and windows. Their major components are the **head**, **jamb** and **sill**. The head is the top, the jambs are the sides and the sill is the bottom. A head detail is created by slicing vertically through the top of the door or window. A jamb detail is created by slicing horizontally through the side. A sill detail is created by slicing vertically through the bottom (an exterior sill's purpose is to drain water away from the door or window and provide support for the jambs). Brand-specific details are available from the manufacturer and may be required, depending on the project.

Figure 9-67 is of an interior door; Figure 9-68 shows its head, jamb and sill details.

Figure 9-67
Interior door in a wood-framed wall with gypsum board finish on both sides.
The cutting planes indicate how the head, jamb and sill details are made.

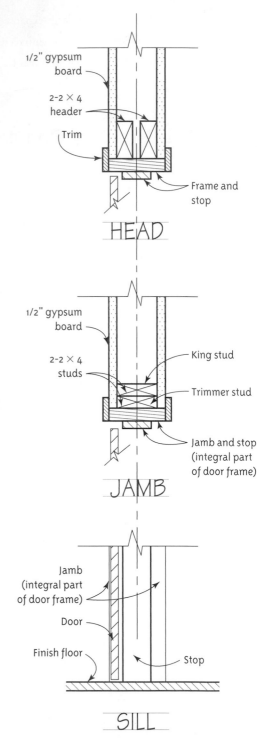

Figure 9-68
Head, jamb and sill details of the door in Figure 9-67.

The wall is wood stud construction with a gypsum board finish on both sides. A **header** (lintel beam) made of two 2 × 4s spans the opening. The door and trim are wood.

Head We're cutting vertically through the top. The two *X*s represent the header; we're looking at two 2 × 4s cut through. The gypsum boards on both sides of the header are the finished wall surfaces. The door frame is attached to the header and the top part of the door and trim is attached on both sides. We also see the top part of the door.

Jamb We're cutting horizontally through the side and looking down. Here the two *X*s represent the two studs on each side of the opening. (The one closest to the door is the trimmer stud; the next one is the king stud). Again, we see the gypsum board, frame trim, and part of the door.

Sill We're making a vertical cut, but from inside the opening. Therefore, even though this is technically a section view, most of what we see is in elevation, and is better described as a threshold drawing. It is standard practice to rotate the jamb drawing 90° and draw the head, jamb, and sill lined up vertically. Scale is usually 1/2" = 1'-0" or 3/4" = 1'-0".

To help visualize the relationship between 2-D and 3-D views, Figure 9-69 presents a cutaway pictorial of a casement window in a vinyl-sided exterior wall, and Figure 9-70 is a manufacturer drawing of its head, jamb, and sill details.

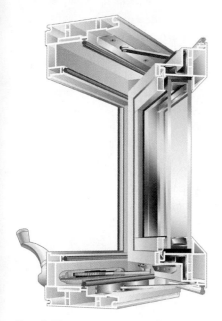

Figure 9-69
Cutaway pictorial of a casement window.

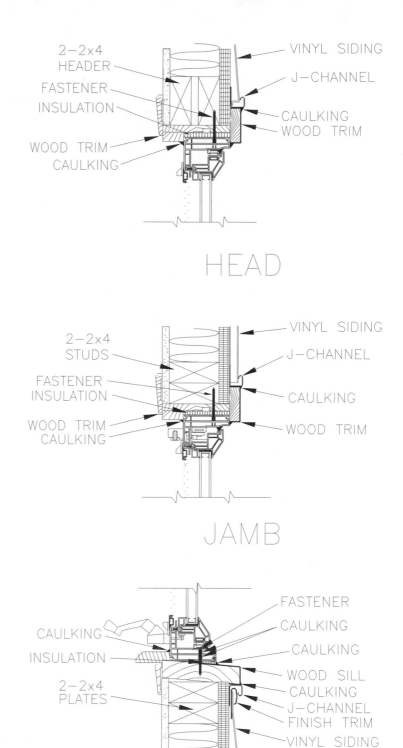

HEAD

2−2x4 HEADER
FASTENER
INSULATION
WOOD TRIM
CAULKING

VINYL SIDING
J−CHANNEL
CAULKING
WOOD TRIM

JAMB

2−2x4 STUDS
FASTENER INSULATION
WOOD TRIM
CAULKING

VINYL SIDING
J−CHANNEL
CAULKING
WOOD TRIM

SILL

CAULKING
INSULATION
2−2x4 PLATES

FASTENER
CAULKING
CAULKING
WOOD SILL
CAULKING
J−CHANNEL
FINISH TRIM
VINYL SIDING

Figure 9-70
Head, jamb and sill details of the casement window in Figure 9-70.

Hardware

Screws, hinges, knobs, screens, and other hardware are usually included in construction detail drawings; know that they all have their own code restrictions. For instance, the finished floor surface on either side of an interior door must be the same elevation as the threshold (the horizontal floor space between the door jambs) or within 1/2" of it. Both sides of a threshold hardware strip must be beveled. Codes require exit door hardware to be easily used and capable of operation with one hand and no special knowledge or skill. Levers, push-handles, and U-shaped pulls meet this description. Locks that would hinder egress do not. All door pulls must be accessibly operating devices, of a certain shape, and installed at a specific height. Closers to make a door self-closing may also be required.

Summary

Doors and windows are important building components that come in many types and sizes. It's important to know the rudiments of their construction so you can draft them properly. Symbols and details tell the reader where they are located, what they look like, and how they operate.

Questions

1. Name three ways doors may operate.
2. Name three types of windows.
3. What is the height of a typical residential swing door?
4. What is a sash?
5. Specify two ways building code requirements affect doors.
6. What are a window head, jamb, and sill?
7. Describe the difference between a flush door and a panel door.
8. What do "left-hand door" and "right-hand door" refer to?
9. What are the four window measurement sizes?
10. What is the difference between a muntin and a mullion?
11. What is a panic bar?
12. What is fenestration?

Suggested Activities

1. Look at some doors in your home or classroom and sketch their elevations and, if possible, their head, jamb, and sill details.
2. Obtain door and window catalogs and study the drawings and descriptions of their products.
3. Go through your home and note which doors are right hand, left hand, right-hand reverse, and left-hand reverse.
4. Take digital pictures of windows in elevation and trace them to familiarize yourself with their construction. Be as detailed as you can with the muntins, and trim.

Internet Resources

http://www.loewen.com
Information about Loewen window products and CAD downloads of their products.

http://www.pella.com
Information about Pella products.

http://www.velux.com
Information about VELUX roof windows, skylights and sun tunnels.

CHAPTER 10

Legends and Schedules

OBJECTIVE

- This chapter discusses legends and schedules: what they are, the components they are used for, and how to draw them.

KEY TERMS

call-outs	key number	legends
keys	marks	schedule.
key letter		

What are Legends and Schedules?

Legends and schedules are information charts (Figure 10-1). Used for particular categories of building components, they present information in an organized manner without cluttering the plan. Interior designers create them for room finishes, windows, doors, accessories, appliances, furniture, fixtures, and equipment. Both legends and schedules are referenced to the floor plan; the difference between them lies in how they are referenced and the type and quantity of information they contain.

Legends are charts that typically explain symbols on a plan view drawing, although sometimes they can explain components (Figure 10-2). For instance, an architectural floor plan with multiple poché symbols may have an accompanying legend that describes each poché. An electrical plan may have a legend for its symbols, telling the reader the item that each electrical symbol represents (Figure 10-3). This information may appear in a tabular grid or in list format.

Schedules are charts that give detailed information about components shown in the plan. They are referenced from the plan via **call-outs** or **marks** (Figure 10-4). These symbols are generic; for instance, a geometric shape such as an ellipse, hexagon, or circle with a letter or number inside is a typical call-out. The number or letter inside the call-out shape tells the reader where on the schedule to look for detailed information.

A schedule is usually in tabular form. The table's rows and columns organize the information lettered within the grid compartments. Sometimes schedules are drawn in pictorial form, and sometimes a tabular form schedule will have a column in which pictures are drawn inside its compartments.

Design practices vary; you may see some legends, such as a wall legend, referenced from the plan via generic markers. **Keys** are a type of legend used for furniture; they, too, often use generic call-outs as stand-ins for the furniture, and these call-outs may be called **key letters** or **key numbers.** Furthermore, some legends contain more information than just a symbol and short text note and hence are schedule-like.

Schedules require their own ID label, just like any other drawing. Legends may or may not have an ID label.

DOOR SCHEDULE

○	SIZE	THK	DOOR MAT'L	DOOR TYPE	FRAME MAT'L	FRAME TYPE	GLASS	LOCKSET	COMMENTS
200	2'8" x 6'8"	1⅜	WD	C	WD	A		CLASSROOM	
201	3'0" x 6'8"	1⅜	H.M.	A	WD	A	1" INSULATED	ENTRY	1'-8" SIDELIGHT
202	3'0" x 6'8"	1⅜	H.M.	B	WD	A		ENTRY	
203	FR 2'4" x 6'8"	1⅜	H.M.	D	WD	B		NONE	BI-PASSING HARDWARE
204	2'6" x 6'8"	1⅜	WD	C	WD	B		PRIVACY	
205	FR 2'6" x 6'8"	1⅜	WD	D	WD	B		PASSAGE R.H. DUMMY TRIM L.H.	
206	2'0" x 6'8"	1⅜	WD	C	WD	B		PRIVACY	
207	(4) 1'0" x 6'8"	1⅜	WD	E	WD	B		NONE	BI-FOLDING HARDWARE
208	2'0" x 6'8"	1⅜	WD	C	WD	B		PRIVACY	
209	2'0" x 6'8"	1⅜	WD	C	WD	B		PRIVACY	
210	2'6" x 6'8"	1⅜	WD	C	WD	B		PRIVACY	
211	2'0" x 6'8"	1⅜	WD	C	WD	B		PASSAGE	
212	2'6" x 6'8"	1⅜	WD	C	WD	B		PRIVACY	
213	2'6" x 6'8"	1⅜	WD	C	WD	B		PASSAGE	

Figure 10-1
This door schedule is a chart that contains detailed information about the building's doors.

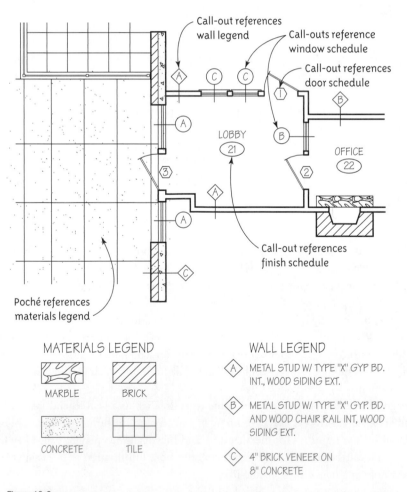

Figure 10-2
This floor plan has generic callouts that reference building components to both legends and schedules.

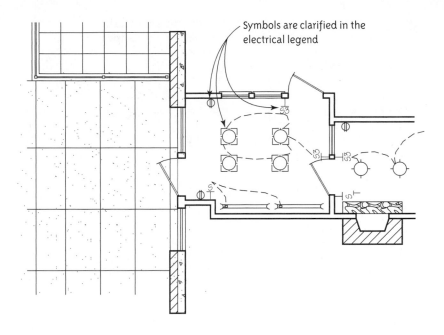

Symbols are clarified in the electrical legend

ELECTRICAL LEGEND	
SYMBOL	DESCRIPTION
⊕	INTERIOR CEILING LIGHT, 100 W INCANDESCENT
▭	INTERIOR FLOURESCENT LIGHT, 100 W
S₃	3-WAY SWITCH
⊡	RECESSED CEILING-MOUNTED LIGHT
⏀	110 V DUPLEX OUTLET
Sₜ	THERMOSTAT

Figure 10-3
An electrical plan and accompanying legend describing what the plan's symbols mean.

Drawing a Call-out Symbol

Circles and hexagons are the most common shapes for window and door call-outs and ellipses are the most common shape for room finish schedules. Triangles or diamonds are typically used for walls. Call-outs may be attached to the component with a line; if not, they should be close enough to it to make obvious what is being referenced. Letter and numbers always read upward, no matter the direction of the call-out. Room finish call-outs are placed at the center of the room.

Center door and window call-outs on those components when possible. However, drafters usually have to work around dimension and other object lines, so use common sense in all call-out placement. Call-outs should not be placed over object lines or other symbols and may be attached to angled or curved lines if this facilitates working around other object lines.

If letters are used for the window call-outs, numbers should be used for the doors (or vice versa) and their call-outs should be the same size, but different shapes. Windows and doors that are exactly the same—in size, type, and material—should have the same call-out. However, any difference requires a different letter or number. Window and door call-outs are sometimes placed in the elevation drawings as well as the plan.

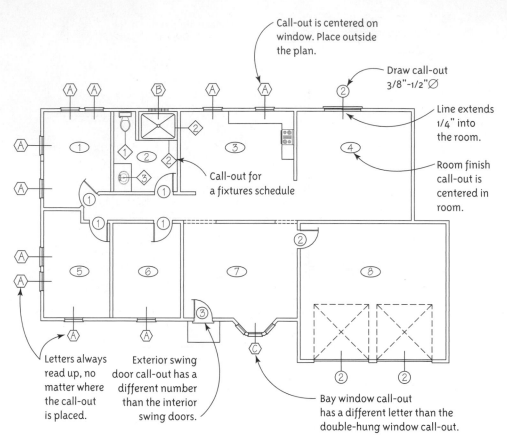

Call-out is centered on window. Place outside the plan.

Draw call-out 3/8"-1/2"⌀

Line extends 1/4" into the room.

Room finish call-out is centered in room.

Call-out for a fixtures schedule

Letters always read up, no matter where the call-out is placed.

Exterior swing door call-out has a different number than the interior swing doors.

Bay window call-out has a different letter than the double-hung window call-out.

Figure 10-4
Floor plan with callouts for the doors, windows, bathroom fixtures and room finishes.

Drawing the Schedule

Schedules have two formats: grid and pictorial. On a tabular grid, information is lettered with 1/8' tall letters inside the grid boxes. Letters should not be placed on the grid lines. Grid size is based on how much information the grid needs to contain; letters should not be crammed in or drawn smaller to fit in an arbitrarily sized grid. Therefore, it is helpful to letter the information first and then draw the grid around it. Sometimes dimensions are written like this: 3° (pronounced 'three-oh') instead of 3'-0', to save space and drafting time. You will occasionally see this done on floor plans, too.

Grid schedules may also be prepared on a computer using software that creates tables, such as Word or Excel, then printed and taped onto the drawing paper or copied onto an adhesive-back plastic film and applied to the drawing sheet. (Be aware that some copy shops will reject such sheets, as adhesive-backed items may damage a copier. This caveat applies to rub-on letters, layout tape, and correction fluid as well.)

A pictorial schedule consists of drawings of each door and window type. Pictorial schedules are used more often for doors than for windows.

The design and the information contained in a schedule vary. While information such as size and type is common to all schedules, other types of information, such as glazing, manufacturer name and model, and head, jamb and sill references may or may not be included.

There is no one correct way to lay out a schedule; it depends on the complexity of the items and the amount of information given. A correctly done schedule is one that is organized and easy to read. Figures 10-5 through 10-10 contain some examples of legends and schedules.

ROOM FINISH SCHEDULE

NO	ROOM NAME	FLOOR: SEALED CONCRETE	FLOOR: CARPET	FLOOR: SHEET VINYL	FLOOR: VINYL COMPOSITION TILE	FLOOR:	FLOOR:	BASE: VINYL BASE	BASE:	BASE:	BASE:	BASE:
300	ENTRY				●			●				
301	LOBBY				●			●				
302	HALLWAY			●				●				
303	WOMEN'S BATHROOM			●				●				
304	MEN'S BATHROOM				●			●				
305	OFFICE				●			●				
306	STORAGE				●			●				
307	LIBRARY	●						●				
308	MECHANICAL				●							
309	HALLWAY				●			●				
310	HALLWAY				●			●				
311	MAINTENANCE	●										
312	GARAGE	●										
313	STORAGE	●										
314	STORAGE	●										
315	OFFICE											
316	CONFERENCE ROOM				●			●				
317	HALLWAY				●			●				
318	OFFICE		●					●				
319	OFFICE		●					●				
320	KITCHEN				●			●				
321	BATHROOM			●				●				
322	UTILITY CLOSET				●			●				

NO	ROOM NAME	WALLS: PAINTED BLOCK	WALLS: EPOXY PAINTED BLOCK	WALLS: PAINTED DRYWALL	WALLS: EXPOSED BRICK	WALLS:	WALLS:
300	ENTRY				●		
301	LOBBY	●		●			
302	HALLWAY	●					
303	WOMEN'S BATHROOM		●				
304	MEN'S BATHROOM		●				
305	OFFICE	●					
306	STORAGE	●					
307	LIBRARY	●					
308	MECHANICAL		●				
309	HALLWAY	●					
310	HALLWAY	●					
311	MAINTENANCE	●					
312	GARAGE	●					
313	STORAGE	●					
314	STORAGE	●					
315	OFFICE	●					
316	CONFERENCE ROOM	●					
317	HALLWAY	●					
318	OFFICE	●					
319	OFFICE	●					
320	KITCHEN		●				
321	BATHROOM		●				
322	UTILITY CLOSET		●				

NO	ROOM NAME	CEILING: EXPOSED STRUCTURE–PNT	CEILING: 2×4 LAY-IN	CEILING: 5/8" DRYWALL SUSPENDED	CEILING: 5/8" DRYWALL ON FRAMING	CEILING:	HEIGHT	NOTES
300	ENTRY				●		8'–0"	
301	LOBBY			●			8'–6"	PAINTED GYP BD
302	HALLWAY			●			8'–6"	
303	WOMEN'S BATHROOM				●		8'–0"	
304	MEN'S BATHROOM				●		8'–0"	
305	OFFICE			●			8'–6"	
306	STORAGE				●		8'–0"	
307	LIBRARY			●			8'–6"	
308	MECHANICAL		●				—	
309	HALLWAY			●			8'–6"	
310	HALLWAY				●		8'–6"	
311	MAINTENANCE				●		9'–0"	
312	GARAGE		●				9'–0"	
313	STORAGE					●	9'–0"	PAINTED GYP BD
314	STORAGE					●	9'–0"	
315	OFFICE			●			9'–0"	
316	CONFERENCE ROOM			●			9'–0"	
317	HALLWAY			●			8'–6"	
318	OFFICE			●			9'–0"	
319	OFFICE			●			9'–0"	
320	KITCHEN				●		8'–6"	
321	BATHROOM				●		8'–0"	
322	UTILITY CLOSET				●		8'–0"	

Figure 10-5
Room finish schedule for an office building.

BATHROOM FIXTURES LEGEND

X	FIXTURE	DESCRIPTION
A	WATERCLOSET	AMERICAN STANDARD "ELLISSE" ELONGATED, SIPHON VET; WATER SAVER TYPE, WITH CHROME SUPPLY & ANGLE STOP. PROVIDE CLOSED FRONT SOLID PLASTIC SEAT WITH COVER.
B	LAVATORY	AMERICAN STANDARD "ELLISSE" VITREOUS CHINA SELF RIMMING "HERITAGE FAUCET WITH POP-UP DRAIN & CRYSTALINE HANDLES. PROVIDE CHROME SUPPLIES, STOPS & P-TRAP.
C	TUB/SHOWER	AMERICAN STANDARD #2265.379 (RIGHT OUTLET) AND #2267.375 (LEFT OUTLET) "BILDOR" RECESS TYPE ENAMELED CAST IRON WITH SLIP RESISTANT SURFACE, AND "HERITAGE" TREE HANDLE BATH/SHOWER COMBINATION WITH SPOUT ADJUSTABLE SPRAY HEAD, WASTE & OVERFLOW WITH P-TRAP.
D	TUB/WHIRLPOOL	AMERICAN STANDARD "FONTANE", ENAMELED CAST IRON, WITH INTEGRAL LUMBAR SUPPORT, BEVELED HEADREST, SLIP RESISTANT SURFACE, GRAB BARS, FACTORY INSTALLED WHIRLPOOL SYSTEM W/SIX MULTI DIRECTIONAL JETS, TWO AIR VOLUME CONTROLS, ANTI-VORTEX SUCTION FITTING, 3/4 HP PUMP AND ELECTRONIC ON/OFF CONTROL WITH LOW WATER CUT-OFF. "HERITAGE" WALL-MOUNTED BATH FILLER WITH 8" SPOUT & INTEGRAL STOPS, PROVIDE WASTE, OVERFLOW 7P-TRAP. "I" UNIT TO BE #130B.473 TUB/SHOWER C.
E	WHIRLPOOL/BATH	AMERICAN STANDARD #2645.302 "ROMA" 51/2 FOOT ACRYLIC, END DRAIN OUTLET, OVAL BATHING WELL WITH SAME WHIRLPOOL AS "D".
F	SHOWER	FIAT 48" × 32" "SIERRA", PRECAST TERRAZZO WITH I" FLANGE ANS S.S. DRAIN BODY, BOTH CAST INTERAL, TERRAZZO TO BE 3000 PSI. WHITE MARBLE CHIP COMPOSITION, PROVIDE AMERICAN STANDARD #1204.319 "HERITAGE" SHOWER FITTING WITH ADJUSTABLE SPRAY HEAD & CRYSTALINE HANDLES, PROVIDE P-TRAP.
G	SHOWER FITTING	AMERICAN STANDARD #1204.320 "HERITAGE" WITH JUSTABLE SPRAY HEAD & CRYSTALINE HANDLES, PROVIDE P-TRAP & S.S. DRAIN BODY & STRAINER.
H	KITCHEN SINK	ELKAY "LUSTERTONE" DOUBLE COMPARTMENT; 30255, 18 GA SELF RIMMING 33" × 22" WITH 4 HOLES FOR AMERICAN STANDARD "HERITAGE" 8" CENTER-SET FAUCET WITH SPRAY, PROVIDE WASTE FITTINGS, SUPPLIES & P-TRAP FOR CONNECTION WITH DISHWASHER AND INSINKERATOR #333, 1/2HP GARBAGE DISPOSAL.
J	WASHER	GUN GRAY OR "CATCH-A-DRIP" WITH TWO 1/2" HOSE VALVES AND 2" DRAIN OUTLET.
K	FLOOR DRAIN	JOSAM BODY WITH 6" ROUND ADJUSTABLE STRAINER W/FLASHING CLAMP DEVICE.
L	AREA DRAIN	JOSAM 12⅝" TOP, CAST IRON BODY & TOP, WITH 5" DEEP GALVANIZED SEDIMENT BUCKET.
M	TRENCH DRAIN	JOSAM 12" WIDE, CAST IRON BODY & TOP, WITH 5" DEEP GALVANIZED SEDIMENT BUCKET. PROVIDE INCLINED BACKWATER VALVE, & DEEP SEAL 4" P-TRAP. TOTAL LENGTH 16'-0".
N	WALL HYDRANT	JOSAM #71050¾" NON-FREEZE TYPE WITH INTEGRAL VACUUM BREAKER.

Figure 10-6
Legend of plumbing components.

BATHROOM ACCESSORIES SCHEDULE

CALL-OUT	A	B	C	D	E	F	G	H	I	J	K	L	M	N
Description	TOWEL DISPENSER / WASTE RECEPTACLE	TOWEL DISPENSER	MIRROR 24"X36"	FOLDING SHELF	2 COMPARTMENT SANITARY DISPOSAL	1 COMPARTMENT SANITARY DISPENSER	SANITARY RECEPTACLE	TOILET PAPER DISPENSER	GRAB BAR	DOUBLE ROBE HOOK	TOWEL PIN	RECESSED SOAP DISPENSER	RECESSED SOAP DISH	CURTAIN ROD
MOUNTING HEIGHT A.F.F.	5'-6"	5'-0"	6'-0"	4'-0"	2'-6"	2'-6"	6'-0"	*	3'-0"	5'-0"	5'-0"	3'-6"	4'-0"	6'-9"

NO.	ROOM NAME	A	B	C	D	E	F	G	H	I	J	K	L	M	N
106	WOMEN'S NO. 2	1			2	1			2	2					
107	MEN'S NO. 2	1							1	2					
108	MEN'S DRESSING RM. NO. 2		1	1					1						
109	WOMEN'S DRESG. RM. NO. 2		1	1					1						
121	TOILET NO. 3		1	1					1						
129	STAR DRESSING RM.		1	1					1		1	1		1	1
130	WOMEN'S DRESSING								1		1	1		1	1
132	TOILET		1	1					1						
133	TOILET		1	1					1						
135	MEN'S DRESSING RM.								1		1	1		1	1
140	STAR DRESSING RM.		1	1					1		1	1		1	1
146	WOMEN'S RM.	2			11	5	1	1	11	2			6		
147	MEN'S RM.	2							4	2			6		
165	MEN'S RM.	1							4	2			3		
167	WOMEN'S RM.	2			4	2		1	3	2			3		
210	WOMEN DRESSING RM.								1	1	1	1		1	1
214	STAR DRESSING		1	1					1	1	1	1		1	1
216	MEN'S DRESSING								1	1	1	1		1	1
243	MEN'S RM.	2							5	2			4		
245	WOMEN'S RM.	2			7	3	1	1	7	2			4		
325	MEN'S RM.	2							5	2			4		
327	WOMEN'S RM.	2			7	3	1	1	7	2			4		

* FIELD VERIFY

Figure 10-7
Schedule of bathroom accessories for a theatre.

WINDOW SCHEDULE

MARK	WIDTH	HEIGHT	MATERIAL	GLASS	TYPE	NOTES
A	3'-0"	3'-0"	METAL-CLAD	3/8" INSULATED	FIXED	
B	3'-0"	3'-0"	D.O.	D.O.	FIXED	
C	D.O.	D.O.	D.O.	D.O.	FIXED	
D	D.O.	D.O.	D.O.	D.O.	DOUBLE-HUNG	
E	D.O.	D.O.	D.O.	D.O.	DOUBLE-HUNG	
F	5'-0"	5'0"	D.O.	1" INSULATED	DOUBLE-HUNG	
G	6'-0"	5'-0"	D.O.	1" INSULATED	DOUBLE-HUNG	
H	6'-0"	6'-8"	D.O.	1" TEMPERED INSULATED	SLIDING GLASS	

Figure 10-8
Window schedule for a residence. *D.O* means *ditto* ("same as above"). Quotation marks are not used as ditto marks in this context.

NO	SIZE	DOOR TYPE	FRAME TYPE	HRDW GROUP	DETAIL REFERENCES			FIRE RATING	REMARKS
					HEAD	JAMB	SILL		
1	PR 3'-0" x 7'-0"	A	1	1	5A6	23A9	12A9 TYP		
2	PR 3'-0" x 7'-0"	A	1	1	5A6	23A9	12A9 TYP		
3	3'0" x 7'-0"	B	3	3	14A9	13A9	——		
4	3'0" x 7'-0"	B	3	3	14A9	13A9	——		
5	3'0" x 7'-0"	C	3	4	14A9	13A9	——		
6	2'0" x 7'-0"	B	3	5	14A9	13A9	——		
7	3'0" x 7'-0"	C	4	5	14A9	13A9 TYP	——		
8	3'0" x 7'-0"	C	4	5	14A9	13A9 TYP	——		
9	3'0" x 7'-0"	D	3	5	14A9	13A9	——	20 min	
10	3'0" x 7'-0"	D	3	2	20A9	19A9	12A9 TYP		INSULATED DOOR & FRAME
11	3'0" x 7'-0"	E	2	1	5A6	24A9/21A9	12A9 TYP		
12	3'0" x 7'-0"	D	3	5	18A9	17A9	——		
13	10'0" x 10'-0"	G	——	——	1A8	4A8/5A8	3A8		
14	FUTURE 3'0" x 7'-0"	——	——	——					PROVIDE KNOCKOUT PANEL

DOOR SCHEDULE

Figure 10-9
Door schedule for a residence.

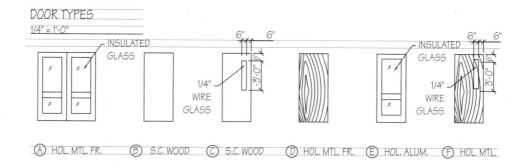

Figure 10-10
Pictorial door schedule for an office building.

Summary

Legends and schedules are charts of information that clarify and explain specific building components. This information is omitted from the floor plan to make both the plan and the information easier to read.

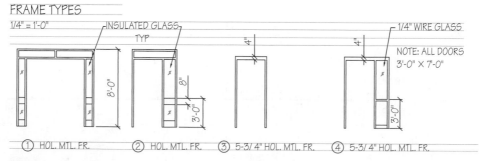

Suggested Activities

1. Find a floor plan in a magazine and create a schedule for its doors and windows.

2. Draft a finish schedule for a floor plan you already have. Draw the symbols in the floor plan that reference the plan to the schedule.

3. Draw some schedule symbols in a floor plan, keying the plan to finishes, doors and windows, and bathroom fixtures.

Questions

1. What is a legend?

2. What is a schedule?

3. Name three building components for which interior designers draw schedules.

4. What is a call-out?

5. How does a designer draw a call-out?

6. Name three pieces of information that you might find on a door schedule.

CHAPTER 11

Stairs and Fireplaces

OBJECTIVE

- This chapter discusses stairs and fireplaces: types, sizes, code requirements, design considerations, and how to draw them. Residential applications are the primary focus of this chapter.

KEY TERMS

area of refuge	guard	ramp
balustrade	headroom	riser
chimney	L	spiral
circular	landing	stairs
circulating heat	handrails	stove
combustion	header	straight run
exit stairs	hearth	surround
fireplace	insert	tread
firebox	mantel	U
firebrick	radiant heat	vent
flue		

Stairs and fireplaces are important building elements, both functionally and visually (Figure 11-1). Architects usually design them, but interior designers also draw them, and may occasionally be asked to select and design them.

Stairs

Stairs are a form of vertical access. Other forms that also transport are elevators and escalators. Stairs may be open, partially enclosed or contained in a well. They may be residential, commercial, indoor, outdoor, built on site, or prefabricated, may be constructed of wood, concrete or metal, and may be left bare or covered with carpet or a combination of materials. They can be a room's focal point and incorporate interior and exterior views, or be strictly utilitarian. However, all stairs share some common features and must meet local code and accessibility requirements. The International Residential Code (IRC) and International Building Code (IBC) are two model codes often utilized for residential stair design, so they are the ones referred to here. A helpful reference for residential requirements is the visual, downloadable interpretation of the IRC found at http://www.stairways.org/. Commercial requirements are more complicated, so you will need to identify and consult the governing codes for each commercial project you do.

Stair Features

Stairs have dozens of parts and vocabulary terms; the following are the ones most frequently used by designers and drafters (Figure 11-2).

Figure 11-1
Stairs can be designed to be works of art, inviting as well as functional. *Courtesy of Bomonite Corporation, Madera, CA.*

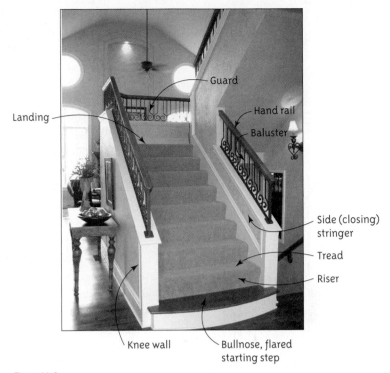

Guard

Landing

Hand rail

Baluster

Side (closing) stringer

Tread

Riser

Knee wall

Bullnose, flared starting step

Figure 11-2
Parts of a stairway. *Courtesy of Willis Construction, Overland Park, KS.*

Balustrade The **balustrade** (Figure 11-3) is an assembly containing the balusters, newel posts and handrail. Balusters are vertical support posts for the handrail. Newel posts (Figure 11-4) are larger than balusters and located at the bottom and top of a staircase and at turns and support

1 1/4"–2 1/2"

Handrail cross-sections

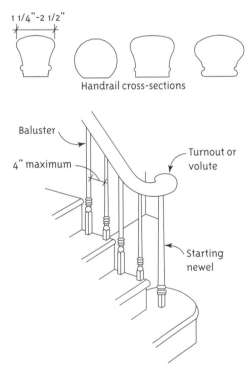

Baluster

4" maximum

Turnout or volute

Starting newel

Figure 11-3
Parts of a balustrade. The outside diameter of a circular handrail's cross-section must be between 1¼"–2½". If the cross section isn't circular, the perimeter must be between 4"–6" with a maximum 2¼" width.

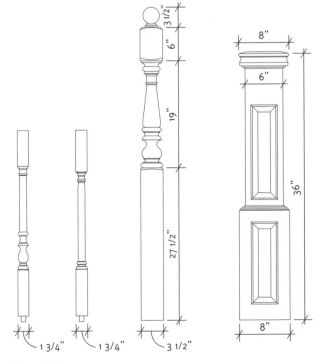

3 1/2"

6"

19"

27 1/2"

1 3/4"

1 3/4"

3 1/2"

8"

6"

36"

8"

Figure 11-4
Newel post styles. *Courtesy of Arcways Inc., Neenah, WI.*

areas; the starting newel is the first one at the foot of the stairs. A **handrail** is the balustrade's horizontal or "rake" member. It sits on top of the balusters, is supported by newel posts, and runs level and continuously along the stairs and landing. A rosette is a small, decorative piece of wood placed where the handrail ties into the wall. A turnout or volute is a handrail fitting on a starting newel that curves away from the stairway.

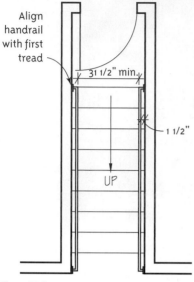

Figure 11-5
Wall rail clearances

Wall Rail This is a handrail that is mounted on a wall (Figure 11-5) and supported by wall-mounted brackets instead of newels. Codes require a wall rail on one side and allow it on both. The clear distance between wall rail and handrail must be 31½" minimum. There must also be a minimum space of 1½" between the handrail and the wall. The handrail must be continuous for the full length of the stair system, start above or beyond the top riser, and continue to a point above the lowest riser.

Headroom Headroom is the clear vertical distance between the tread and the ceiling, measured linearly along a sloped plane (Figure 11-6). Minimum ceiling and headroom heights are set by the codes. The minimum headroom allowed is 6'-8". The bottom of any ceiling-mounted item, such as an exit light or ductwork, must be at that 6'-8" height.

Pitch Also called rake angle, pitch is the angle of the staircase, calculated by dividing its rise (height) by its run (length).

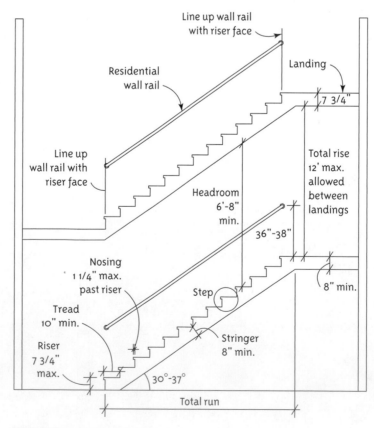

Figure 11-6
Residential staircase clearances and dimensions.

Rise Rise is vertical distance. Unit rise is riser height: the height of one step as measured from the top of one tread to the top of the next. Total rise is the height from finished floor to finished floor.

Run Run is horizontal distance. Unit run is tread depth: the depth of one step as measured from the nosing of one tread to the nosing of the next (or, in the absence of a nosing, from the face of one riser to the face of the next). Total run is measured from the face of the first riser to the face of the last.

Stairs, stairway, staircase, stairwell, flight All of these terms denote a series of steps or flights of steps used for passing from one floor or landing to another. A stair system consists of two or more stringers, multiple steps, possibly one or more landings, and a guardrail or handrail. A box stair is enclosed by walls on both sides. A knee wall is a short wall that holds the balusters. An open or freestanding stair is not enclosed by any walls and is open underneath.

Landing A **landing** is a level rest area on a staircase. Each flight of stairs must have one landing at the top and bottom, and if the flight rises more than 12 feet, it must have an intermediate one as well, which need not necessarily be in the center. A landing's width is the width of the stairs and its length must be at least the stair width (a minimum of 38" in residences) measured in the direction of travel. Note that if a door is at the head of the stairs, it cannot swing over them (Figure 11-7), and when fully open, it cannot project into the clear exit width more than 7". Additionally, a door cannot at any open position impede on more than half the required exit width.

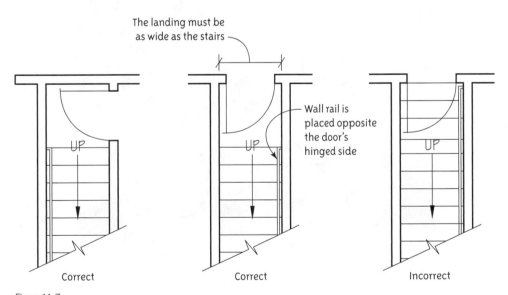

Figure 11-7
Correct and incorrect doors/stairs placement. A door may not swing over stairs.

Stringer This is a diagonal structural support for the steps (Figure 11-8). It can be a return tread stringer (Figure 11-9) which means the stairs have a bullnose finish along the front and the open side edge. Alternatively, it can be a closed stringer (Figure 11-10), where treads are not seen from either side; instead, they fit into a rout, which is a groove in the stringer.

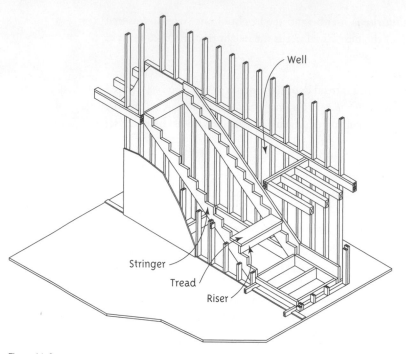

Figure 11-8
Stairwell construction.

Figure 11-9
Return tread stringer; note the bullnose, open-to-one-side treads.

Figure 11-10
A free-standing, circular stairway. Both edges of the treads are housed in a closed stringer.

Step A step is one of a series of structures within a stair system that consist of a riser and a tread (Figure 11-11). A **riser** is the vertical board at the front of the step. Risers are usually required on interior stairs. A **tread** is the horizontal board at the top of the step. A bullnose tread has a radius-finished edge and is often used as a starting step, an extra-large first step designed for appearance purposes. If the tread is less than 11" deep it needs a nosing, a part that protrudes 1¼" beyond the riser face. Within any flight of stairs, the difference between the largest and smallest riser or tread cannot exceed 3/8".

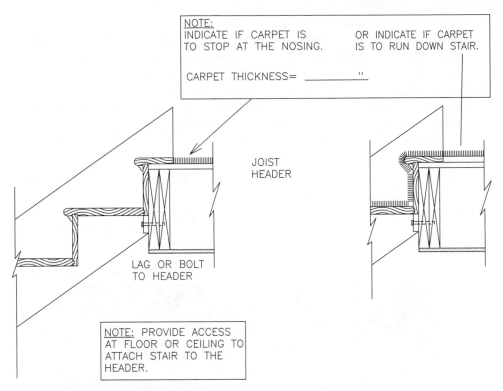

NOTE:
INDICATE IF CARPET IS   OR INDICATE IF CARPET
TO STOP AT THE NOSING.   IS TO RUN DOWN STAIR.

CARPET THICKNESS= _____ "

JOIST
HEADER

LAG OR BOLT
TO HEADER

NOTE: PROVIDE ACCESS
AT FLOOR OR CEILING TO
ATTACH STAIR TO THE
HEADER.

Figure 11-11
Step construction detail. *Courtesy of Arcways, Inc., Neenah, WI.*

Well opening This is a hole made in an upper floor for stairway placement.

Handrails and Guards **Handrails** help people steady themselves. They usually consist of a single rail installed at a specified height.

Guards prevent people from falling over the edge of a staircase or balcony and are required on any elevation over 30" where there is no adjacent wall. A handrail and **guard** are often combined as one assembly (Figure 11-12); the guard portion comprises the vertical balusters below

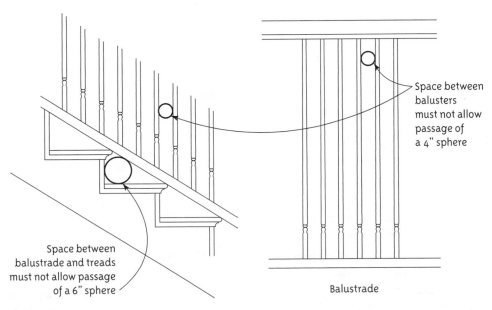

Space between
balusters
must not allow
passage of
a 4" sphere

Space between
balustrade and treads
must not allow passage
of a 6" sphere

Balustrade

Figure 11-12
IRC requirements for baluster spacing. A sphere is referenced instead of a circle because a sphere is the same size from any direction, as opposed to a flat, linear measurement.

the handrail. (You may sometimes encounter the term "guardrail." This is an outdated term used in old building codes). Handrails and guards are code requirements whether the stairs are open or enclosed. A handrail or guard must be constructed so a 4" diameter sphere cannot pass between balusters or through any opening in the baluster; a 6" sphere may not pass between the bottom rail and the steps; and an 8" sphere is allowed above 34".

Guards must extend from a code-mandated height (usually 3'-0 to 3'-6" high) down to a specified distance above the floor. All stairways with three or more risers must have a handrail that is continuous for the entire stair length (Additional intermediate handrails may be needed on wide stairs; see Figure 11-13). The rail is placed on the open side of the stairs. Its height will vary between 2'-6" and 3'-2" above the tread, depending on the situation. Handrails must be 1½" from the wall and cannot extend more than 3½" into the required stair width. The maximum handrail size is 1¼" to 2" for circular ones; the maximum cross-section for other shapes may be as large as 2¼". Horizontal ladder-style baluster designs cannot be used.

Figure 11-13
(Commercial) stairs outside the theatre at Johnson County Community College. Wide stairs require at least one intermediate handrail. Handrails can be plain or ornamental but should be smooth and sturdy. Note the level parts at the top and bottom. Codes also require specific weight limits to be supported.

Non-residential vs. Residential Code Compliance

While many design features for both residential and commercial stairs are the same, clearances and sizes may vary. Common examples are:

- Handrails on residence landings must be between 34" –38" high, measured from tread top to handrail top, while handrails on commercial building landings must be a maximum of 42" high.

- Handrails on residential stairs typically end in newel posts or turnouts, whereas the level part of a handrail on commercial stairs extends past the stairs. Sometimes it can turn back into the wall or into a newel post (Figure 11-14).

- Stairs must be at least 36" wide in residences and 44" wide in commercial buildings with occupancy over 50. Residential tread depth can be a minimum of 10" with a maximum riser height of 7¾". Commercial stairs riser height varies from 4" to 7" high (service stair risers may be 8" high, which is too steep for other uses) and commercial tread width varies from 11" to 14." The exact riser and tread dimensions are a function of the type of staircase and the total rise.

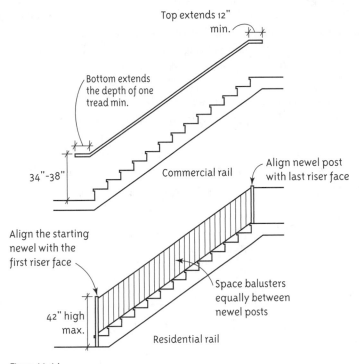

Top extends 12" min.

Bottom extends the depth of one tread min.

34"-38"

Commercial rail

Align newel post with last riser face

Align the starting newel with the first riser face

Space balusters equally between newel posts

42" high max.

Residential rail

Figure 11-14
Commercial vs. residential rail.

Stair Types

Different stair types accommodate different building designs. Stairs can be straight-run, U, L, winder, circular or spiral. All types are found in both residential and commercial construction.

Straight run Straight-run stairs have no turns (Figures 11-15, 11-16). They may have a landing. Straight-run stairs require a long, open space. Fire codes usually restrict total rise in a

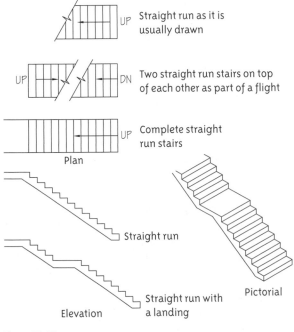

Straight run as it is usually drawn

Two straight run stairs on top of each other as part of a flight

Complete straight run stairs

Plan

Straight run

Straight run with a landing

Elevation

Pictorial

Figure 11-15
Straight-run stairs.

Figure 11-16
Straight-run, enclosed stair with a wall rail on both sides.

straight run for fire exit stairs to 11'-0". About half of the complete stair is drawn on each floor plan, followed by a diagonal break line. When drawing a straight-run stair, place an arrow in the center of the plan view. It should point upward and its note should say "up" on all flights except for the top, where the arrow will be reversed and the note will say "down."

L L-shaped stairs have a landing and a turn and are used when there is not enough space for a straight-run (Figure 11-17). They may be equal leg (meaning the turn is at the midpoint), or one leg may be longer than the other one. When there is insufficient room for a landing, winder treads may be used. These can be trapezoidal (pie-shaped) if there is not enough room for regular-size treads. The midpoint of the winder treads must be equal to the tread width of the regular steps. L-shaped stairs are suitable for residences and low-traffic areas, but cannot be fire exit stairs.

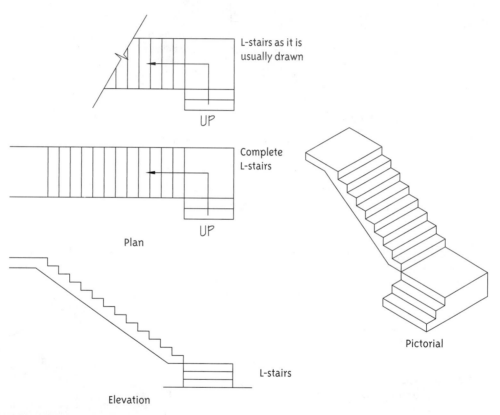

L-stairs as it is usually drawn

UP

Complete L-stairs

UP

Plan

Pictorial

L-stairs

Elevation

Figure 11-17
L stairs.

Winder A winder stair is an L stair where the L-shaped turn is made with trapezoidal treads due to lack of space for a landing (Figure 11-18). The trapezoidal treads must have a minimum width of 10" at a point no more than 11" from the side of the treads. The narrowest portion of a winder stairway may not be less than 6" wide. A winder tread has non-parallel edges and is used to make a directional change on a winder stair case without the use of a landing.

U *or* **Scissors** A U-shaped or scissors stair consists of two parallel flights (Figures 11-19, 11-20). This configuration is used when the space for a straight-run isn't available. A scissors or narrow U has a small space between flights, a wide U has a well hole, or large space.

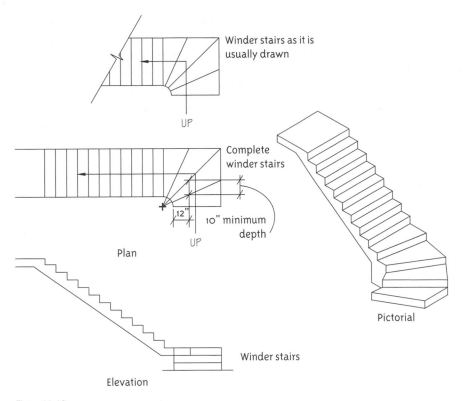

Winder stairs as it is usually drawn

UP

Complete winder stairs

12"

10" minimum depth

UP

Plan

Winder stairs

Elevation

Figure 11-18
Winder stairs.

Pictorial

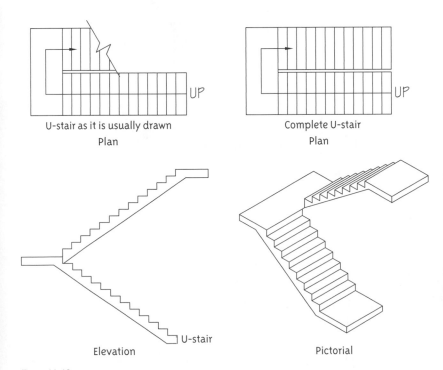

U-stair as it is usually drawn
Plan

UP

Complete U-stair
Plan

UP

Elevation

U-stair

Pictorial

Figure 11-19
Narrow U stairs.

Figure 11-20
U-stairs with curved landings.

Spiral A **spiral** stair rises in a circle above a center point (Figures 11-21, 11-22) and is used where there is little horizontal space available; it should be avoided otherwise, as it is not a very safe stair, and it is very difficult to move furniture and other large objects up or down a spiral stair. Most spiral stairs are premade in a factory and shipped to the site. Spiral stair treads must provide a clear walking width of 26" measured from the outer edge of the center column to the inner edge of the handrail. A 7½" run must be provided within 11" of the tread's narrowest part. No riser may exceed 9½" high, and all risers must be equal. A headroom of 6-'6" is allowed for a spiral staircase. It has circular treads, which are wider on one side than the other. (These treads are also used on circular staircases.)

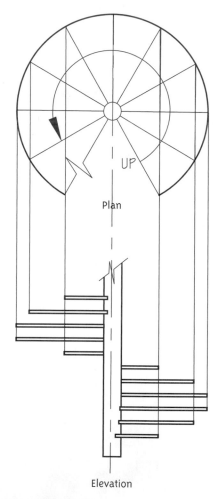

Figure 11-21
Spiral stairs.

Figure 11-22
Prefabricated spiral stairs. *Courtesy of Arcways, Inc., Neenah, WI.*

Circular Also called a curved stair, a **circular** stair requires a lot of horizontal space. Its steps are trapezoidal and rise along an irregular curve or arc. Circular stairways must have a minimum tread depth of 11" measured at a point not more than 11" from the narrow edge. The depth cannot be less than 6" at any point.

These are the main types of stairs. However, a stairs can take on other configurations or combine characteristics of different types. Examples include T, Z, O, triangle-shaped, and wide U' s with a return flight in the middle. Figures 11-23–11-28 show different stair configurations.

Exit stairs Commercial buildings must include **exit stairs**, which consists of the stairs, a protected enclosure of fire-rated walls, and any doors. Depending on occupancy classification,

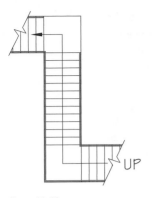

Figure 11-23
Z or Double L stairs.

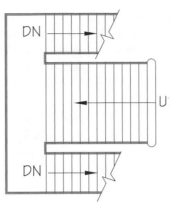

Figure 11-24
Wide U with a middle return flight.

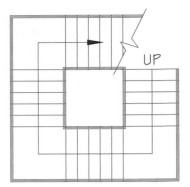

Figure 11-25
Square stairs.

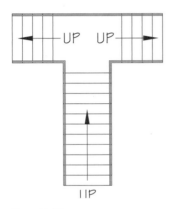

Figure 11-26
T stairs.

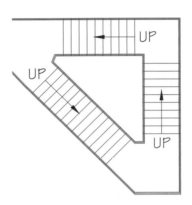

Figure 11-27
Triangle stairs.

Figure 11-28
A staircase that's a combination of straight run and wide-U, and incorporates accessories.

number of occupants, use of the stair, and tread dimensions, the codes allow most stair types to be part of the exit access. However, codes disallow use of some stair types as a means of egress. The doors of an exit stair must swing in the direction of the exit discharge. Therefore, all doors must swing into the stairway except at the ground level, where the door must swing toward the exit discharge or public way.

Exit stairs may also require an **area of refuge**, which is a place where people can safely wait for emergency assistance (Figure 11-29). Areas of refuge are usually provided next to, or inside, exit stairwells or at elevator lobbies. An area of refuge located next to an exit stair must be wheelchair accessible and the number of wheelchair spaces that must be provided is determined by the floor's occupant load, with the most common requirement being one space for every 200 occupants.

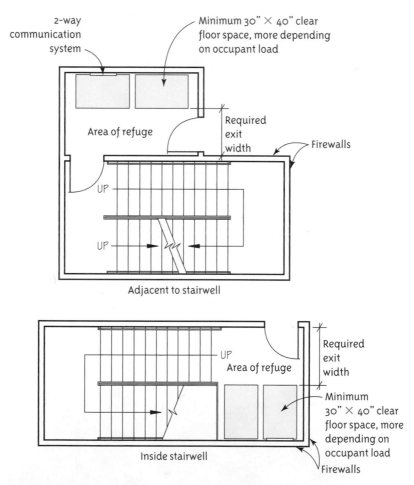

Figure 11-29
Some stairwells must have areas of refuge in case of fire.

Exterior Stairs

Exterior stairs are usually designed with smaller riser heights and wider treads than interior stairs; a popular proportion is a 6" riser paired with a 11" tread (Figures 11-30, 11-31). There should be a landing every 16 risers on continuous stairs. There is no riser requirement for wood stairs, and wood exterior stair treads are thicker than those of interior steps. Usually the same material used on the deck or porch is used for treads, and a non-skid material can be used to cover them.

Note that the rise on a concrete stair is drawn at a 10° angle. This angle is not labeled, because the exact as-built angle will usually be based on what forms the construction crew has available at the job site. However, it cannot exceed 30°.

Figure 11-30
Exterior stairs are shallower and wider than interior stairs.

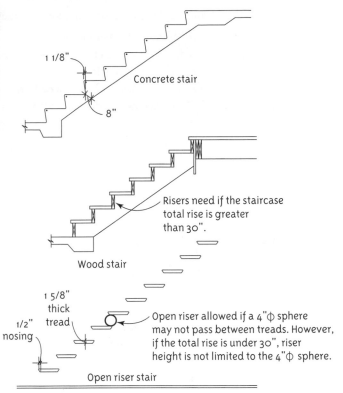

1 1/8"

Concrete stair

8"

Risers need if the staircase total rise is greater than 30".

Wood stair

1 5/8" thick tread

1/2" nosing

Open riser allowed if a 4"φ sphere may not pass between treads. However, if the total rise is under 30", riser height is not limited to the 4"φ sphere.

Open riser stair

Figure 11-31
Concrete, wood, and open-riser exterior stairs. Basement and exterior wood steps rarely require risers.

Stair Design

Here are some considerations for stair design.

- Consider the staircase's potential use when deciding how wide to make it. Minimum code requirements result in steep, narrow stairs. Will the occupants frequently need to transport furniture and appliances on the stairs? Will there be much traffic or opposing traffic? If so, avoid narrow stairs. Appearance-wise, stairs that are wider than what is required by code are more attractive and easier to use (Figure 11-32).

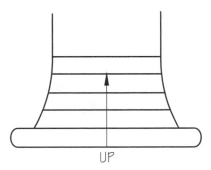

UP

Figure 11-32
A flared starting step makes a stairs more attractive.

- Use several short landings instead of one large one to provide more resting places and make a more visually appealing staircase. Placing windows at the landing incorporates exterior views, and landings can be strategically placed to overlook living or kitchen spaces for nice interior views. Such landings also are convenient places for plants and decorative pieces. Space under a landing can be made accessible for storage by placing a door under the upper flight.

- Avoid single steps to sunken rooms, as they are tripping hazards. Use at least two steps.
- Select handrails that complement the staircase. They should be continuous, and a user who runs a hand along the railing should not hit mounting hardware or brackets. An elaborate newel post and interesting balusters can dress up a knee wall. Newel posts and balusters can be custom-designed or ordered from a manufacturer. There are many catalogs available containing details and drawings of railings, ornamental posts, balusters, and trim.
- A well-designed staircase follows these formulae:

$$\text{Two risers plus 1 tread} = 24\text{"}-25\text{"}$$
$$\text{Riser} \times \text{tread} = -72\text{"}-77\text{"}$$
$$\text{Riser} + \text{tread} = 17\text{"}-18\text{"}$$

Angle should be $30°-37°$.

Calculating How Many Stairs Are Needed

Here are steps to calculate how many risers and treads are needed, and the height and width of each.

1. *Know the total rise;* that is, the distance from finished floor to finished floor. If you have drawings available, obtain the distance from their dimensions. Otherwise, you will need to do some fieldwork to get this dimension. For this example, we'll use a total rise of 11'-6".

   It's easiest to work with this number in inches, so convert it. Convert by multiplying $11' \times 12"$ (because there are 12 inches in a foot) and adding the remaining 6". This gives a total rise of 138".

2. *Calculate the height and number of risers needed.* To do this, assume an ideal riser height of 7". The final height may be off somewhat, but 7" is a good place to start. Divide the total rise of 138" by 7". This gives us 19.71 risers. We need an even number—either 18 or 19 stairs. Let's see which number yields us a riser height closest to 7".

$$138/18 = 7.6"$$
$$138/19 = 7.26"$$

   Therefore, 19 risers is a more desirable number to use. Each one will be 7.26" tall.

3. *Calculate the width and number of treads needed.* You will need one tread less than the number of risers, because the last tread is on the second floor. Therefore, there will be 18 treads. Using the $R + R + T = 25$ formula, plug in the riser height and solve for T.

$$7.26 + 7.26 + T = 25$$
$$T = 25 - 7.26 - 7.26$$
$$T = 10.48$$

   This is close to the ideal tread width of 11", so the combination of this riser height and tread width is acceptable. To verify this, plug these figures into one of the formulae given earlier.

   A total run length of 188.64" is needed (18 treads $\times$ 10.48" wide each) for a straight-run stairs. If that run is not available, you will need to use a U or other, less space-consuming configuration. Remember that if the total rise (height) is greater than 11', a landing is needed, increasing the run length.

Creating a Grid to Draw the Stairs

Once you know the number of risers and treads, the plan view of the stairs is easy to draw. The elevation is a bit harder to draw, but using a grid with squares for the riser and tread dimensions makes the task easier. Here are steps to drawing a grid for the riser and treads referenced above.

1. Draw a rectangle, making the length the total run available and the height the total rise (Figure 11-33).

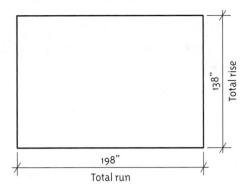

Figure 11-33
Drawing a stairs grid: step 1.

2. Divide one of the vertical lines into the number of risers needed (19). Do this by selecting a scale—any scale—where the 0 can be placed at one corner and the 19 at the other corner (Figure 11-34).

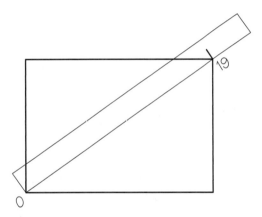

Figure 11-34
Drawing a stairs grid: step 2.

3. Mark off all 19 increments and project these increments horizontally to the vertical lines (Figure 11-35).

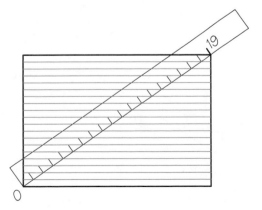

Figure 11-35
Drawing a stairs grid: step 3.

4. Now divide one of the horizontal lines into the number of treads needed (18). Do this by placing the zero where shown and the 18 anywhere on the vertical line. The lines from step 3 have been removed for clarity (Figure 11-36).

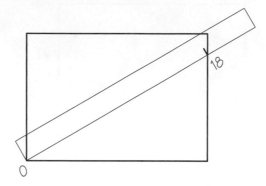

Figure 11-36
Drawing a stairs grid: step 4.

5. Mark off all 18 increments and project these increments vertically to the horizonal lines (Figure 11-37).

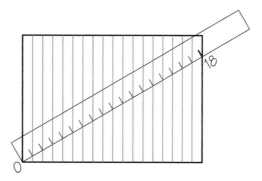

Figure 11-37
Drawing a stairs grid: step 5.

6. Darken the risers and treads (Figure 11-38).

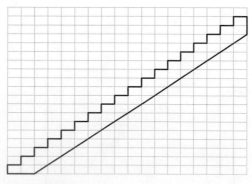

Figure 11-38
Drawing a stairs grid: step 6.

Figures 11-39-11-42 show examples of commercial stairs in plan and section.

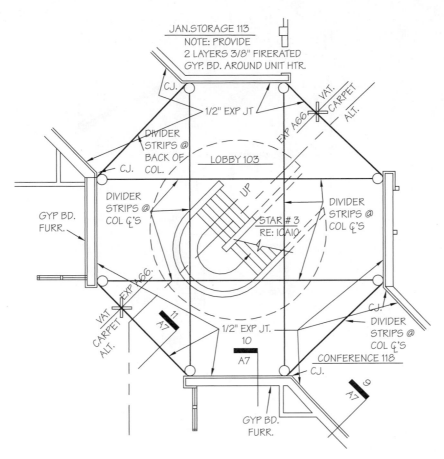

Figure 11-39
First floor of a U staircase in a set or working drawings. *Courtesy of Edwin Korff, AIA, Prairie Village, KS.*

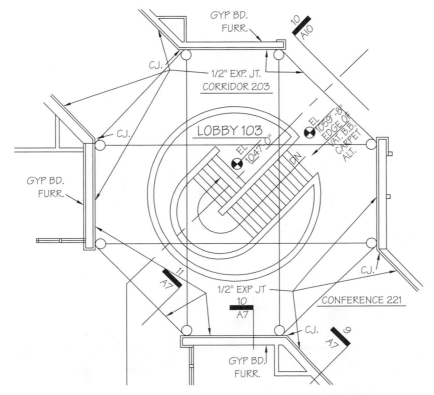

Figure 11-40
Second floor of a U staircase in a set of working drawings. *Courtesy of Edwin Korff, AIA, Prairie Village, KS.*

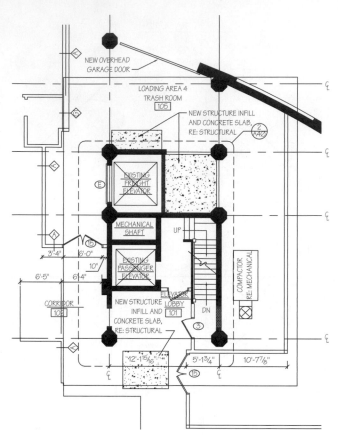

Figure 11-41
First floor of a U staircase in a set of working drawings. *Courtesy of h+p architecture, Chicago, IL.*

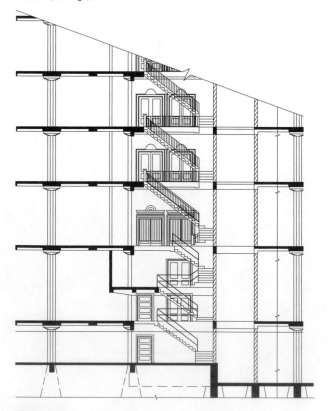

Figure 11-42
Section through the U staircase in Figure 41. *Courtesy of h+p architecture, Chicago, IL.*

A **ramp** is a sloped surface whose purpose is to make a building accessible (Figure 11-43). It should have a nonslip surface and be protected from rain, snow, and ice where possible. The IRC-recommended slope for both residential and commercial ramps is 1:8 (one vertical unit for every 8 horizontal units), with 1:12 being the maximum. Thirty feet is the maximum length for a ramp section. Ramps longer than 30' are built in sections, each separated by a landing the same width as the ramp and at least 5' long. Landings are required at the top and bottom, and their length must take into account any adjacent doors. An entry platform should extend 18" beyond the handle side of the door to facilitate wheelchair use.

A ramp's minimum allowed width is 31½", as measured to the inside of the handrail. 36" is needed for wheelchair accessibility, with 4'-0" recommended.

Any ramp that exceeds a 1:12 ratio must have at least one handrail, the height being between 34"– 38". Guards 36" high are needed when there is no adjacent wall and the ramp's overall rise is greater than 30".

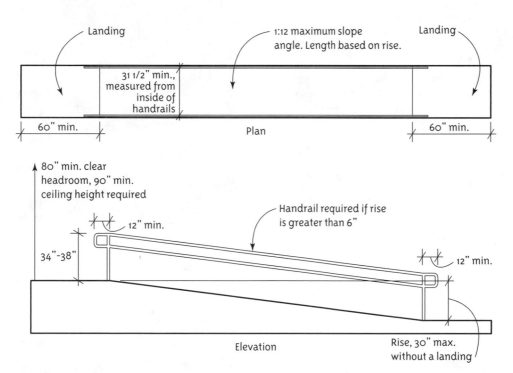

Figure 11-43
Clearances for an interior or exterior ramp.

Fireplaces and Stoves

A **fireplace** is a framed opening in a chimney that holds an open fire. Fuel is placed in a combustion chamber and ignited, drawing oxygen from the room or outdoors to supply the fire. The flame produces heat, which warms the room, and gas and smoke, which exit through a chimney or vent. A **stove** is a free-standing external fireplace in a metal container. It expels gas and smoke through a chimney or vent.

Modern central heating has made the fireplace technically obsolete. However, it remains an enduring building component because of the symbolic and decorative atmosphere it provides. A fireplace is often a room's focal point. Fireplaces appear in kitchens, living rooms, dining rooms, home offices, bedrooms, and bathrooms, and in commercial spaces such as hotel lobbies.

A fireplace can be site-built, prefabricated, rectangular, round, floor level, eye level, indoor, outdoor, decorative, or functional. It can be made of brick, tile, stone, stucco, or metal, and can be a traditional one-face design or a piece of contemporary art. It may have a few simple mechanical parts or it may utilize thermostats, oxygen sensors, remote control ignition, and technologies for heat output and circulation. A fireplace might be a simple opening with no mantel or decoration besides its material texture or it may have an overmantel, a metal hood, and built-in log storage. It can burn wood, natural gas, propane, coal, oil, electricity, or a combination of fuels. Despite these differences, all fireplaces share some common features and must adhere to building codes for proper build and installation. Figures 11-44 through 11-50 show different types of fireplaces.

Figure 11-44
The ornamentation of this traditional masonry fireplace comes from the texture of its materials. *Courtesy of Pella Windows, Pella, IA.*

Figure 11-45
A "chiminea" stove warms up an outdoor living room. *Courtesy of Empire Comfort Systems, Belleville, IL.*

Figure 11-46
A Heat & Glo "Cyclone" electric fireplace is designed like a piece of wall art. *Courtesy of hearthnhome.com.*

Figure 11-47
A Heat & Glo prefabricated, 3-sided gas fireplace. *Courtesy of hearthnhome.com.*

Figure 11-48
A prefabricated fireplace arrives as a complete assembly ready to be installed in the wall. *Courtesy of Empire Comfort Systems, Belleville, IL.*

Figure 11-49
Prefabricated fireplace with an overmantel. It is built flush with the wall and built-in bookcases. *Courtesy of Empire Comfort Systems, Belleville, IL.*

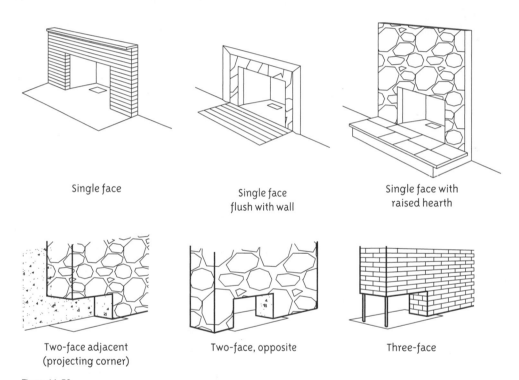

Single face

Single face flush with wall

Single face with raised hearth

Two-face adjacent (projecting corner)

Two-face, opposite

Three-face

Figure 11-50
Different fireplace styles.

Features

Chimney A **chimney** is a vertical, freestanding structure that carries smoke and gas out of the room (Figures 51, 52). Building codes prohibit a fireplace or chimney from direct contact with any of the building's framing. There must be a 2" clearance, and that space must be filled with a non-combustible material. Inside the chimney is a **flue**, a vertical path to the roof. The flue must be lined with clay tile and surrounded by at least 4" of masonry on all sides, and its area should be at least 1/10th the area of the fireplace opening. (The areas of multi-face fireplaces are added together.) Each fireplace needs its own flue, but one chimney can hold multiple flues. Inside the flue is the damper, an adjustable metal piece that regulates airflow and prevents downdrafts. It is placed 6" or 8" above the top of the fireplace opening.

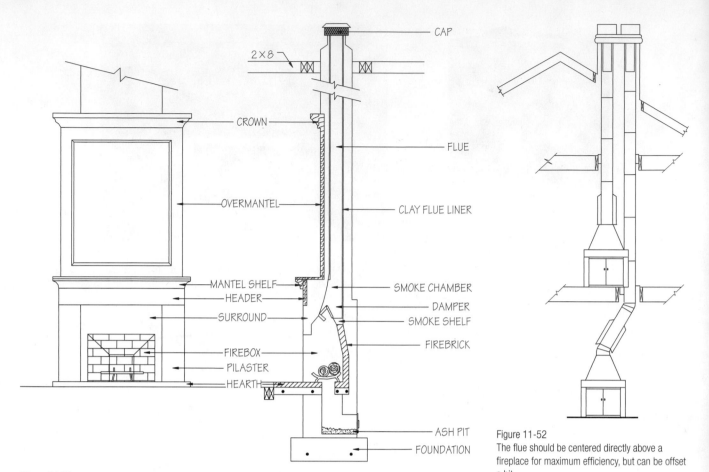

2×8

CAP

CROWN

FLUE

OVERMANTEL

CLAY FLUE LINER

MANTEL SHELF
HEADER
SURROUND

SMOKE CHAMBER
DAMPER
SMOKE SHELF

FIREBOX
PILASTER
HEARTH

FIREBRICK

ASH PIT
FOUNDATION

Figure 11-51
Fireplace elevation and section.

Figure 11-52
The flue should be centered directly above a
fireplace for maximum efficiency, but can be offset
a bit.

The flue also contains a smoke shelf, which deflects cold downdrafts back up, and a smoke chamber, which is the area just above the smoke shelf and damper. It is pyramid-shaped with a vertical back. The chimney cap keeps objects from falling into the chimney and sparks from landing on the roof. The chimney must extend at least 2' above any roof surface that comes within 10' of the chimney (Figure 11-53).

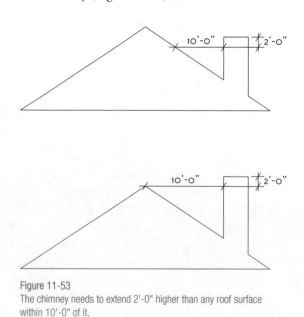

10'-0"   2'-0"

10'-0"   2'-0"

Figure 11-53
The chimney needs to extend 2'-0" higher than any roof surface
within 10'-0" of it.

Here are some common terms associated with fireplaces.

Header Sometimes called the facing or, when decorated, the entablature, the header is the horizontal component over the fireplace opening. It is supported by vertical columns called pilasters.

Firebox The **firebox** is the combustion chamber where the fire is contained. It is lined with firebrick and its shape is designed to direct gas and smoke up the chimney. Its size must be correctly proportioned because if it is too shallow, smoke will pour into the room, and if it is too deep, not enough heat will reflect out. The back and side walls of the firebox should be at least 8" thick. An ash pit is a $5" \times 8"$ hole below the firebox where ashes from the burning wood fall. The cleanout is an $11" \times 11"$ door that lets the user access the ashes for removal. The firebox may feature built-in andirons, a grate, or even an indoor charcoal grill.

Firebrick **Firebrick** is heat-tempered brick. It is used inside a firebox.

Surround A **surround** can be either the immediate border of the face around the firebox opening or a non-combustible/masonry decorative frame around the whole fireplace. Any wood trim should be at least 8" from the firebox opening.

Opening The **opening** is the rectangular recess in the surround.

Hearth The hearth is the floor of the fireplace. It can be flush with the room floor or above it. Its purpose is to protect the rest of the floor from sparks.

Mantel The **mantel** is the whole frame surrounding a fireplace, but the term *mantel* is often used to describe just the shelf over the fireplace. A mantelpiece is the frame around a fireplace plus the overmantel, the decorative panel above the mantelshelf on which a decorative accessory can be hung.

Vent A **vent** is an opening that draws air into the fire and allows **combustion** byproducts to escape (Figure 11-54). Gas fireplaces are categorized by their vent types. Natural-draft vents use a chimney pipe similar to that used on furnaces and gas-powered water heaters. Direct

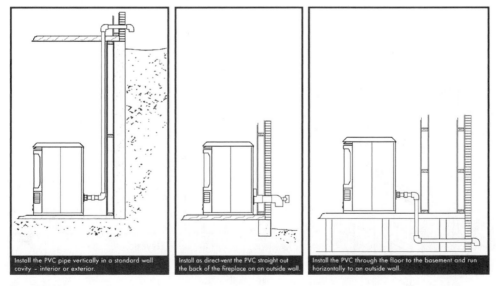

Figure 11-54
Three examples of direct venting a stove through a wall. Stoves can be installed in existing fireplaces as an insert or as a zero-clearance system with a wood cabinet mantel or as a freestanding stove. *Courtesy of Empire Comfort Systems, Belleville, IL.*

vents open directly through an exterior wall. They draw cold air for combustion from outside, but do not let warmed air escape from inside. They do not require a full chimney, so a fireplace that utilizes this vent type is suitable for apartments, condominiums, or unusual locations such as under a desktop or window. Unvented or vent-free fireplaces draw combustion air from inside the home and burn so efficiently that there is no need for a vent system.

Fireplace Styles

Site-built A traditional wood-burning fireplace is built on a masonry foundation and is usually rectangular with one, two, or three open faces (Figure 11-54). Its function is primarily aesthetic, as it is an inefficient heat source. In fact, a fireplace may actually increase drafts by drawing room air through the fireplace opening and up the chimney, along with most of the heat generated by the fire. However, by designing a fireplace with correct proportions and glass doors, energy inefficiencies can be somewhat minimized.

Figure 11-55 is a cutaway through a typical site-built fireplace. Figures 11-56 and 11-57 show a typical site-built fireplace and the steps to draw it in plan. Figures 11-58 and 11-59 are drawings of different site-built fireplace styles. The red lines show how the views orthographically relate to each other (the 45° angle line shown on two of them projects information from the top view to the side view, which in this case is the section view). The letters correlate to the fireplace dimension charts in Figures 11-60, 11-61, and 11-62.

Prefabricated Over 70 percent of the fireplaces installed in new construction today are prefabricated and burn both wood and gas. Prefabricated fireplaces are also called zero-clearance fireplaces, since they can be installed in wood frame walls within an inch of combustible materials. They can mimic the styles of traditional site-built fireplaces or look completely

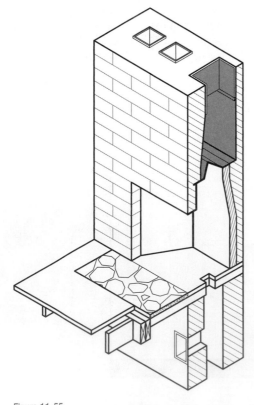

Figure 11-55
Cutaway view of a masonry fireplace and chimney.

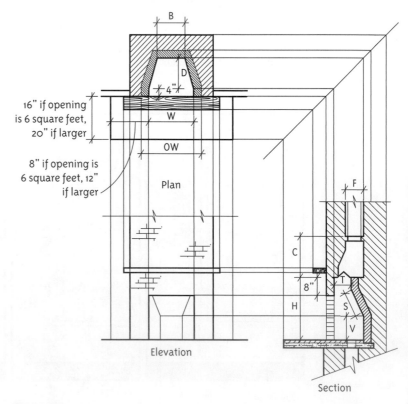

Figure 11-56
Single face site-built masonry fireplace.

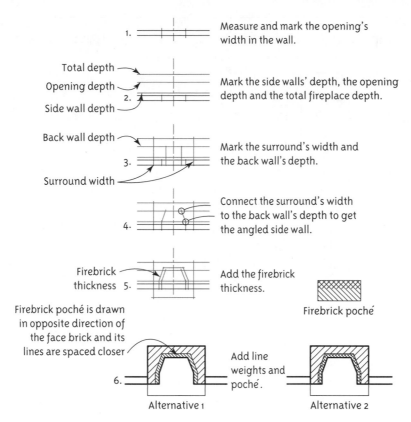

1. Measure and mark the opening's width in the wall.

Total depth
Opening depth
2.
Side wall depth

Mark the side walls' depth, the opening depth and the total fireplace depth.

Back wall depth
3.
Surround width

Mark the surround's width and the back wall's depth.

4. Connect the surround's width to the back wall's depth to get the angled side wall.

Firebrick thickness 5.
Add the firebrick thickness.

Firebrick poché

Firebrick poché is drawn in opposite direction of the face brick and its lines are spaced closer

6. Add line weights and poché.

Alternative 1   Alternative 2

Figure 11-57
Steps for drawing a single-face fireplace.

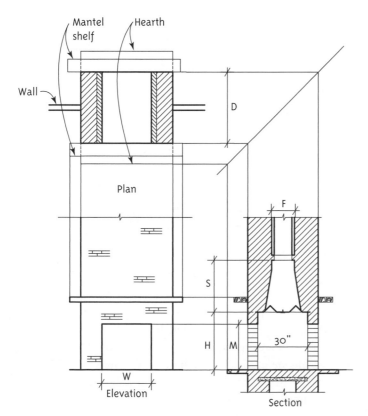

Figure 11-58
Two-face opposite masonry fireplace.

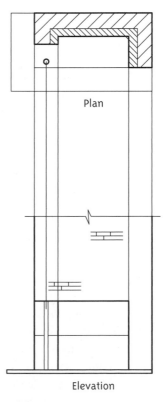

Figure 11-59
Two-face adjacent fireplace.

DESIGN DATA FOR SINGLE-FACE FIREPLACES

WIDTH	HEIGHT	DEPTH	BACK WALL	VERTICAL BACK	SLOPED BACK	THROAT	OVERALL WIDTH	OVERALL DEPTH	SMOKE CHAMBER	RECTANGULAR FLUE
W	H	D	B	V	S	T	OW	OD	C	F
24"	24"	16"	11"	14"	15"	8-3/4"	32"	20"	19"	8-1/2" x 8-1/2"
26"	24"	16"	13"	14"	15"	8-3/4"	34"	20"	21"	8-1/2" x 8-1/2"
28"	24"	16"	15"	14"	15"	8-3/4"	36"	20"	21"	8-1/2" x 13"
30"	29"	16"	17"	14"	18"	8-3/4"	38"	20"	24"	8-1/2" x 13"
32"	29"	16"	19"	14"	21"	8-3/4"	40"	20"	24"	8-1/2" x 13"
36"	29"	16"	23"	14"	21"	8-3/4"	44"	20"	27"	13" x 13"
40"	29"	16"	27"	14"	21"	8-3/4"	48"	20"	29"	13" x 13"
42"	32"	16"	29"	14"	23"	8-3/4"	50"	20"	32"	13" x 13"
48"	32"	18"	33"	14"	23"	8-3/4"	56"	22"	37"	13" x 13"
54"	37"	20"	37"	16"	27"	13"	68"	24"	45"	13" x 18"
60"	37"	22"	42"	16"	27"	13"	72"	27"	45"	13" x 18"
60"	40"	22"	42"	16"	29"	13"	72"	27"	45"	18" x 18"
72"	40"	22"	54"	16"	29"	13"	84"	27"	56"	18" x 18"
84"	40"	24"	64"	20"	26"	13"	96"	29"	67"	20" x 20"
96"	40"	24"	76"	20"	26"	13"	108"	29"	75"	24" x 24"

Figure 11-60
Design data for a single-face fireplace.

DESIGN DATA FOR TWO-FACE OPPOSITE FIREPLACES

WIDTH	MOUTH	HEIGHT	SMOKE CHAMBER	FLUE
W	M	H	S	F
28"	24"	35"	19"	13" x 13"
32"	29"	35"	21"	13" x 18"
36"	29"	35"	21"	13" x 18"
40"	29"	35"	27"	18" x 18"
48"	32"	37"	32"	18" x 18"

Figure 11-61
Design data for a two-face opposite fireplace.

different. Prefabricated fireplaces are relatively lightweight, easier to install than site-built ones, and designed to be more energy efficient; hence they are preferred for new construction. They have a metal shell and a brick-lined firebox, and are insulated. Combustion air enters from outdoors through a direct vent; room air is warmed as it circulates through the firebox and is blown back into the room through warm air outlets. Some models have vents that pipe room air past the firebox so it can be heated and then returned to the room. You can obtain specific models' construction drawings from their manufacturers (Figure 11-63).

TYPE	OPENING WIDTH (W)	OPENING HEIGHT (H)	OPENING DEPTH (D)	FLUE SIZE
DESIGN DATA FOR OTHER FIREPLACES				
SINGLE-FACE	28"	24"	20"	
	30"	24"	20"	12" x 12"
	30"	26"	20"	
	36"	26"	20"	
	36"	28"	22"	12" x 16"
	40"	28"	22"	
	48"	32"	25"	16" x 16"
TWO-FACE ADJACENT	34"	27"	23"	
	39"	27"	23"	16" x 16"
	46"	27"	23"	
	52"	30"	27"	
THREE-FACE	39"	21"	30"	16" x 16"
	46"	21"	30"	
	52"	21"	34"	20" x 20"

Figure 11-62
Design data for other style fireplaces.

Insert An **insert** is a heating unit that fits inside an existing fireplace to converts it into an efficient zone heater. It can burn wood, gas, or pellets. Inserts utilize the existing chimney. Most inserts have blowers to circulate heat.

Design Considerations

Here are some considerations for fireplace selection and design.

Purpose Know if the primary goal is to provide an aesthetic focal point, a zone heater, or both. You will need to research the many hearth products available to decide which styles, materials, colors, and venting options best fit the overall building design.

Placement A fireplace can occupy an interior wall, an exterior wall, or a corner. Older homes usually have chimneys placed in the center so that heat cannot escape directly outdoors. Newer homes tend to have exterior-wall chimneys, which save floor space. Chimney position is important because even though a chimney can enhance the exterior appearance of a home, a frame structure will be weakened by a masonry fireplace and chimney on an exterior wall. Conversely, a masonry structure will be strengthened by it. The chimney should not emerge at a location close to higher-elevation roofs, a consideration on multilevel homes. A fireplace should have a prominent position in a room and away from major traffic routes.

If a location other than a wall is wanted, a stove, which is a prefabricated, freestanding fireplace, is an option. A gas stove's venting options enable it to be installed almost anywhere, such as in the middle of a room, under a desktop or window, or within an existing fireplace. Stoves are made of steel, stone, or cast iron and are available in different styles, finishes, and colors. Most modern stoves are airtight and allow the amount of combustion air to be controlled, making them highly efficient. They burn cleanly because they are built to generate secondary

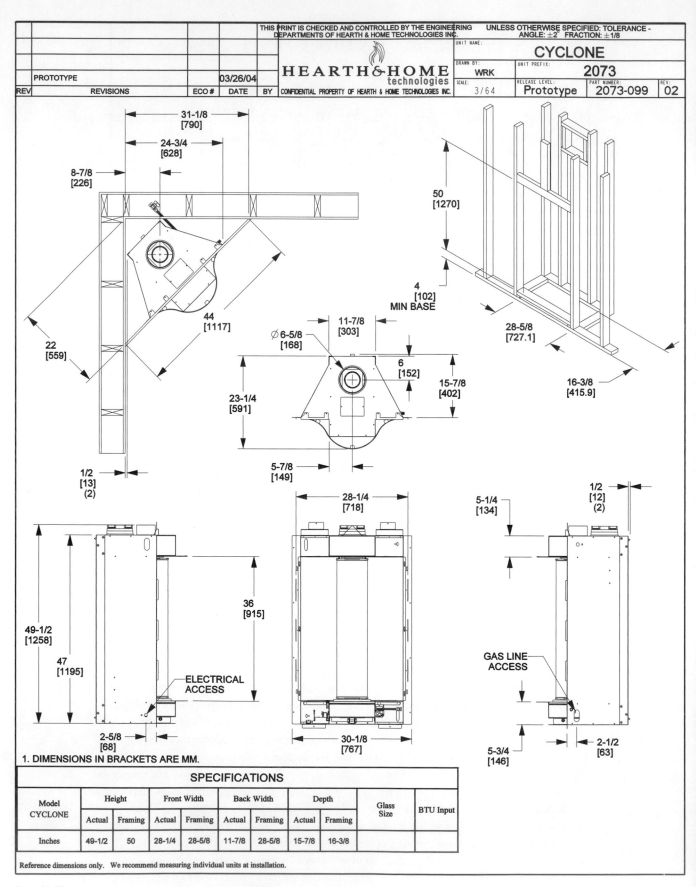

1. DIMENSIONS IN BRACKETS ARE MM.

SPECIFICATIONS

Model CYCLONE	Height		Front Width		Back Width		Depth		Glass Size	BTU Input
	Actual	Framing	Actual	Framing	Actual	Framing	Actual	Framing		
Inches	49-1/2	50	28-1/4	28-5/8	11-7/8	28-5/8	15-7/8	16-3/8		

Reference dimensions only. We recommend measuring individual units at installation.

Figure 11-63
Manufacturer drawings for the Cyclone prefabricated gas fireplace. *Courtesy of hearthnhome.com.*

combustion, which burns excess gases; in doing this, they turn more fuel into heat. There are catalytic, non-catalytic, and pellet fired stoves. A catalytic stove uses a catalytic combustor to reburn combustion gases at low temperatures. Non-catalytic stoves burn cleanly and efficiently by heating the incoming combustion air and circulating combustion gases. Pellet stoves are sophisticated, efficient appliances that utilize electronic controls, blowers, and heat exchangers to produce heat. They produce either **radiant heat**, which emanates from the stove, or **circulating heat**, which is distributed throughout a room via airflow.

Firebox opening size The firebox opening size is important for appearance and operation (Figure 11-64). It cannot be randomly chosen; if it is too small, the fireplace will not produce enough heat. If it is too large, the fire can overheat the room. Floor plan and climate are factors, but design guidelines generally call for the opening to be 1/30th of the area of small rooms and 1/65 of the area of large rooms. A common size for a single-face fireplace is 36" wide by 26" high. Small rooms should have small fireplace openings; large rooms should have large openings. A rectangle whose width is greater than its height will have more pleasing proportions than a square, A raised hearth, usually 11"–16" above the floor, will brings the opening nearer to eye level and make the fireplace appear larger.

SUGGESTED WIDTH OF FIREPLACE OPENINGS IN RELATION TO ROOM SIZE

ROOM SIZE	PLACED IN SHORT WALL	PLACED IN LONG WALL
10' x 14'	24"	24"-32"
12' x 16'	28"-36"	32"-36"
12' x 20'	32"-36"	36"-40"
12' x 24'	32"-36"	36"-48"
14' x 28'	32"-40"	40"-48"
16' x 30'	36"-40"	48"-60"
20' x 36'	40"-48"	48"-72"

Figure 11-64
Chart of suggested fireplace widths relative to room size. After determining a fireplace's width, follow design guidelines for the rest of its proportions. Small variations to meet any restrictions of its materials, such as its masonry course sizes can be made.

Openings that are 6 square feet in area require hearths that extend at least 16" into the room and 8" past the opening's sides. Openings that have more square footage require a hearth that extends 20" into the room and 12" on each side. The inner hearth is the floor inside the fireplace. It is made of a non-combustible material such as brick, stone, or tile outside the fireplace and firebrick inside.

Available Fuel Options Making a fire can be as simple as pushing a button or as labor-intensive as stacking cordwood, building, tending, and cleaning up. If the latter is not feasible, fireplaces that burn wood pellets, fire logs, natural gas, propane, oil, coal, and electricity are options. Fire logs are made of sawdust and wax, ignite quickly and leave little ash. Pellets are made of compressed sawdust, also leave little ash, and produce very hot fire and burns longer and more efficiently than cordwood. They come in 40-pound bags and are poured into a hopper that feeds into the stove. Coal burns longer than wood and provides a clean, even, controllable heat with no visible smoke. Some coal stoves can also burn wood. Oil offers powerful zone heating at the same cost as natural gas. Natural gas and propane (a byproduct of natural gas production that is most commonly used for barbeque grills) offer the most flexibility in design and placement. Seldom-used fireplaces that are equipped to use natural gas or propane can benefit from a set of gas logs.

An electric fireplace has a simulated wood fire and does not require a chimney or venting system. It has a built-in heater controlled by a switch and can be installed in an existing fireplace, making it an option for apartments, townhouses, offices, hotel lobbies, and hotel rooms. A masonry heater creates a small, hot fire. Its heat is absorbed by a huge mass of masonry and slowly released into the room. It only has to be started once or twice a day and is appropriate for cold climates. A masonry heater requires a heavy foundation.

Summary

Stairs and fireplaces are important building components. Many styles and materials for each are available. When choosing and placing them, you must consider technical considerations and building codes along with aesthetic, functional, and design criteria.

Suggested Activities

1. Measure an existing space for total rise, then calculate the number and size of steps needed to clear it.

2. Visit a new-construction site and examine some stairways as they are being built. Note how treads and risers are housed inside the stringers.

3. Do an Internet search for stair manufacturers and look at pictures, vocabulary, and information about their different products.

4. Go to buildings with interesting staircases and fireplaces and sketch plan and elevation views of them.

5. Obtain manufacturer catalogs for different fireplace types and examine their design data, photographs, and drawings.

Questions

1. Name four stair types.
2. What are treads, risers, and nosing?
3. What is the maximum IRC indoor riser height and tread length?
4. What is the height at the top of a handrail?
5. When are handrails required?
6. Name three fireplace styles.
7. How far should the hearth extend from a fireplace that has a six-foot-square opening?
8. What is the difference between a fireplace and a stove?
9. What is a vent?
10. What should the width, height, and depth of the firebox be for a fireplace in a 10' × 12' room?

Internet Resources

http://www.arcways.com.
Stair manufacturer. Its website has lots of good photos.

http://www.stairways.org/
Association of stairway manufacturers. Its website has information on stairs, including a visual interpretation of the IRC (International Residential Code) as it applies to stairs.

http://hearthnhome.com/
Fireplace manufacturer. Its website contains lots of information and drawings.

CHAPTER 12

Drafting Interior Elevations and Sections

OBJECTIVE

- This chapter discusses what interior elevations and sections are and how to draft them.

KEY TERMS

callout	longitudinal	slab
case goods	millwork	standard overlay
detail	molding	stile
elevation	partial	tambour
exterior elevation	rail	transverse
full overlay	raised panel	trim
inset	rake	typical detail
interior elevation	recessed panel	vaulted
lipped	reveal	wall section

What Is an Interior Elevation?

An **elevation** is a vertical, or height, drawing of a wall (Figure 12-1). It assumes the viewer is standing up and looking straight ahead. An **exterior elevation** is of a wall outside the building; an **interior elevation** is of a wall inside the building or of the interior side of an exterior wall. Its purpose is to show built-in and attached items on the walls, such as doors, windows, fireplaces, shelving, cabinetry, trim, molding, wainscots, thermostats, outlets, and switches with their heights. Scale is usually 1/4" = 1'-0".

An interior elevation is created by projecting features down from the floor plan, from interior corner to interior corner and finish floor to finish ceiling (Figure 12-2). Wall, floor, and ceiling thicknesses are usually not included in an interior elevation as they are not relevant to its purpose. Vertical heights and distances from the floor are dimensioned, as are any horizontal distances that do not fit on the floor plan. It is referenced from the plan with a **callout**, a symbol that links the two drawings. This symbol contains an arrow that points to the wall that is being drawn in elevation. The viewer is positioned at the arrow's apex. Everything in front of this point is drawn, so it is important to strategically place it. Different locations in the same room may yield different elevations (Figure 12-3).

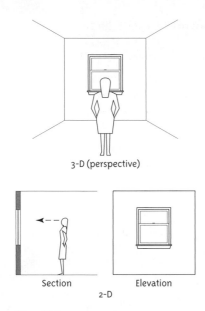

3-D (perspective)

Section   Elevation

2-D

Figure 12-1
Interior elevations and sections are height drawings of walls.

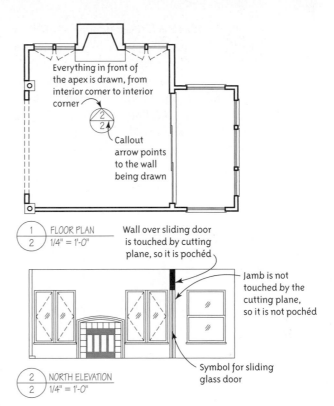

Everything in front of the apex is drawn, from interior corner to interior corner

Callout arrow points to the wall being drawn

(1/2) FLOOR PLAN
1/4" = 1'-0"

Wall over sliding door is touched by cutting plane, so it is pochéd

Jamb is not touched by the cutting plane, so it is not pochéd

Symbol for sliding glass door

(2/2) NORTH ELEVATION
1/4" = 1'-0"

Figure 12-2
Floor plan, callout , and interior elevation.

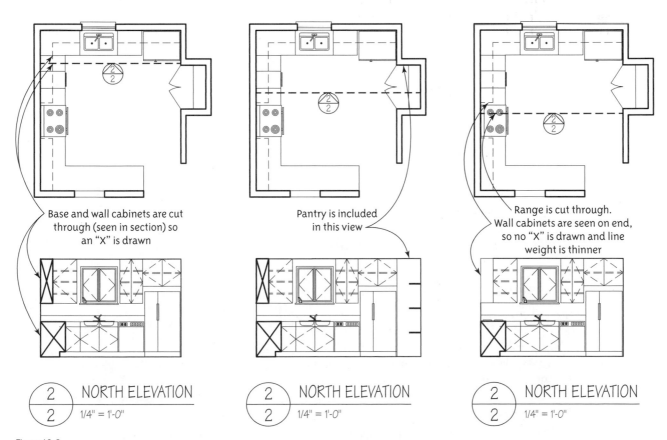

Base and wall cabinets are cut through (seen in section) so an "X" is drawn

Pantry is included in this view

Range is cut through. Wall cabinets are seen on end, so no "X" is drawn and line weight is thinner

(2/2) NORTH ELEVATION
1/4" = 1'-0"

(2/2) NORTH ELEVATION
1/4" = 1'-0"

(2/2) NORTH ELEVATION
1/4" = 1'-0"

Figure 12-3
The location of the callout determines what is drawn on the elevation. Everything in front of the arrow apex is projected down. Nothing behind the arrow apex is drawn.

Figure 12-4 is a kitchen floor plan. (Wall and base cabinet divisions have been left out for clarity.) Let's draw an interior elevation of one of its walls.

1. Orient the floor plan so the wall of interest is at the top, with the callout arrow pointing up to it (Figure 12-4). If the wall is at the bottom with the arrow pointing down, projecting the plan's lines down will yield a "flipped" view of that wall; that is, the items that should be on the viewer's left will appear on the right in the elevation drawing, and the items that should be on the viewer's right will appear on the left. Next, align a horizontal line on the plan with the parallel bar and tape the paper down (Figure 12-5). The steps are shown on Figure 12-6.

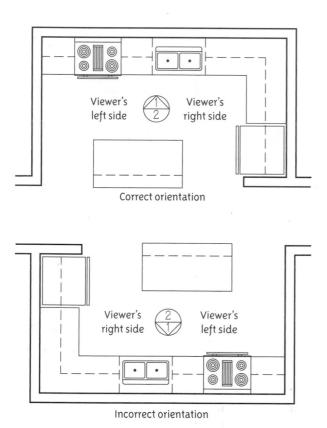

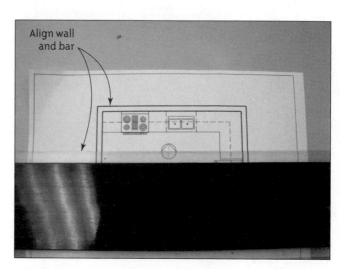

Figure 12-5
Align a wall on the plan with the parallel bar before taping the paper to the board.

Figure 12-4
Orient the floor plan with the callout arrow pointing up. Placing the arrow down, as in the incorrect orientation example, and projecting down will result in an elevation where items that should be on the viewer's right will appear on the left, and vice versa.

2. Draw the back wall. To do this, project construction lines from the floor plan's interior corners down. Draw a ground line, and then measure up the desired height for the ceiling line. The resulting rectangle is the outline of the wall. (Alternatively, if a detail drawing for this wall exists, it can be taped in the side view location and its floor and ceiling lines can be projected over.)

3. Project the base and wall cabinets down, then measure and mark their heights, obtaining correct heights from industry references or product literature. If the elevation is of an existing space, measure the cabinets (and other features) for their heights. Add division lines between them.

4. Darken object lines and add line weights and details. The dashed lines on the cabinet doors in the construction elevation shown in Figure 12-7 indicate the side they are

hinged on (the apex points to the hinges). Note that the base and wall cabinets on the right side have a thicker line weight and an "X" through them. This is because this small portion of the elevation appears in section, as it is cut at the location of the callout arrow.

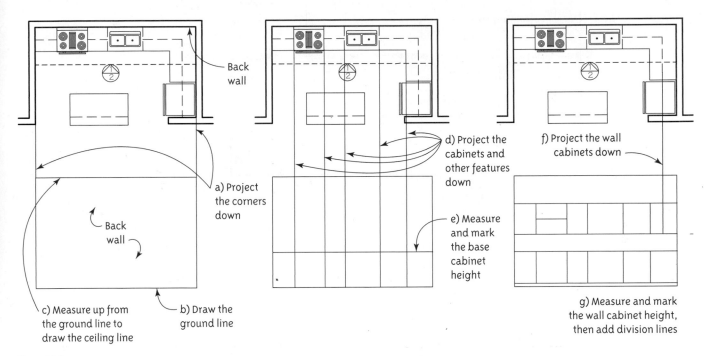

Figure 12-6
Constructing the elevation.

If the elevation is drawn for construction purposes it will be a simple line drawing containing only the information needed to build it. An elevation for presentation purposes may have different details, such as door handles instead of hinge symbols and people and accessories to give it eye appeal and a human, recognizable scale (Figure 12-7).

Sheet Layout While orthographic views of small objects such as furniture can be displayed on one sheet of paper, with edges aligned for easy reading, larger architectural drawings don't lend themselves to this. Typically, floor plans are drawn on one sheet, elevations on another, and sections on yet another. Like smaller objects, they are drawn from view to view on a single sheet of paper when being constructed. However, they are laid out on separate sheets for final tracing. When tracing the final drawings, draw elevations and sections right side up. They should never be drawn perpendicular to other drawings on a sheet. Cross-reference all drawings with callout symbols and ID labels.

Dimensioning An elevation is dimensioned for the heights of important features. As with a floor plan, it may be dimensioned to architectural construction (Figure 12-8) or cabinet industry standards (Figure 12-9). See Chapter 8 for information on dimensioning floor plans.

What Is a Section?

A section is a cut through a building or a portion of a building. There are different types of section drawings: full, partial, and detail.

A **full** section is a vertical cut through the entire length or width of the building and from foundation to roof (Figure 12-10). It's named by the direction of the slice; a **transverse** section

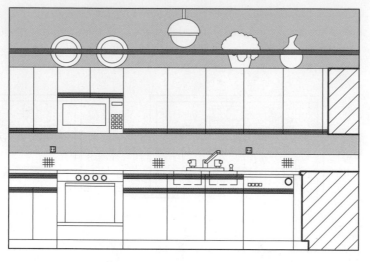

1 / 2 KITCHEN ELEVATION
1/4" = 1'-0"

Presentation

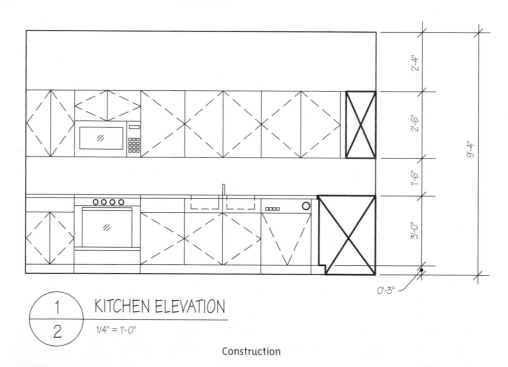

1 / 2 KITCHEN ELEVATION
1/4" = 1'-0"

Construction

Figure 12-7
Finish the elevation in a style appropriate to construction or presentation purposes.

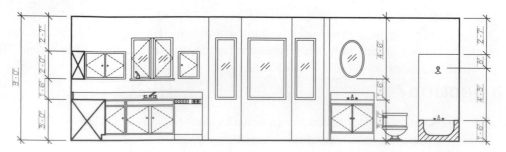

Figure 12-8
Elevation dimensioned to architectural construction standards.

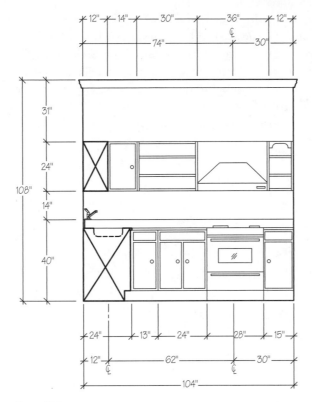

Figure 12-9
Elevation dimensioned to cabinet industry standards.

is perpendicular to the roof ridge and a **longitudinal** section is parallel to it. Its purpose is to show heights, major vertical elements (such as chimneys, stairs, or a sloped roof), spatial relationships, framing, thicknesses of floors walls and ceilings, and details of construction that are not visible on an elevation. Interior ornamentation and trim are not included unless the drawing is a "section-elevation" and is meant to be drawn.

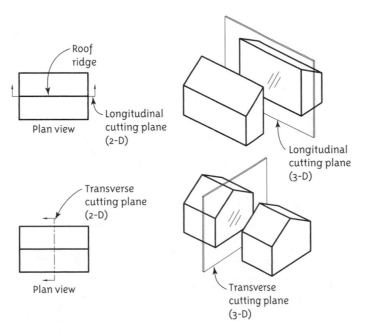

Figure 12-10
Transverse and longitudinal sections through a house with a gable roof.

If the building has multiple levels, rooms will appear stacked on top of each other. Parts of a full section may be encircled with a heavy, dashed line that has an attached symbol telling the reader where to look for a larger-scale drawing. Scale varies; if the section's purpose is to show shape and be a reference for enlargement details, 1/8" = 1'-0" is suitable. If it has to show pochés, studs, insulation, finishes, and how they all go together, then you will need to use a 1/4" = 1'-0" scale at minimum. Figures 12-11, 12-12, and 12-13 show a floor plan with cutting plane, a section, and an exterior elevation of a cottage.

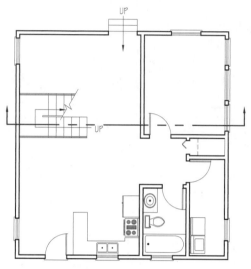

Figure 12-11
Floor plan of a cottage with a cutting plane symbol showing where a section is taken.

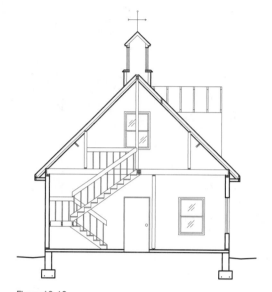

Figure 12-12
Section through the cottage.

Figure 12-13
Exterior elevation of the cottage.

Here are steps for drawing a longitudinal section through the house below (Figures 12-14, 12-15). We will draw a simple one appropriate for referencing enlargement details.

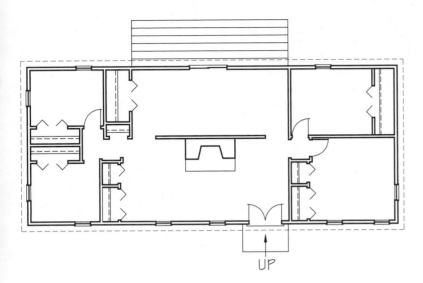

Figure 12-14
Floor plan.

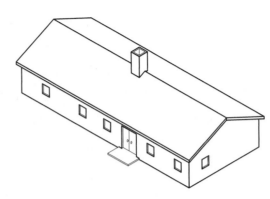

Figure 12-15
Pictorial view of the house in Figure 12-14.

1. Choose where to place the cutting plane (Figure 12-16). As with the elevation, this must be chosen carefully, because some locations will be more informative than others. The section symbol is drawn precisely where the slice is made, with arrows pointing in the direction the viewer is looking. Avoid slice locations where long lengths of walls or large items are cut, or locations that will result in a confusing or uninformative picture. Where possible, include windows and doors in the slice location. Orient the paper with the arrows pointing up; if arrows point down, projecting the plan's features downward will result in an incorrect drawing. Align a horizontal line on the plan with the parallel bar and tape the paper down.

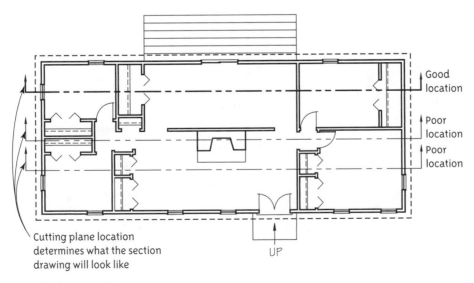

Figure 12-16
Good and poor cutting plane locations.

2. Project the walls down (Figure 12-17). Note that unlike the elevation, exterior wall thickness is projected down and included in the section drawing. Draw a line for the floor, and from there measure and mark ceiling height. Project windows and doors down, then measure and mark their heights.

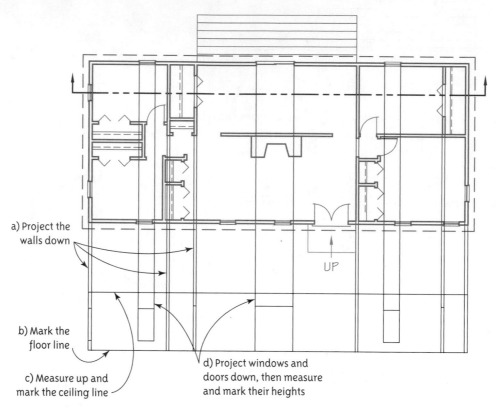

a) Project the walls down

b) Mark the floor line

c) Measure up and mark the ceiling line

d) Project windows and doors down, then measure and mark their heights

UP

Figure 12-17
Project walls, windows, and doors down from the plan to construct a section drawing.

3. Poché the walls that are cut through. A small-scale section whose purpose is chiefly to show special relationships and reference other drawings does not need material poché symbols. Instead, poché the walls solid black, placing window symbols where any windows are cut. When cutting through door openings (Figure 12-18), wall jambs that are looked at, not touched, by the cutting plane are left open (un-pochéd).

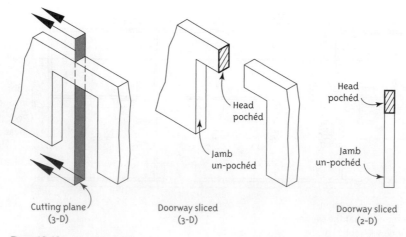

Cutting plane (3-D)

Head pochéd

Jamb un-pochéd

Doorway sliced (3-D)

Head pochéd

Jamb un-pochéd

Doorway sliced (2-D)

Figure 12-18
When you cut through a doorway, the door head is cut, hence it is pochéd. The jamb is not touched by the cutting plane, so it is left open (un-pochéd).

In this example we're cutting through a doorway, so the wall is un-poché from floor to door head. However, the wall *above* the door head is poché, since the cutting plane touches it. Add floor thickness and foundation information, if available and needed. Add the ground line; this will be hard-lined if paved, drawn freehand if not. Project the closet shelves and rods down, and then poché them, the floor and any walls that require it. Add details to the windows and doors. Project the roof overhang down for reference in the next step (Figure 12-19).

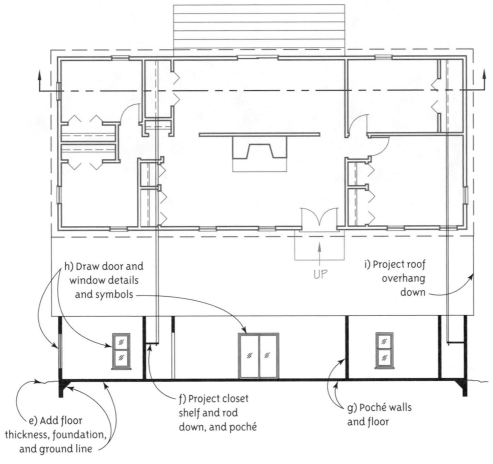

Figure 12-19
Adding detail to the section drawing.

4. This house has a gable roof, so the ceiling is **vaulted,** meaning it angles up on one or both sides (Figure 12-20). Draw the ceiling's height at the slice location in the floor plan. An existing interior or exterior elevation of a side (gable) wall is needed to graphically determine this.

Figure 12-21 shows the relationship between the location of the cutting plane in plan and the height of the ceiling in the section drawing. To determine this height, set the dividers tool to mark distance "X" on the plan. Then transfer distance "X" to the section drawing. Draw a vertical line from "X" until it intersects the roof. That vertical line represents the house's floor-to-ceiling height at the cutting plane location. Project that height horizontally to the section view. (Note: When drawing sloped ceilings in elevation, draw the ceiling's height at the callout symbol's location). Add the roof overhang and thickness.

The ceiling's slope is not visible in this view because the cut is parallel to the roof ridge. This particular view of the ceiling is the rake view, **rake** being the inclined edge of a sloped roof over a wall. To see the ceiling's slope, you will need a transverse section (Figure 12-22).

Figure 12-20
A vaulted ceiling. *Courtesy of Pella Windows and Doors, Pella, IA.*

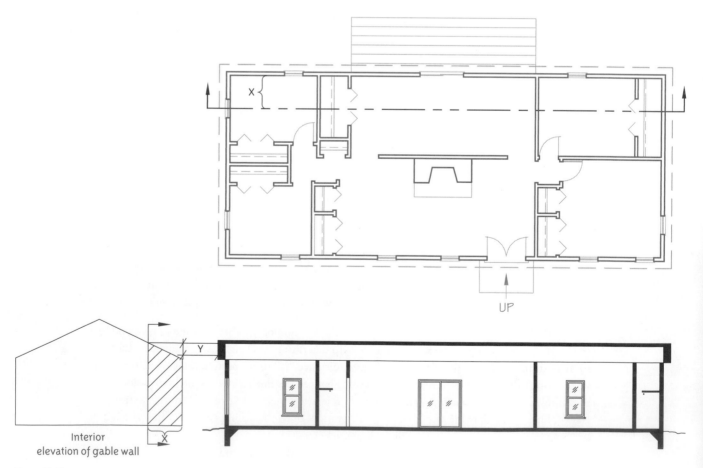

Interior
elevation of gable wall

Figure 12-21
The ceiling height drawn depends on the cutting plane location.

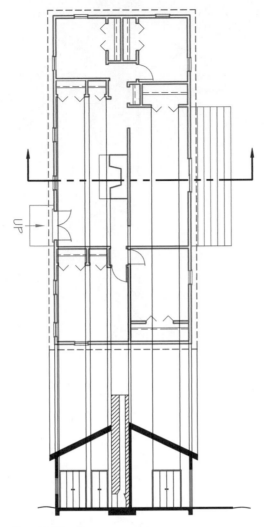

Figure 12-22
Transverse section.

Wall Sections and Details

Wall section A **wall section** is a vertical slice through a wall from footing to roof (Figure 12-23). Its purpose is to show how the wall is put together. Wall sections look different depending upon what they are made of and if they are interior or exterior. Figures 12-24, 12-25, and 12-26 are some examples. Scale is between 1/4" = 1'-0" to 1" = 1'-0".

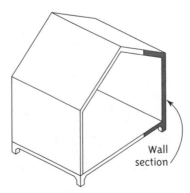

Figure 12-23
A wall section is made with a vertical cutting plane from foundation to roof.

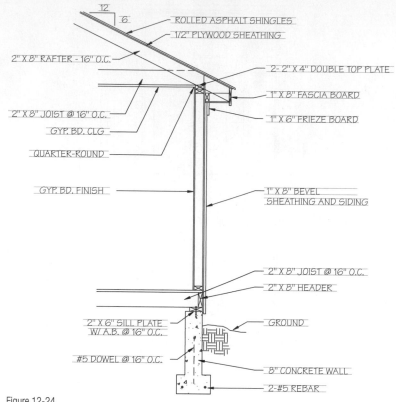

Figure 12-24
Wall section (wood stud).

Labels for Figure 12-24:
- ROLLED ASPHALT SHINGLES
- 1/2" PLYWOOD SHEATHING
- 2" X 8" RAFTER - 16" O.C.
- 2- 2" X 4" DOUBLE TOP PLATE
- 1" X 8" FASCIA BOARD
- 2" X 8" JOIST @ 16" O.C.
- 1" X 6" FRIEZE BOARD
- GYP. BD. CLG.
- QUARTER-ROUND
- GYP. BD. FINISH
- 1" X 8" BEVEL SHEATHING AND SIDING
- 2" X 8" JOIST @ 16" O.C.
- 2" X 8" HEADER
- 2" X 6" SILL PLATE W/ A.B. @ 16" O.C.
- GROUND
- #5 DOWEL @ 16" O.C.
- 8" CONCRETE WALL
- 2-#5 REBAR

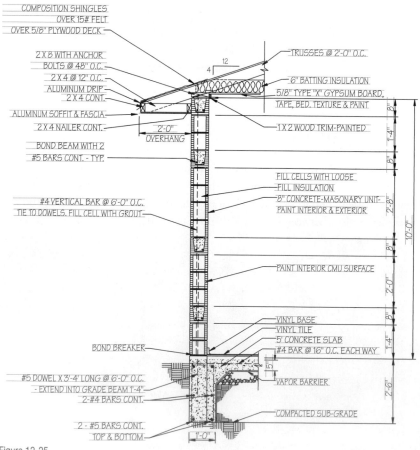

Figure 12-25
Wall section (concrete block).

Labels for Figure 12-25:
- COMPOSITION SHINGLES OVER 15# FELT OVER 5/8" PLYWOOD DECK
- 2 X 8 WITH ANCHOR BOLTS @ 48" O.C.
- 2 X 4 @ 12" O.C.
- ALUMINUM DRIP
- 2 X 4 CONT.
- ALUMINUM SOFFIT & FASCIA
- 2 X 4 NAILER CONT.
- BOND BEAM WITH 2 #5 BARS CONT. - TYP.
- #4 VERTICAL BAR @ 6'-0" O.C. TIE TO DOWELS. FILL CELL WITH GROUT
- BOND BREAKER
- #5 DOWEL X 3'-4' LONG @ 6'-0" O.C. - EXTEND INTO GRADE BEAM 1'-4"
- 2-#4 BARS CONT.
- 2 - #5 BARS CONT. TOP & BOTTOM
- TRUSSES @ 2'-0" O.C.
- 6" BATTING INSULATION
- 5/8" TYPE "X" GYPSUM BOARD, TAPE, BED, TEXTURE & PAINT
- 1 X 2 WOOD TRIM-PAINTED
- 2'-0" OVERHANG
- FILL CELLS WITH LOOSE FILL INSULATION
- 8" CONCRETE-MASONARY UNIT-PAINT INTERIOR & EXTERIOR
- PAINT INTERIOR CMU SURFACE
- VINYL BASE
- VINYL TILE
- 5' CONCRETE SLAB
- #4 BAR @ 16" O.C. EACH WAY
- VAPOR BARRIER
- COMPACTED SUB-GRADE

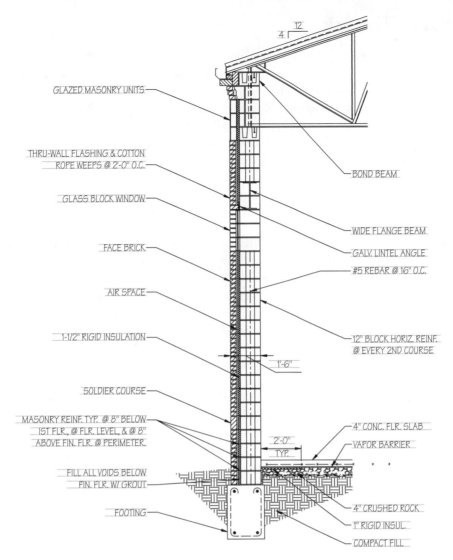

GLAZED MASONRY UNITS

THRU-WALL FLASHING & COTTON ROPE WEEPS @ 2'-0" O.C.

GLASS BLOCK WINDOW

FACE BRICK

AIR SPACE

1-1/2" RIGID INSULATION

SOLDIER COURSE

MASONRY REINF. TYP. @ 8" BELOW 1ST FLR., @ FLR. LEVEL, & @ 8" ABOVE FIN. FLR. @ PERIMETER.

FILL ALL VOIDS BELOW FIN. FLR. W/ GROUT

FOOTING

BOND BEAM

WIDE FLANGE BEAM

GALV. LINTEL ANGLE

#5 REBAR @ 16" O.C.

12" BLOCK HORIZ. REINF. @ EVERY 2ND COURSE

1'-6"

4" CONC. FLR. SLAB

VAPOR BARRIER

2'-0" TYP.

4" CRUSHED ROCK

1" RIGID INSUL.

COMPACT FILL

Figure 12-26
Wall section (brick veneer on concrete block).

The poché inside the sections represents the material the wall is made of. Charts of common pochés were shown in Chapter 8 (Figures 8-14 and 8-15). Looking at them, you can see that many materials are pochéd differently in section and elevation. This is to make drawings easier to read. For instance, look at the wall section in Figure 12-26. At the glass block window the brick and concrete block are not pochéd like they are in the rest of the wall. This is because, as discussed earlier in this chapter, the section slice touches the wall but not the jamb. So the wall material is pochéd in section and the jamb in elevation.

Partial section A **partial** section is a vertical cut through a small portion such as one wall. A **detail** is a vertical or horizontal cut made through an even smaller portion, such as a window, doors, stairs, level change, or anything else where multiple pieces join. Its purpose is to show how the pieces go together. The amount of details needed depends on the building's complexity, number of floors, and changes in the materials and methods of construction. **Typical details,** abbreviated TYP, show repetitive features; for instance, if all exterior walls are built the same, one detail called "typical wall detail" represents all walls even though the cutting plane on the floor plan is only placed through one wall. Designers often have typical detail libraries consisting of drawings or AutoCAD blocks (templates) that they can use on multiple projects. Details are drawn at scales between 1/2" = 1'-0" and full size.

Architectural Millwork

Interior designers may be asked to design and draft architectural **millwork**. Also called *woodwork,* this is an umbrella term for finished wood items manufactured in a lumber mill (although polyurethane, paper veneers over substrate, foams, and MDF products used for the same purpose are also referred to as millwork). Millwork includes any piece placed inside or outside a home, such as windows, doors, shutters, columns, pediments, and cabinets. Millwork pieces commonly drawn by interior designers include case goods, cabinets and cabinet doors, front doors, interior moldings, and trim. Since such millwork is frequently included in elevation and section drawings, it is discussed here.

Cabinets

Cabinets are a type of **case goods**. This refers to any non-upholstered piece of furniture used to store, or case, household goods. Bookcases, desks, and dressers are other examples of case goods, since they are designed for storage.

Cabinet choices include framed and unframed (also called frameless) (Figure 12-27), full or standard overlay, and concealed or exposed hinge. To tell whether a cabinet is framed or unframed, open the door and check for **rails** (horizontal framing pieces) and **stiles** (vertical framing pieces). Framed cabinets have rails and stiles; unframed ones do not. **Overlay** is the amount of front frame covered by the door and drawer. The overlay is usually half an inch, with one inch of the face frame exposed. The **reveal** is the part of the front frame that is exposed. Cabinet widths were illustrated in Chapter 6; Figures 12-28 and 12-29 show illustrations of heights and Figure 12-30 shows a section through a typical wall and base cabinet.

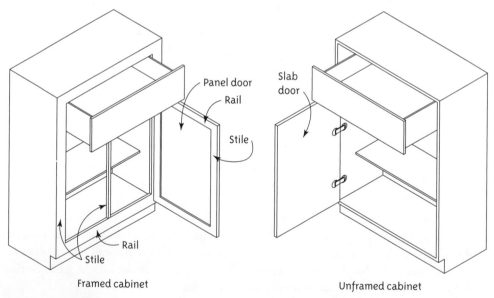

Figure 12-27
Framed and unframed cabinets.

Cabinet door types refer to how the door fits the front of the cabinet box. The types are tambour, inset, lipped, full overlay, and standard overlay (Figure 12-31).

Tambour A **tambour** door has its own frame and can be used on both framed and frameless cabinets. A tambour door is made of many separate pieces attached to a flexible backing sheet. It is installed in a track that allows it to slide around a corner or roll up. It is useful for appliance garages or in any cabinet where function calls for the door to remain open without getting in the way. Roll-up tambours can be installed under straight or corner cabinets.

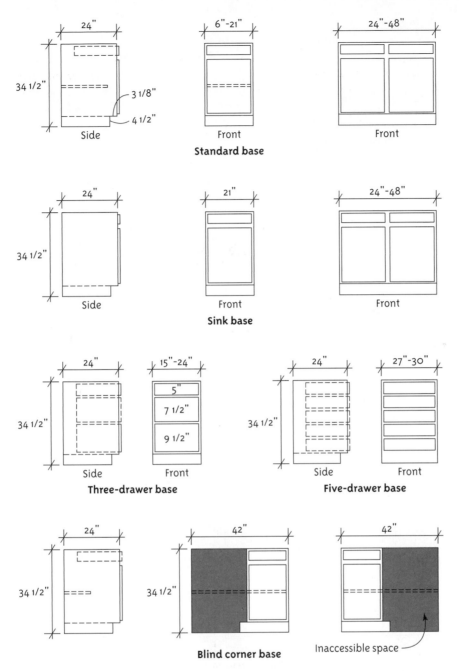

Figure 12-28
Different cabinet types and their sizes in elevation. Base cabinets are usually 36" high, which includes a 4"–4½" toe space and 1"–1½" thick countertop. Width is between 9–60" (in 3" increments), and depth is 24". Wall cabinets are usually between 12"–33" high, with their width between 9"–60" (in 3" increments) and a depth of 12"–13". A sit-down work surface should be 30" above the floor.

Inset An **inset** door sits within the face frame and is flush with the front edges of the cabinet box. Inset doors are most often used to achieve a formal colonial or a rustic farmhouse look. It requires door pulls or knobs to open the doors and drawers, and its hinges are exposed. Inset doors are designed to sit within the rails and stiles of the cabinet frame.

Lipped A **lipped** door has a rabbet (groove) cut all the way around the door on the back edge. The rabbet allows part of the door to go back into the cabinet and leaves the remaining part resting on the cabinet or face frame. Because the door sits tight in the frame, it must have

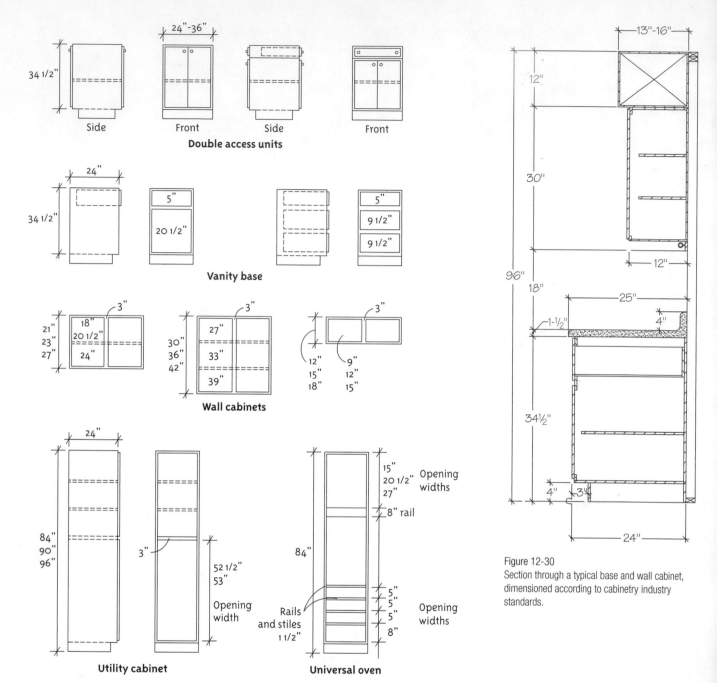

Figure 12-29
Different cabinet types and their sizes in elevation.

Figure 12-30
Section through a typical base and wall cabinet, dimensioned according to cabinetry industry standards.

a knob or pull. When viewed from the front, a lipped door appears to be an overlay door. Only when the door or drawer is opened can it be seen that it is a lipped door.

Full Overlay Door A **full overlay** door is the one option available for frameless cabinets. There is 1/8" or less of cabinet frame showing around each door and drawer head and between the doors. The doors and drawer fronts are mounted to rest against the face frame. Hinges are concealed and a pull or knob is necessary for opening. The look presented is seamless and modern.

Standard Overlay Door A **standard overlay** door, also called traditional, is the most common option for framed cabinets. The face frame has a full reveal around the door and drawer perimeters. The reveal is typically 1", which allows enough clearance between the doors and drawer for operational access without pulls or knobs.

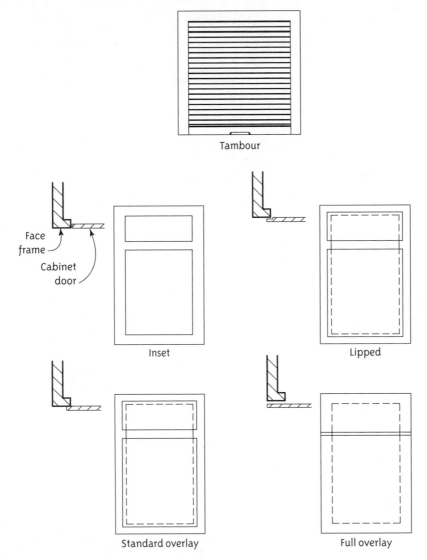

Tambour

Face frame

Cabinet door

Inset

Lipped

Standard overlay

Full overlay

Figure 12-31
Tambour, inset, lipped, full overlay and standard overlay doors on framed cabinets. The hidden lines representing the cabinet frame are typically not shown in an elevation drawing.

Cabinet door style This refers to the appearance of the door. There are two major door styles, slab and panel (Figure 12-32). Both have hundreds of variations; a few are shown in Figure 12-33.

Slab door Also called flat panel or frameless, a **slab** door appears made of one solid piece. It has no raised or recessed profile.

Recessed panel door Also called framed, a **recessed panel** door looks like a picture frame; a wood frame around the edge surrounds a panel in the middle. The vertical sections of the frame are the stiles, and the horizontal sections are rails. The frame is typically 1/2" to 3/4" thick. It starts as a flat piece of wood, and then the frame is placed around the perimeter. The panels can be smooth, grooved, or decorated.

Raised Panel Also called *solid*, a **raised panel** door is made of solid or veneered wood that protrudes forward.

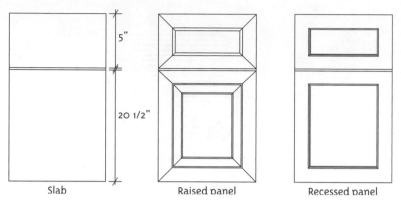

Figure 12-32
Slab panel cabinet door style.

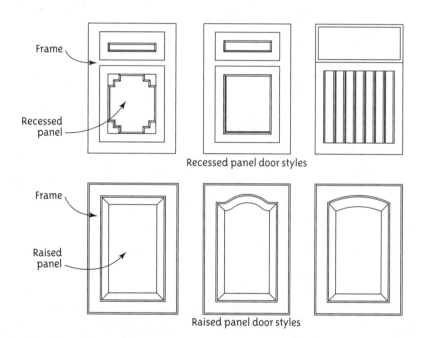

Figure 12-33
Examples of raised and recessed panel door variations.

Architectural woodwork also includes interior moldings and trim. **Moldings** are decorative lengths of wood that hide joints at intersections of the floor, wall, and ceiling of interior walls. Cornice moldings (Figure 12-34) are placed where the wall and ceiling meet; base moldings are placed where the wall and floor meet.

Trim is an umbrella term for different types of finish millwork (Figures 12-35 and 12-36). It may be lengths of wood that run around a door, on the floor and ceiling, between ornamental items on a fireplace mantel, or are built in, and as such, is also called casing or molding. It is further categorized as standing or running. Standing refers to short lengths of trim, such as around a door, and running refers to longer lengths, such as a baseboard. Hundreds of designs and cross-sections of such trim exist, such as ogee, egg-and-dart, and dentil.

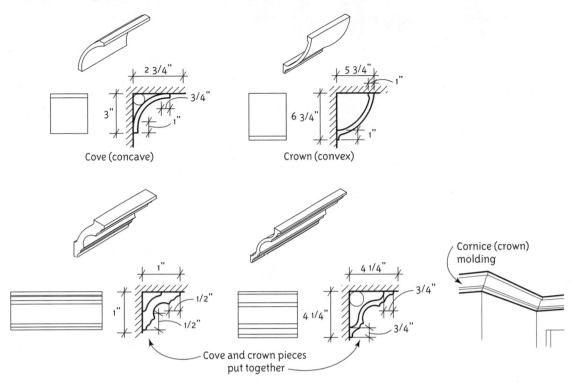

Cove (concave)

Crown (convex)

Cornice (crown)
molding

Cove and crown pieces
put together

Figure 12-34
Cornice moldings can be crown (convex), cove (concave), or a mixture to create an elegant look.

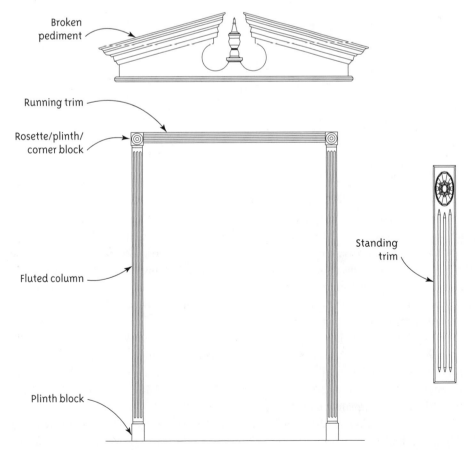

Broken
pediment

Running trim

Rosette/plinth/
corner block

Fluted column

Standing
trim

Plinth block

Figure 12-35
Examples of decorative trim pieces.

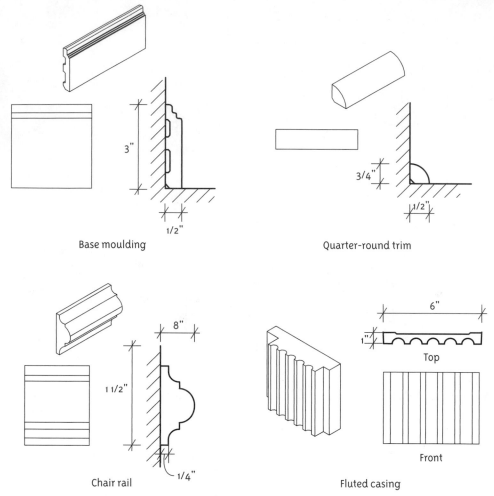

Base moulding

Quarter-round trim

Chair rail

Fluted casing

Figure 12-36
Common trim pieces.

Trim is also referred to as architectural ornamentation or detailing, as it may be purely decorative. Examples include onlays, chair rails, corbels, medallions, cornices, wainscots, dados, plate rails, rosettes, and plinth blocks. Onlays are decorative wood carvings. Chair rail is a type of trim usually placed 36" above the floor, named so because its purpose is to protect walls from chair backs. Corbels are decorative brackets. Medallions are round decorative pieces. A cornice, also called a cove or crown, is trim placed where a vertical surface meets a horizontal one (such as at the intersection of the wall and ceiling). A wainscot is a protective surface on the lower three or four feet of an interior wall, of a different material from the rest of the wall. A pediment is a surface used ornamentally over doors and windows. A dado is a heavily decorated wainscot; it may refer to just the middle section of ornamental paneling or the whole decoration. A dado joint is a rectangular groove cut for a woodworking joint; a dado rail is a rail at the top of a dado. A plate rail is a rail or narrow shelf along the upper part of a wall for holding plates or ornaments. Rosettes and plinth blocks are decorative square blocks placed on both sides of a door. (Rosettes are upper blocks with a carved rose or geometric form inside, and plinths are lower, plainer blocks.)

Figures 12-37 through 12-42 are generic drawings of common items that may come in handy when drafting elevations. Consult manufacturer catalogs for specific sizes and features.

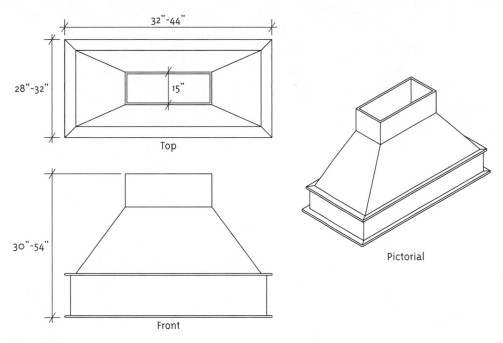

Figure 12-37
Range hood. It's depth determines the height at which it is located. A hood 17" or less deep should be placed a minimum of 20" above the range top. A hood 18" or more deep should be placed a minimum of 24" above the range top.

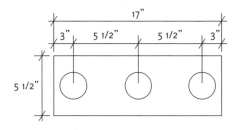

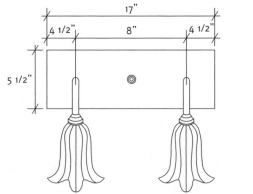

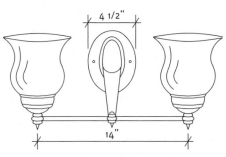

Figure 12-38
Bathroom light fixtures.

Figure 12-39
Front door designs.

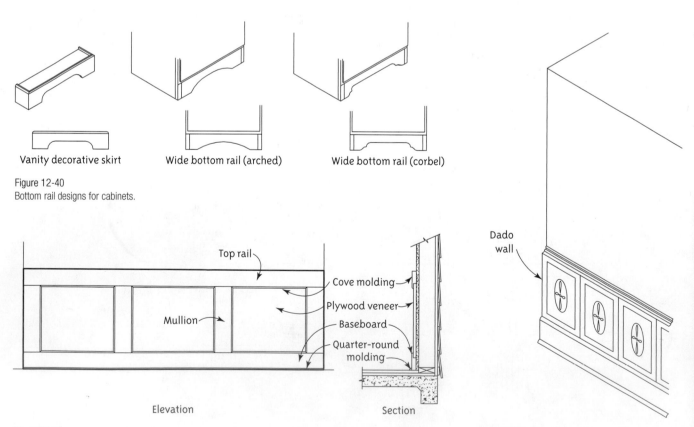

Vanity decorative skirt

Wide bottom rail (arched)

Wide bottom rail (corbel)

Figure 12-40
Bottom rail designs for cabinets.

Top rail

Cove molding

Plywood veneer

Mullion

Baseboard

Quarter-round molding

Elevation

Section

Figure 12-41
Paneled wainscot.

Dado wall

Figure 12-42
Dado wainscot.

Summary

Elevations and sections are orthographic drawings that give heights and details of features shown on the floor plan. They include everything that is hung on the wall. Details show intricate features of construction. With dimensions, they are paired with floor plans to completely describe a design.

Suggested Activities

1. Browse a manufacturer's catalog for cabinet door styles.
2. Do an Internet search for "architectural millwork" to see what different items are available.
3. Measure and draw interior elevations of walls at home and on campus. Draw elevations with both flat and non-flat ceilings.
4. Obtain a floor plan from a home décor magazine and select a cutting plane location. Overlay it with tracing paper and project lines down from it to sketch a section. Since magazine floor plans are not to scale, estimate appropriate heights.

Questions

1. What is an interior elevation?
2. What is the difference between a transverse and longitudinal section?
3. How does the cutting plane location in a room with a vaulted ceiling affect the elevation drawing?
4. How should an interior elevation's callout be oriented in the floor plan?
5. What is a wall section?
6. What is a detail?
7. What is architectural woodwork?
8. What are case goods?
9. What are cabinet rails and stiles?
10. What is the difference between cabinet door type and style?
11. Describe the difference between a full and standard overlay cabinet door.
12. What is architectural millwork?
13. What is casing?

Further Reading and Internet References

Burden, E. (2003). *Entourage: A Tracing File and Color Sourcebook* (4th ed.). New York: McGraw-Hill.

http://www.woodweb.com Contains woodworking information and forums.

CHAPTER 13

Other Construction Drawings

OBJECTIVE

- This chapter gives an overview of building systems and the drawings that represent them, specifically demolition, plumbing, heating, ventilation and air conditioning, and site plans.

KEY TERMS

air conditioner	GFI	register
appliance	ground	riser diagram
benchmark	heat pump	schedule
boiler	hose bib	schematic
branch	humidifier	septic tank
circuit	inlet	service panel
clean-out	lamp	site plan
convection	mechanical plan	stack switch
conduction	main	sump pump
contour line	meter	thermostat
current	outlet	trap
demolition	pipe	valve
duct	plumbing plan	volt
electrical plan	power	water heater
electricity	power/telephone/	water pipes
equipment plan	data/VOIP plan	watt
fixture	property line	well
footprint	radiant hydronic	zone
forced-air	radiation	
furnace	reflected ceiling plan	

Knowing how a building's utility systems work enables a "whole building" approach to design. You may not draw the plumbing or mechanical systems, but the ability to interpret these drawings will enable you to select appropriate fixtures and appliances and ensure that others involved in the project acknowledge them. For instance, you need to select a tub with a drain on the end that will connect to existing pipes. Verification of electrical conduit and pipe location is needed before placing holes in walls to make new doors or windows. Overlaying electrical and HVAC drawings onto floor plans will help ensure that outlets, thermostats, gas and water lines, and other service items are placed appropriately (Figure 13-1).

Figure 13-1
This shower utilizes water, gas, and electronics, all of which are shown on their respective plans. *Courtesy of Kohler Plumbing, Kohler, WI.*

Electrical Plans

The **electrical,** or lighting, **plan** is usually prepared by an electrical engineer. The National Electrical Code, NFPA-70, governs its design. It shows lights, switches, circuits, outlets, and, sometimes, dimensions (Figures 13-2, 13-3).

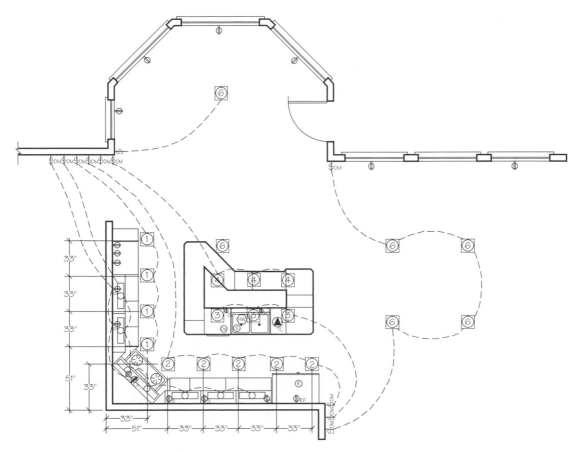

Figure 13-2
Electrical plan for a kitchen.

APPLIANCE AND ELECTRICAL LEGEND

SYMBOL	SPECIAL PURPOSE OUTLETS	VOLTAGE	MANUFACTURER
TC	1. TRASH COMPACTOR	110 VOLTS	KITCHEN AID
HW	2. HOT WATER DISPENSER	110 VOLTS	KITCHEN AID
GD	3. INSINKERATOR GARBAGE DISPOSAL	110 VOLTS	INSINKERATOR
DW	4. KITCHEN AID DISHWASHER	110 VOLTS	KITCHEN AID
CT	5. JENN-AIR COOKTOP	GAS	JENN-AIR
RF	6. KITCHEN AID REFRIGERATOR	110 VOLTS	KITCHEN AID
MW	7. GE MICROWAVE	110 VOLTS	GE
CO	8. VIKING CONVECTION OVEN	220 VOLTS	VIKING
WD	9. VIKING WARMING DRAWER	110 VOLTS	VIKING

LIGHT FIXTURE LEGEND

SYMBOL	LIGHT	WATTAGE
1	RECESSED LIGHT	150 WATTS
2	RECESSED LIGHT	150 WATTS
3	RECESSED LIGHT	150 WATTS
4	RECESSED LIGHT	150 WATTS
5	RECESSED LIGHT	150 WATTS
6	RECESSED LIGHT	150 WATTS
S	FLUORESCENT LIGHT / FLUORESCENT LIGHT WITH SWITCH	
$DM $	DIMMER SWITCH / SINGLE-POLE SWITCH	
SW	DUPLEX OUTLET WITH SWITCH	

Figure 13-3
Legend for the plan in Figure 13-2.

Some background on electricity and its delivery is needed to make a discussion of electrical plans more meaningful.

Electricity and its Delivery

Electricity is the flow of electrical **power**. This essentially consists of *electrons* (negatively charged particles) that are put to work. It is generated at a power plant using fuel. Conventional energy sources include coal, oil, and nuclear raw materials. "Green" fuels such as solar, wind, and combustible gas from landfills and decomposing corn stalks are also used. The electricity generated is transmitted under great pressure, called *voltage,* across high-voltage power lines to local utilities. The utilities then deliver that electricity to homes and businesses via a local power grid that runs either above ground on poles or underground over cables. Voltage is stepped down at power stations and transformers along the way. Electricity enters the building at a greatly reduced pressure through a meter and distribution panel and, still under pressure, is distributed throughout via feeder and/or branch circuits (Figures 13-4, 13-5, 13-6).

Electrical Terms

The flow of electrons is called **current**, which is measured in *amps*. Electricity is converted from energy to power, and then into heat when consumed. It is measured in **watts** for appliances and *horsepower* for cars. Since household electrical consumption is relatively high, its unit of measure is the *kilowatt* (1000 watts), abbreviated kW. The total amount of electrical energy used is measured in *kilowatt-hours* (kWh). A kWh is the work performed by one kilowatt in one hour and is what electricity's price is based on. A 1000-watt light bulb operating for one hour uses one kWh.

A typical incandescent light bulb uses 75 watts; a computer needs 60 watts to turn on, a TV needs 250 watts, and a hairdryer needs 1 kW. A person climbing stairs works at about 200 watts; an athlete can work at up to 900 watts.

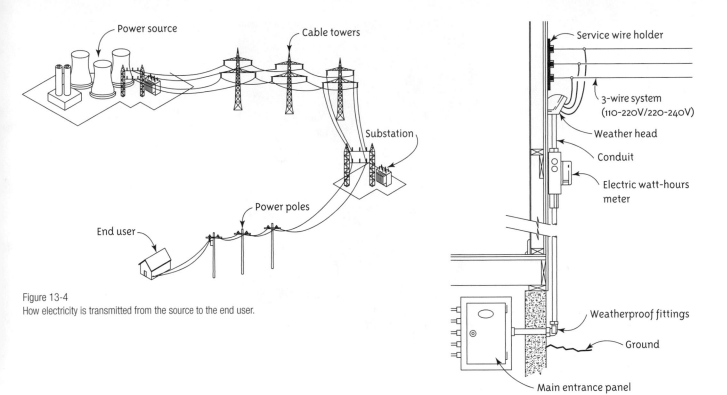

Figure 13-4
How electricity is transmitted from the source to the end user.

Figure 13-5
How electricity enters the house.

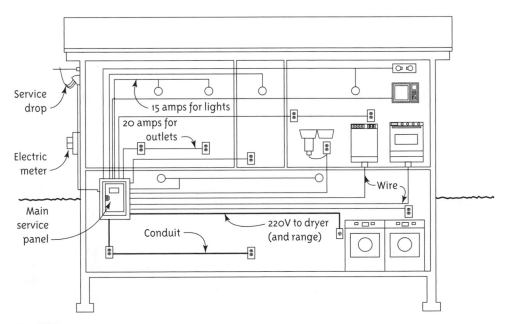

Figure 13-6
How electricity is delivered within the house.

Electric meter Also called a watt-hour meter, an electric **meter** (Figure 13-7) measures the amount of electrical energy the customer uses, which in turn tells the power company how much to charge. There are two types of meters. One has a row of small dials on its face, with each dial recording kilowatt-hours (Figure 13-7). For example, if a 100-watt bulb burns for 10 hours, the meter will record 1 kWh (10 × 100 = 1,000 watt-hours, or 1 kWh).

Figure 13-7
Residential electric watt-hours meter.

HOW TO READ AN ELECTRIC METER

Each dial registers a certain number of kilowatt-hours of electrical energy. From right to left the dial records or measures individual kilowatt-hours from 1 to 10; the next counts from 10 to 100 kilowatt-hours; the third counts up to 1,000; the fourth counts up to 10,000; and the dial at the extreme left counts kWhs up to 100,000. The second type has numerals in slots on the meter face, similar to a car odometer, instead of dials. It's read from left to right, and the numbers show total electrical consumption.

Service Panel   Also called the *main* or *disconnect panel,* the **service panel** (Figure 13-8) receives electricity into the house from the service entrance or meter box and distributes it throughout via branch circuits. It consists of a large metal box inside which are *circuit breaker*s or *fuses.* A circuit breaker is a safety device that disconnects the live wire if the current is too large. It can be reset by toggling it back on, unlike a fuse, which is an older technology protective device that melts when the circuit is overloaded. Circuit breakers and fuses protect conductors and appliances. The service panel can be disconnected from the interior wiring system to shut it off. A tripped circuit breaker or blown fuse generally means there are too many appliances plugged into that circuit or that an appliance is malfunctioning. Service panels are usually located inside the house.

Three wires from the meter enter this box. Two are attached to the tops of a pair of exposed copper *buses,* or bars. These two are the *live,* or *hot* wires and provide 110V and 220V power. The third is the *neutral* or *ground* wire. It is attached to a separate bus in the main box and is connected to the ground by a heavy solid copper wire clamped to an underground bar or plate.

Figure 13-8
Residential service panel.

Circuit   A **circuit** includes the conductor, switching device(s), the load, and the outlet for that load. There are two types of circuits: *feeder* and **branch**. Both emanate from the service panel. Feeder circuits (not found in all houses) use thick cables that travel from the main entrance panel to smaller distribution panels called *sub-panels*. These redistribute power and are located in remote parts of a house or in outbuildings, inside and outside. *Branch circuits* run from the service panel or a sub-panel and are routed to outlets and switches. They include *lighting circuits,* which supply permanently installed light fixtures, *small appliance circuits,* which supply areas where multiple 110V appliances are likely to be used; and *dedicated* circuits; which supply large 220V and other appliances that require *dedicated* (their own) circuits. Such appliances include sump pumps, garbage disposals, dishwashers, washers and dryers, furnaces, water heaters, refrigerators, and microwaves. This is why you will see some outlets on electrical plans specifically marked for the appliance planned for that location, such as DW for dishwasher or GD for garbage disposal.

Voltage Voltage is the pressure that forces electrons through a wire. A **volt** is a unit of electric potential. Standard voltages in the United States are 110V and 220V.

Current Current is the flow of electrons through a wire, measured in *amperes* (amps).

Appliance **Appliance** is a general term for any item powered through a plug and a flexible cable.

Watt A watt is a unit of power, calculated as energy per unit of time. It measures the work done by electricity. Appliances are rated in watts. When an appliance is turned on, it draws current from the electrical grid or a battery.

Lamp **Lamp** is the technical term for light bulb. There are many different kinds of lamps, but the most common are *incandescent* and *fluorescent*. An incandescent bulb has a filament (very thin wire) that emits light when current is run though it. A *fluorescent* bulb is typically a long, thin tube with an inner phosphor coating that emits light when a current runs through it; some are small and shaped like filament lamps.

Fixture A light **fixture** consists of a lamp, a reflector for directing the light, an opening, a housing, and connection to a power source. It may also contain a *ballast* to regulate current. There are many different types of light fixtures; Figure 13-9 shows four.

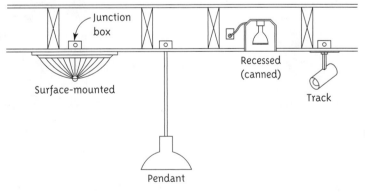

Figure 13-9
Different types of light fixtures.

Conduit A conduit is a hollow tube that holds *conductors,* or wires.

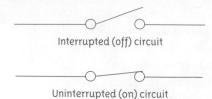

Interrupted (off) circuit

Uninterrupted (on) circuit

Figure 13-10
Symbolic representation of a switch in a circuit.
When the switch is closed, the lights are on,
because there is an uninterrupted flow of
electricity. When the switch is open, the lights are
off, because the flow has been interrupted.

Switch A **switch** is an electrical device that opens and closes the circuit in an electrical circuit and is operated by the user. When the circuit is open, current cannot flow, meaning the lights or appliances (the loads) are off (Figure 13-10).

There are many different kinds of switches. The *hand switch,* operated by a person to open and close a circuit, is most relevant to building design. Types of hand switches include the *toggle,* an angled, two-position lever used for lights (Figure 13-11), and *push-button* switches, which are two-position momentary devices activated with a press-and-release button, and are used for doorbells and garage door openers. A *dimmer* switch (Figure 13-12) allows light to be adjusted to a specific brightness by turning a knob. A *timer* switch usually controls indoor lights, outdoor security lights, post and porch lights, heat lights, pool motors, or landscape lighting and can be programmed to switch on and off automatically. A *joystick* switch is operated by a lever that moves in multiple directions, activating different contact mechanisms depending on which way the lever is pushed. It's typically used for crane and robot controls and video games. *Single-pole, three-way,* and *four-way* switches are toggles that operate lights. Single pole means one switch operates one light; three-way means two switches operate one light (e.g., switches placed at opposite ends of a hall); and four-way means three switches operate one light.

The electrician marks
where the top of the
junction box is to be
placed, and what kind
of switch (in this case an S3)

Figure 13-11
Toggle switch. The electrician marks the junction box height on the stud,
along with what type of switch to be installed (in this case an S3).

Figure 13-12
One type of dimmer switch.

Three-way switches are placed in all large rooms that have two exits, at hallway and staircase ends, and in garages. Four-way switches are placed in rooms with more than two exits. This is so the user doesn't enter a dark room and doesn't have to backtrack. Switches are placed on the door latch side, 48" above the floor or 30" to 40" above the floor for wheelchair users and no closer than 2½" from the trim. Switches should never be placed near bathtubs or showers.

Convenience Outlet Also called a *receptacle,* a convenience **outlet** is a connection device integral to a circuit that allows electricity to be siphoned off for appliances or lights via plug-in cords. Most outlets are duplex; single outlets are found on circuits dedicated to specific appliances. A *split-wire* outlet is an outlet where one contact is always hot and the other is activated with a switch. (The user might plug a lamp in the contact that can be turned on and off, and plug a clock in the other.) It is recommended that general purpose (not dedicated) outlets be placed 6' apart, 12" to 18" above the floor, no further than 6' from a room corner (unless a door or built-in item occupies that space), and on any wall between doors.

Ground fault circuit interrupter (GFI) A red push button on an outlet indicates the presence of this device (Figure 13-13). A **GFI** a safety feature installed in the circuit that monitors the amount of current going to the appliance and compares it to the current coming back. As long as the two are equal, it allows electricity to flow. However, if the amount of returning current returning is less than it should be, the GFI will trip, or open, the circuit. This is based on the premise that if the current is not traveling via the wire it must be traveling somewhere else, such as through a person (which is how electrocution takes place). The NEC regulates the outlets on which GFIs are placed; they must be on kitchen countertop outlets that are within 6' of the sink, on all outdoor and bathroom outlets, and on all readily accessible, non-dedicated garage, crawlspace, and unfinished basement outlets.

Figure 13-13
GFI (also called GFIC) duplex outlet.

In kitchens, outlets are GFIC and should be placed 6" above the countertop (48" above the floor line) and every 4' of wall space. Hall outlets should be placed every 15'. A split-wire outlet should be provided in each room. Bathrooms outlets are also GFIC, there should be at least two, and at least one should be placed above each countertop or vanity. Outlets shouldn't be in close proximity to phone jacks, as that causes line interference, and each room should have at least three outlets. Each fixture, device, or appliance in the plan needs an outlet, and an outdoor weatherproof one should be placed on each side of a house. Outside outlets are also needed for decorative and security lighting, and entry and garage doors.

Ground A **ground** is an electrical connection to the earth. Its purpose is to terminate electrical and lightening protection systems. The part directly in contact with the earth can be a buried copper rod or plate, a connection to buried metal water piping, or a complex system of buried rods and wires.

Special Outlets Types of special outlets include phone jacks, TV antenna outlets (cable, satellite or aerial), built-in outlets for home entertainment or theater speakers, and burglar and fire alarm systems.

Automated Systems These low-voltage, electronically controlled climate control timers, lights, alarms, locks, appliances, and smoke and fire detectors. Installed in so-called "smart homes," they can be programmed to turn those features on or off as desired.

Drawing the Electrical Plan

Draw the electrical plan on a floor plan that shows walls, major appliances, and built-in features only (no room dimensioning, section symbols, or other details). Show the location of switches, outlets, phone jacks, circuit wiring, conduit size, lights, security systems, distribution panels, fire alarms, and communication devices. Identify outlets as 220V, split wire, weatherproof, or other special use. Symbols for automated systems may also be included (or drawn on a separate electronics plan).

Align alphanumeric symbols (such as S3) with the walls, just like you would align dimensions on an architectural construction drawing. While there isn't a specific size required for any symbol, lights are usually drawn larger than switches and outlets, with switches and outlets drawn between 3/16"−1/4" long, and the circle for surface-mounted and recessed lights drawn with a diameter between 1/4"−5/16" in diameter (Figure 13-14).

After the symbols are drawn, curved, hard lines (drawn with a flexible or irregular curve) between the switches and devices show what operates what—for example, which toggle switches operate which lights and outlets, and which push-button switches operate which door chimes or door openers. The ends of the circuit lines should touch the switches and devices. The line's endpoint should touch the "S" portion of the switch symbol, should touch a quadrant on the outlet symbol, and can touch anywhere on a round surface-mounted or canned light symbol. Draw the endpoint to the ballast inside a fluorescent light symbol.

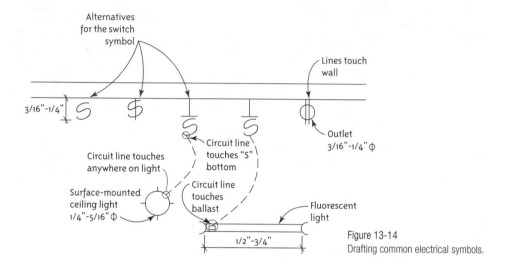

Figure 13-14
Drafting common electrical symbols.

Depending on office standards, the circuit lines may be drawn as short, evenly spaced, evenly sized dashes or drawn solid. They are never drawn horizontal and vertical. Sometimes dashed lines mean bare wire and solid lines mean wire housed inside conduit. Note that the lines don't reflect the positions of the wires themselves; they just show the control link. Often, the electrician on a residential job will decide exactly where to run the wire. Figure 13-15 shows good and poor ways of drawing circuits. Figures 13-16, 13-17, and 13-18 are legends of electrical symbols.

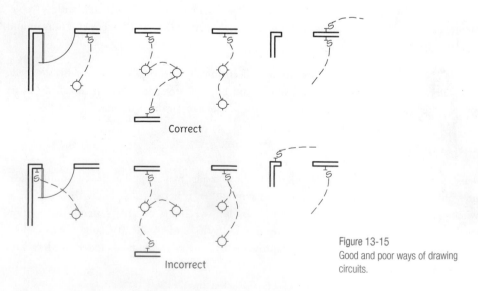

Figure 13-15
Good and poor ways of drawing circuits.

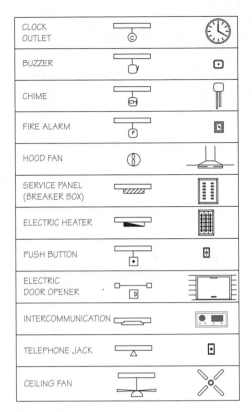

Figure 13-16
Electrical plan symbols legend.

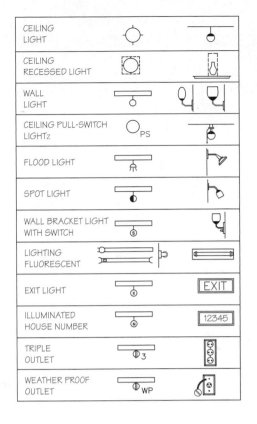

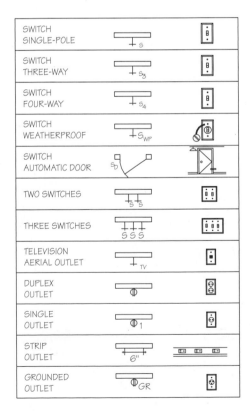

Figure 13-17
Electrical plan symbols legend.

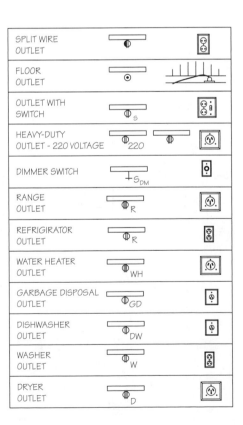

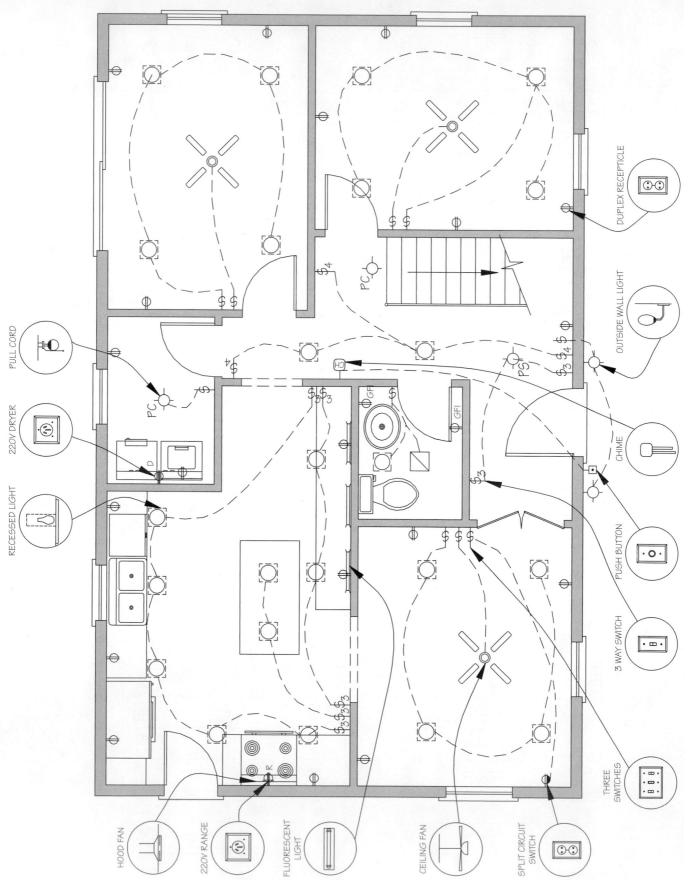

DUPLEX RECEPTICLE

OUTSIDE WALL LIGHT

CHIME

PUSH BUTTON

3 WAY SWITCH

THREE SWITCHES

PULL CORD

220V DRYER

RECESSED LIGHT

HOOD FAN

220V RANGE

FLUORESCENT LIGHT

CEILING FAN

SPLIT CIRCUIT SWITCH

Figure 13-18
Pictorial legend of electrical plan symbols.

Figures 13-19, 13-20, 13-21, 13-22, and 13-23 show examples of electrical plans.

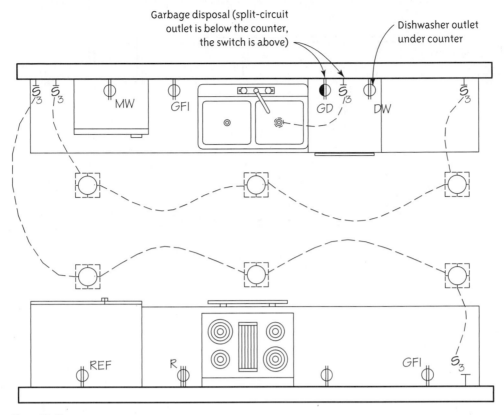

Garbage disposal (split-circuit outlet is below the counter, the switch is above)

Dishwasher outlet under counter

Figure 13-19
Electrical plan for a galley kitchen.

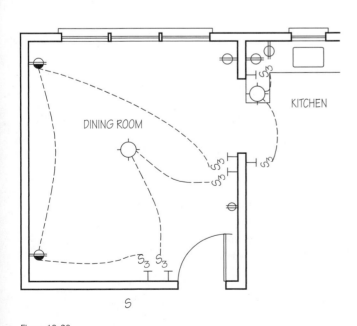

Figure 13-20
Electrical plan for a dining room.

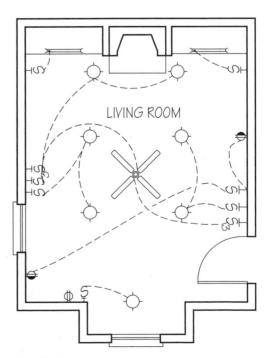

LIVING ROOM

Figure 13-21
Electrical plan for a living room.

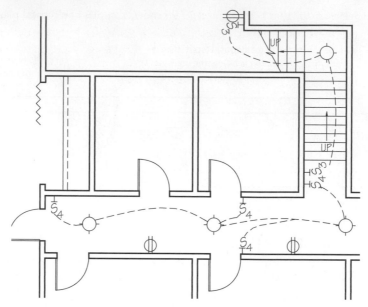

Figure 13-22
Electrical plan for a hall and stairs.

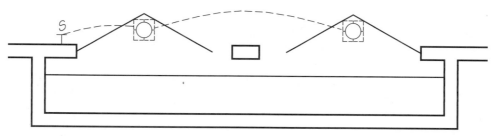

Figure 13-23
Electrical plan for a closet.

Power/Telephone/Data/VOIP Plans

Interior designers may prepare a **power/telephone/data/VOIP plan**. This shows where electrical outlets, phone jacks, data ports, computers, video equipment, and communications systems are located. Some of this information may overlap with the electrical plan or be combined with it. But if the project is large and complex, a separate plan for this information is usually prepared.

Power Plan Terms

Data Port This is a receptacle that allows telecommunications devices such as fax machines, modems, and USB hubs to be easily plugged in.

Digital Subscriber Line A digital subscriber line (DSL), also called a *T-1 line,* is a phone or data connection that is separate from a regular Internet connection. It transmits data over phone lines without interfering with voice service. It is possible to transfer 1.544 MB of data per second over such a line, which is used by businesses and Internet service providers. DSL or T-1 lines connect to even faster T-3 lines, which are the Internet's backbone.

Telephony Telephony is a general term for the use of equipment to provide voice communication over distances, specifically by connecting telephones to each other. *Voice over Internet protocol (VoIP)* uses a broadband Internet connection instead of an ordinary phone line to make phone calls by converting a voice signal from the telephone into a digital signal that travels over the Internet. You can make VoIP calls directly from a computer, from a special VoIP phone, from a traditional phone with an adapter, or via a wireless connection. While VoIP typically relies on an Internet connection, it can also utilize regular telephone lines and DSL connections. Residential VoIP service generally only uses an Internet connection, but some businesses have dedicated point-to-point T1 connections.

Communications Systems These include intercoms, alarms, videophones, closed-circuit video cameras, doorbells, chimes, music, audio devices with remote speakers, and TV cables. Sound and video are converted to electric current, travel through a wire, and are converted back to sound and video at the other end.

Reflected Ceiling Plan

A **reflected ceiling plan** (RCP) is a view of the ceiling as if it were reflected onto a mirror that is flat on the floor (Figure 13-24). Another way to think of it is a plan drawn from the perspective of someone who is floating above and parallel to the ceiling, looking straight down on it, and then projecting the ceiling straight down onto the floor. This view enables the RCP to be overlaid directly onto the floor plan with the same orientation and scale. It shows walls and partitions that extend to and above the ceiling.

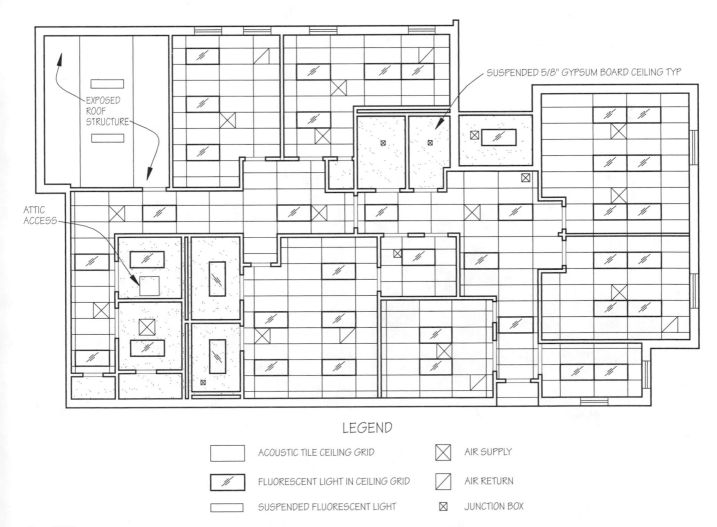

LEGEND

☐ ACOUSTIC TILE CEILING GRID	☒ AIR SUPPLY
☐ FLUORESCENT LIGHT IN CEILING GRID	◿ AIR RETURN
☐ SUSPENDED FLUORESCENT LIGHT	⊠ JUNCTION BOX

Figure 13-24
Reflected ceiling plan of a small office building. It has a combination of suspended acoustic tile and gypsum board ceiling.

RCPs also show ceiling materials (Figures 13-25, 13-26), molding, ornamentation (Figure 13-27), exposed structural elements (Figure 13-28), HVAC, soffits, exposed beams, skylights, building grid lines, notes for ceiling heights, changes in ceiling heights, light fixture locations, exit lights, sprinkler heads, air diffusers and vents, access panels, speakers, and anything else that is on or touches the ceiling. Specific items whose placement isn't clear are dimensioned. For instance, lights in a suspended, gridded ceiling system can be located by counting the number of tiles and/or grid lines, but lights in a gypsum board ceiling cannot, and hence are dimensioned to their centers.

Figure 13-25
Suspended acoustic tile ceiling. It is commonly used in commercial buildings and residential basements. It consists of a grid suspended from the roof with wires; panels and lights are dropped into the grid.

Figure 13-26
Insulation may be placed on top of the tiles in a suspended ceiling. *Courtesy of CertainTeed, Valley Forge, PA.*

Figure 13-27
Rich ornamentation is described on a reflected ceiling plan.

Figure 13-28
Ceiling structure is described on a reflected ceiling plan.

Although some items like lights and air vents are shown on engineer-prepared drawings, these items should also be shown on the reflected ceiling plan. This makes it easier to coordinate all items and ensure that the designer completely understands what the ceiling will look like. RCPs also contain section symbols that direct the reader to ceiling details.

When drawing a suspended ceiling grid (Figure 13-29), first draw a centerline down the room and center the grid on it. That way, the less-than-full-size tiles can line the room's perimeter.

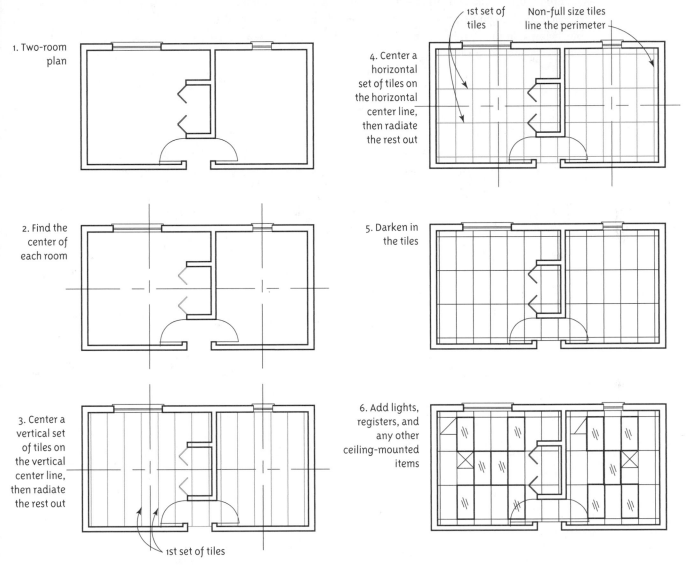

1. Two-room plan

2. Find the center of each room

3. Center a vertical set of tiles on the vertical center line, then radiate the rest out

1st set of tiles

4. Center a horizontal set of tiles on the horizontal center line, then radiate the rest out

1st set of tiles

Non-full size tiles line the perimeter

5. Darken in the tiles

6. Add lights, registers, and any other ceiling-mounted items

Figure 13-29
Steps for drawing a suspended acoustic tile reflected ceiling plan. The center wall is assumed to go to the roof, so plans are constructed separately for both spaces. An area with partition walls might have one continuous plan.

Climate Control Plans

Climate control systems deliver temperate, clean air through a building. They regulate heat, coolness, freshness, humidity, and temperature, and replace treated air that gets lost through the building's enclosures. Their design is regulated by the Uniform Mechanical Code and the drawings that describe them are **mechanical plans** (Figure 13-30) and **equipment plans.** A mechanical plan shows heating, ventilation, and air conditioning systems, hence is also called the *HVAC* plan. It shows the furnace, *air conditioner, water heater, ducts, filters, humidifiers,* pipes, *control devices, outlets, registers,* and *vents.* An equipment plan shows equipment used for heating and cooling only. The equipment plan varies depending on the building. For instance, in a factory or large commercial building there are pumps and valves that control systems other than heating. A restaurant contains pumps and motors for dumbwaiters, and a waste (sump) pump. Hotels and large buildings usually have elevator motors with hydraulic pumps. A mechanical engineer prepares both plans; as an interior designer you will not draw these, but being able to interpret them is useful, as they must be coordinated with the structural, plumbing, and electrical plans.

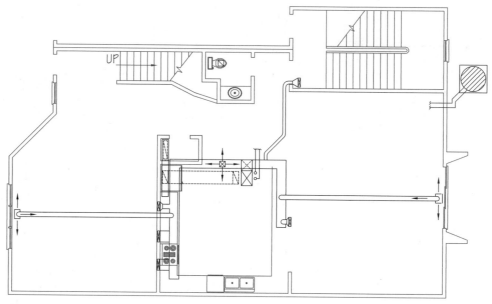

Figure 13-30
Mechanical (HVAC) plan.

Climate Control Terms

Natural Gas This is the energy source for gas-fueled furnaces. It is composed primarily of methane, found in geological formations deep underground. Natural gas is extracted from gas wells, transported through pipelines, treated at a refinery, then stored in underground facilities or moved across the country in large-diameter, high-pressure transmission pipes to border stations, where the pressure is reduced before delivery to the community.

Furnace A **furnace** is an appliance that produces heat (Figure 13-31). Most furnaces are fueled by natural gas, as it is a cheaper energy source than electricity and more convenient than oil, coal, or wood. A furnace mixes gas with air and ignites it with a pilot light or electronic device. The combustion chamber fire warms a heat exchanger. Carbon monoxide gas is produced in this process and is vented outside via a flue that runs through the roof or wall. A blower forces the heated air circulating around the heat exchanger out of the furnace, through the ductwork, out the registers, and into the rooms (Figure 13-32). Separate return air ducts carry the cooler room air back to the furnace where it is reheated. Ducts are insulated to contain heat loss.

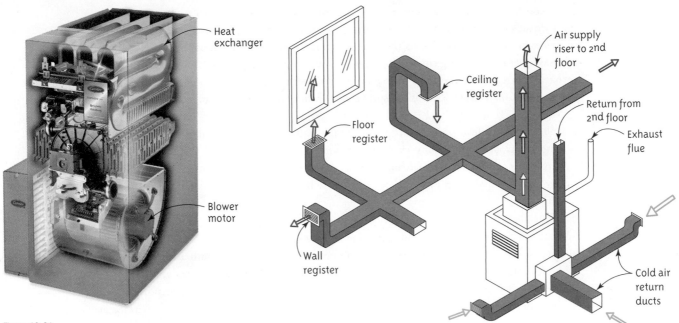

Figure 13-31
Cutaway of furnace. *Courtesy of the Carrier Corporation.*

Figure 13-32
Forced-air heating.

Air Conditioner An **air conditioner** is an appliance that cools, filters, and dehumidifies air (Figure 13-33). It's powered by electricity. Central air conditioners cool an entire building and room air conditioners cool single rooms. A refrigerant (chemical that produces a refrigerating effect) circulates through copper tubing that runs between an outdoor coil called a *condenser* and an indoor coil called an *evaporator,* transferring heat from indoors to outdoors. When cold refrigerant circulates through the evaporator, it absorbs heat from room air blown across it. While absorbing, it vaporizes and travels through the tubing to the condenser, dissipating the heat into the outdoor air during the process. The refrigerant then cools, becomes liquid, and returns to the evaporator.

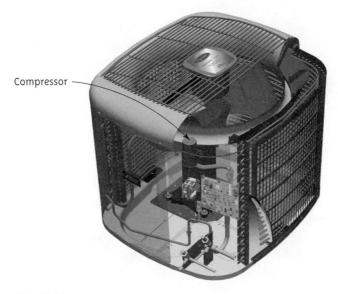

Figure 13-33
Cutaway of the outside unit (compressor) portion of an air conditioner. *Courtesy of the Carrier Corporation.*

A central air conditioner is usually used in combination with a forced-air furnace. It uses the furnace's blower to draw room air into the unit via ductwork. The furnace's air filter removes dust, hair, and lint. When the room air moves past a chilled indoor evaporator coil it releases heat. The resulting cold air travels on to the plenum, a large metal box at the top of the furnace, where it's channeled to the air supply ductwork and returned to the rooms. A drain carries away *condensate,* which is moisture created when warm air cools and condenses.

Heat pump A **heat pump** is an appliance that heats and cools a building. It uses air conditioning principals to pump heat from the air, earth, or groundwater. Heat is removed from an area to cool it or added to warm it. Since it is an electrical appliance, it's clean and doesn't require a chimney. The main unit is located outside, hence it doesn't need much interior space.

Thermostat A **thermostat** is a device that automatically regulates temperature in a furnace or air conditioner via sensors and activating switches (Figure 13-34). Some are mechanical dial types; others are electronic and programmable. Most can control both a furnace and an air conditioner. However, not all thermostats work with all types of furnaces and heaters; a thermostat that works on a forced-air system may not work with a heat pump. Each heating or cooling unit needs its own thermostat. Thermostat location is important because a thermostat measures room temperature and, based on this measurement, activates the furnace. A thermostat should be placed on an interior wall where the temperature is even. If it is placed in a sunny spot, in a drafty area, or near a lamp, its performance will be compromised.

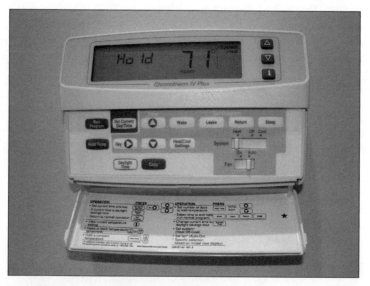

Figure 13-34
Programmable thermostat controls both heating and air conditioning.

Zone A **zone** is a specific area heated or cooled by one unit. Some homes and commercial buildings have multiple zones. As many zones as are needed can be provided with forced air, electric radiant, or hydronic systems.

Hydronic A heating or cooling system that transfers heat via a circulating fluid (e.g., water or vapor) inside *pipes* is called hydronic. Hot water or steam is moved from boiler to radiators, baseboard units, or radiant panels.

Pipes **Pipes** are copper tubes (older homes have galvanized steel), round in cross-section, that serve as the distribution method in *hydronic* systems.

Humidifier A **humidifier** is an appliance that adds moisture to the house (Figure 13-35). Too much humidity causes mold and mildew, too little causes dry air and static electricity. Both take their toll on interior finishes and furniture. A central-air humidifier is attached to the furnace, but freestanding tabletop models are also available.

Figure 13-35
This residential humidifier is attached to the house's main supply duct.

Ducts, Ductwork **Ducts** are the distribution and return-air path in a forced-air system; large quantities of air for heating and cooling are moved through them (Figure 13-36). Ducts are made of sheet metal and can be rectangular or round in cross-section. The two major ductwork types in forced-air systems are *radial* and *extended plenum*. In a radial system, round ducts radiate out in all directions from the furnace. In an extended plenum system, a large rectangular duct (plenum) is the main supply and round ducts extend from it to each register (Figure 13-37). The extended plenum system is most common. The round duct it uses is typically 6" or 8" in diameter (the larger size is needed when the system is used for both heating and cooling). A vertical duct designed to fit between the wall studs is called a *wall stack* and is usually 12" × 3¼".

Figure 13-36
Ductwork carries warmed and cooled air throughout this commercial building.
Courtesy of CertainTeed, Valley Forge, PA.

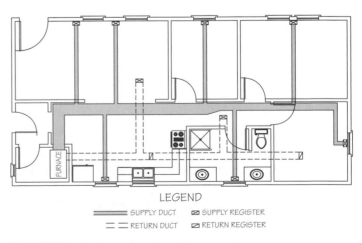

LEGEND
▬▬▬ SUPPLY DUCT   ⊠ SUPPLY REGISTER
═ ═ RETURN DUCT   ⊠ RETURN REGISTER

Figure 13-37
Plan of an extended plenum duct system.

Register A **register** is an outlet in a forced-air system through which treated air is returned to a room (Figure 13-38). It can be on the floor, wall, or ceiling. In a hydronic or electric radiant system, the outlet is usually a baseboard unit. **Inlets**, or return registers, are cold air returns and are a component of forced-air systems. They receive air that needs to be returned to the furnace or air conditioning coil. They are not needed for hydronic or electric radiant systems. Registers and inlets are available in different sizes and are generally placed around the perimeter of a building to provide uniform heating or cooling by concentrating the conditioned air where it is needed most, along the outside walls. There should be at least one in each room, hall, and stairwell. Rooms over 180 square feet or which have more than 15' of exterior wall need two or more registers and inlets.

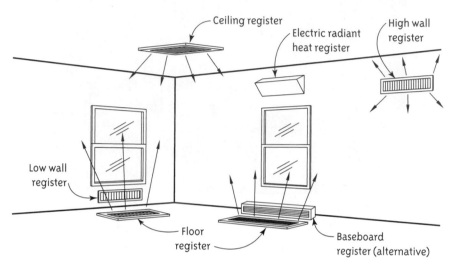

Figure 13-38
Types of registers.

Conductor A conductor is a material that transfers heat. An *insulator* is a material that resists heat transfer. In buildings, conductors are wires and insulators are insulating materials.

How Warmed and Cool Air Travels

There are three ways air travels: via **radiation, convection,** and **conduction** (Figure 13-39). In radiation, heat travels as waves through space. Radiant heat travels in a straight line, penetrating and warming walls and roofs. With convection, an energy source heats the air around it, the warmed air rises, cool air moves in to replace it, and a current is formed. Convection heat travels through liquids and gases. In conduction, heat moves through a solid material. The denser the material, the better it conducts heat.

Types of Climate Control Delivery Systems

The most common climate control delivery systems are *forced-air, radiant hydronic,* and *heat pump.* They can be fueled by electricity, natural gas, oil, propane, kerosene, wood, coal, or solar heat.

Forced-air Heat System Also called *central air,* **forced-air** heat systems are usually installed in new construction. A furnace draws room air through ductwork and returns the warmed air to the rooms back via more ductwork. This system can be combined with an

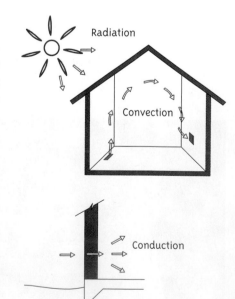

Figure 13-39
Types of heat travel.

air conditioner, humidifier, and air filter. Small, electric forced air heaters may be installed for *zone heating* of one room or small area. They have their own thermostats, draw air from the room, run it over heating elements, and send it back into the room. These fit into walls or along baseboards.

Radiant Hydronic Heat System A **radiant hydronic** heat system uses hot water or steam from a **boiler** to distribute heat through the house (gas, electricity, or wood can also be used). The boiler heats water and a pump circulates it through pipes to floor registers, radiators, or baseboard units that radiate this heat to each room, then back to the boiler. Radiant heat does not warm a room; instead, it warms people and items in the room. This is why a buildings that uses it can feel warmer than one that uses forced-air heat, even if the temperature is the same as with a forced-air system. There are no fans or ducts, which increases energy efficiency because there's no chilled moving air or energy loss during transportation. However, radiant heat cannot be combined with air conditioning, humidifiers, or air filters.

Flexible Tubes and Panels Heat System A flexible tubes and panels heat system is a radiant system that consists of panels or flexible tubes (Figure 13-40) installed in the floor, walls, or ceiling. It absorbs heat from a fuel source (electric or hydronic) and transfers it via radiation and convection to the room. Tubes can be woven into fabric mats or installed in concrete. In a hydronic system water is heated by a boiler or water heater; with an electric system, heating elements warm the surfaces. These systems can heat a whole house or provide zone comfort (usually in kitchens and baths). They are completely hidden from sight.

Figure 13-40
Embedded radiant heat coils. They can be hydronic or electric. These free-form cables are typically embedded in a 1½" thick slab of concrete. The ability of the framing to support the slab's weight must be considered, and adjustments made to window and door heights for the slab's extra thickness.

Reading the Mechanical Plan

Walls and built-in architectural features are outlined. The rest of the drawing is strictly mechanical items. The type, size, and location of all heating, cooling, ventilating, humidification, and air cleaning equipment, pipes, ducts, registers, inlets, baseboard units, thermostats, and other control devices are shown. Unlike architectural floor plans where many symbols resemble the items they represent, such as sinks and toilets, mechanical drawings are **schematic,** meaning they utilize simple line types and symbols that do not resemble the physical appearance of the items represented (Figures 13-41, 13-42).

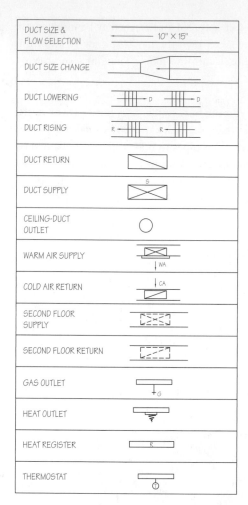

DUCT SIZE & FLOW SELECTION	10" X 15"
DUCT SIZE CHANGE	
DUCT LOWERING	D → D →
DUCT RISING	R → R →
DUCT RETURN	
DUCT SUPPLY	S
CEILING-DUCT OUTLET	
WARM AIR SUPPLY	WA
COLD AIR RETURN	CA
SECOND FLOOR SUPPLY	
SECOND FLOOR RETURN	
GAS OUTLET	G
HEAT OUTLET	
HEAT REGISTER	R
THERMOSTAT	
RADIATOR	RAD
CONVECTOR	CONV
ROOM AIRCONDITIONER	RAC
FURNACE	FURN
FUEL-OIL TANK	OIL
HUMIDISTAT	H
HEAT PUMP	HP
PUMP	
CEILING DUCT OUTLET	
FORCED CONVECTION	
HYDRONIC-RADIANT PANEL COIL	
HOT WATER HEATING RETURN	
HOT WATER HEATING SUPPLY	
HUMIDIFICATION LINE	H
MEDIUM-PRESSURE STEAM	

Figure 13-41
Symbols on a mechanical plan.

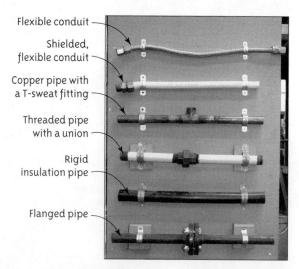

Flexible conduit

Shielded, flexible conduit

Copper pipe with a T-sweat fitting

Threaded pipe with a union

Rigid insulation pipe

Flanged pipe

Figure 13-42
Examples of the pipes and fittings that mechanical symbols and plans schematically represent.

Ducts are drawn to scale. Horizontal ducts are shown by outlining their position; vertical ducts are shown with diagonal lines. Air return ducts are drawn with a hidden line. Pipes are indicated with single lines and are not drawn to scale. Heating units are drawn connected to their supply ducts or hot water mains. This plan is usually drawn at the 1/4"−1'-0" scale.

Plumbing Plans

Plumbing systems deliver fresh water to a building and remove dirty water and waste. Their design is guided by the Uniform Plumbing Code. The drawings that describe them are the **plumbing plan** (Figure 13-43) and **riser diagram**. They show how water is supplied, distributed, and controlled. Sometimes they are combined with HVAC plans. They are prepared by a mechanical engineer; but as with other systems, the ability to interpret them ensures that a watchful eye can be kept on details. Since vertical plumbing pipes must be installed when floor slabs are poured, checking the drawings will ensure that stacks passing through those slabs are precisely located to align with partitions and fixtures. Knowledge of common plumbing terms and how water is delivered and discharged is useful when reading them.

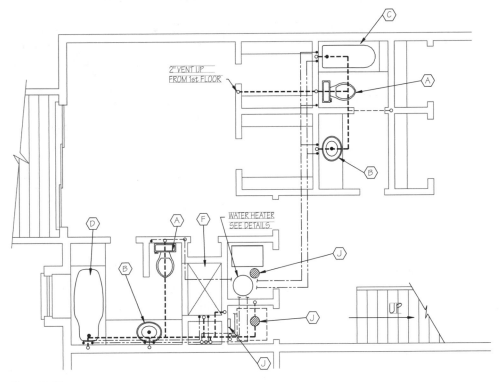

Figure 13-43
Plumbing plan.

Plumbing Terms

Water Heater A **water heater** is an appliance that heats and stores water (Figure 13-44). The cold water supply pipe connects to the top of the water heater. A dip tube carries the cold water to the tank's bottom so cold water doesn't get poured on top of the hot water. The outbound hot water pipe takes the water from the tank's top, ensuring that when hot water is drawn out, it is at full temperature. As the hot water is drawn, cold water enters the tank to replace it. After all the hot water is used, cold water is delivered; running out of hot water halfway through use can be a problem with small tanks. In the center of the water heater is a metal *sacrificial anode,* whose purpose is to corrode and thereby prevent corrosion of the tank. The corrosive action of the hot water attacks the anode, which extends the life of the tank. The tank also has a layer of glass inside for corrosion protection. There is a drain near the bottom of the tank to which the water and any accumulating sediment is directed.

Gas- and electric-fueled water heaters are both common. An electric heater's advantage is that it does not require venting of combustion gases. However, because water is heated continuously, whether it is in use or not, electric heaters are more expensive to operate and do not heat cold water as quickly as gas. Tankless heaters (Figure 13-45) are now available that heat water instantaneously on demand, cutting energy costs.

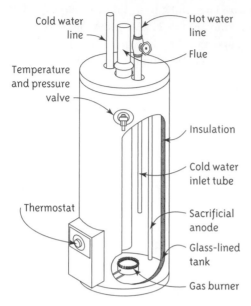

Cold water line

Hot water line

Flue

Temperature and pressure valve

Insulation

Cold water inlet tube

Thermostat

Sacrificial anode

Glass-lined tank

Gas burner

Figure 13-44
Cutaway of a water heater.

Figure 13-45
Tankless water heater. *Courtesy of Stiebel-Eltron, Chico, CA.*

PEX This is the abbreviation for cross-linked polyethylene. A thin, flexible plastic (Figure 13-46), it doesn't require as many joint fittings as other types of pipe material. It may be installed on a *manifold* (Figure 13-47), which is a control center for all hot and cold water lines. Individual fixtures can be shut off here. It is used for water lines and in hydronic radiant heating systems.

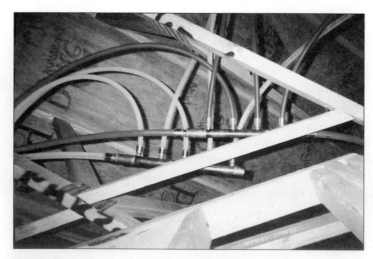

Figure 13-46
PEX piping.

Figure 13-47
Manifold. This enables all water fixtures to be individually shut off from one location. *Courtesy of Viega, Bedford, MA.*

Water Pipes Also called *lines*, **water pipes** are tubes that emanate from the water heater and run throughout the building. They can be made of copper, plastic, brass, iron, steel, or PEX. If a fixture includes both hot and cold water outlets, the hot water line and valve are located to the left of the cold water line and valve. Hot water pipes are insulated to contain heat loss. Placement of hot water pipes on exterior walls in cold climates should be avoided, because this makes them vulnerable to freezing. A joint is the connection location where separate pieces of pipe are attached to one another and a *fitting* is a threaded piece that holds the pipes together.

Fixture A **fixture** is a plumbing device that is permanently attached to a building, such as a sink, toilet or shower (Figure 13-48), and which draws freshwater and discharges wastewater. Codes require minimum clearance and location dimensions for fixtures, and manufacturers specify rough opening sizes.

Figure 13-48
Plumbing fixtures carefully laid out in an elegant bathroom. *Courtesy of Kohler Plumbing, Kohler, WI.*

Trap A trap is a curved or S-shaped section of pipe under a drain. Every fixture has one, but normally they are only visible under sinks. Water flows from the basin with enough force to go through the trap and out through the drainpipe, but enough remains in the trap to form a seal that prevents sewer gas from backing up into the house. There are different kinds of traps; bathtubs, lavatories, and showers have P traps, house mains have a U-trap. Older homes may have drum and S traps, which are no longer allowed by code. Bathtubs have traps that seal against sewer gas and collect hair and dirt. Some kitchen sinks have grease traps. Because grease and hair are generally the causes of drain clogs, traps have *clean-outs*. Toilets are self-trapped; that is, they have built-in traps that are not accessible for unclogging (Figure 13-49).

Valve   A **valve** is a mechanical device that controls the flow of water or gas. A *shut-off valve* or *stop* shuts off the water supply. The house main has one to shut off water to the whole house, and each individual fixture has one. Since all fixtures have a free-flowing supply of water, individual shut-offs enable a fixture to be worked on without interrupting the whole building's water supply.

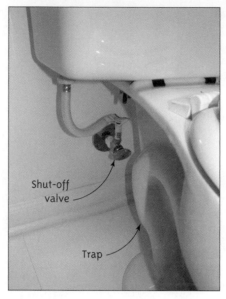

Figure 13-49
All water fixtures need a trap and a shutoff valve.

Stack   A **stack** is a vertical pipe. A *soil stack* transports waste (Figure 13-50). The *vent stack* is the portion of the soil pipe above the highest branch pipe intersection and rises at least 6" above the roof. It allows sewer gases to escape and equalizes the system air pressure. If there were no air coming in from the vents, wastewater could not flow out. All fixtures are *wet-vented* for a short distance, meaning that the portion of the branch pipe between the trap and the vent serves as a vent pipe.

Figure 13-50
Stack and branch pipes remove wastes from plumbing fixtures.

Stack wall Also called a *wet* or *chase* wall, a stack wall is a wall wide enough to contain large stack pipes (Figure 13-51) They are nominally at least 6" wide.

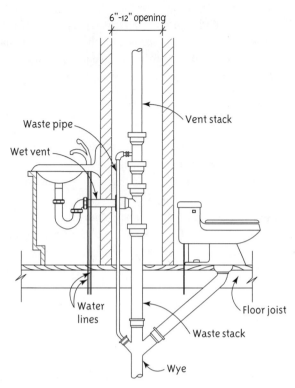

Figure 13-51
A stack wall. Also called a chase wall or wet wall, it is a thick wall behind plumbing fixtures to accommodate the pipes.

Clean-out A cleanout is a plugged hole in a pipe or trap that allows access to the inside for unclogging (Figure 13-52). Cleanouts are at the base of all stacks and wherever pipes make sharp turns.

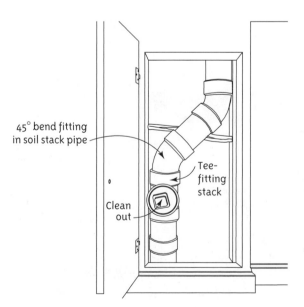

Figure 13-52
Clean-out valve in a soil stack. It's often found in a home's basement.

Well A **well** is a water source obtained by digging into an *aquifer,* which is an underground layer of porous, water-bearing rock. Groundwater is stored in the spaces between the rocks and extracted.

Main A **main** is an underground pipe used to transport potable (drinkable) water from a water utility to individual consumers. The *house main* is the supply pipe that enters the house and from which the branch pipes emanate. It has a valve to shut off the water supply to the entire building and may have a water softener and various filters attached.

Hose Bib A **hose bib** is an outdoor faucet.

Septic tank A **septic tank** is a watertight, underground tank. Solid waste settles to the bottom, is converted to liquid by bacteria, and then flows up out the top into drainage fields through perforated pipes 12" to 16" underground that are spread out over a wide area.

Sump Pump A **sump pump** is a pump that drains basement water accumulated from ground water or bad plumbing (Figure 13-53). It is placed in a concrete pit in the basement, called a *sump,* that is lower than the basement floor. When water reaches a certain level in the sump, the pump automatically activates and pumps the water out of the sump and into the house drainpipe.

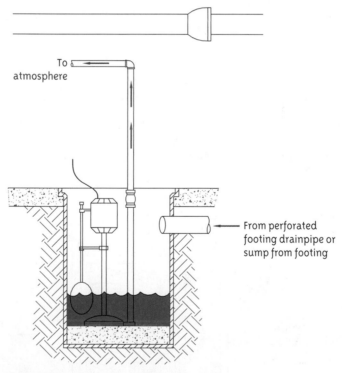

Figure 13-53
Sump pump. Water is directed to a pit below the basement floor, then pumped up to the house drain for removal.

How Fresh Water Is Delivered

A building's water supply begins at a river, lake, or *well,* from which it is routed to a public utility. Some municipalities store water in high towers to provide large volumes of water in a short time for firefighter use and for peak volume times or emergencies, such as electric pump failure. These towers are built high enough for gravity to provide the pressure needed to push the water through the pipes. A water tower typically holds enough water to serve the community's needs for one day. Some towers are whimsically designed and some are even designated as historic landmarks (Figure 13-54).

Most people obtain their water from a utility, but some use their own well. The utility filters and treats the water and pumps it through an underground *main* to the house. After going through the meter, some water is routed to a *hose bib* or sprinkler system. The rest enters the house through the *building main,* may be passed through a softener and some filters, is branched into cold and hot water mains, and finally is routed to a *water heater* and on to appliances, sinks, water closets (toilets), lavatories (sinks), tubs, filters, and showers. Hot and cold water lines run parallel to each other, usually 6" apart, and the water is under pressure the whole distance from the municipal source to the last fixture in the home (Figures 13-55).

Figure 13-54
Water tower in Collinsville, IL.

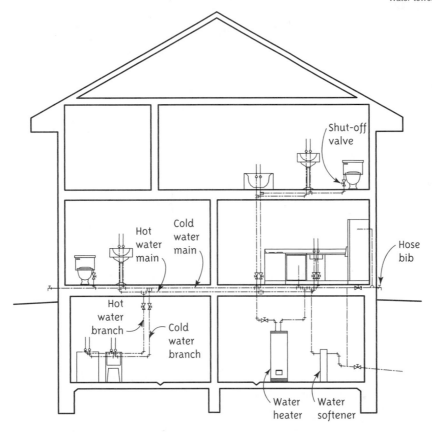

Figure 13-55
Hot and cold water lines.

How Waste Water Is Discharged

Dirty water, called *wastewater* or *soil* water, is discharged via a system that utilizes gravity instead of pressure to push the contents through the pipes. Water leaving the house either goes into a *septic tank* where it seeps into the ground or goes through sewer pipes to a treatment plant, where it is cleaned and sent as *reclaimed wastewater* into streams or for agricultural and municipal irrigation.

There are two types of waste pipes: *stacks* and *branches*. Branches are sloping horizontal pipes that carry waste from each fixture to the stacks, or vertical pipes. Waste stacks carry soil to the house sewer line. They are empty except when water is flushed through them. All waste pipes slope down, usually 1/4" per foot. They are larger than water pipes. All stacks have *clean-outs* at their base.

Waste water flow starts at a fixture. Each fixture has a **trap** to prevent sewer gas backflow into the branch lines. A house trap for the entire system is built in the sewer line outside the house. **Clean-outs** are built into the house's main sewer drain (Figures 13-56, 13-57).

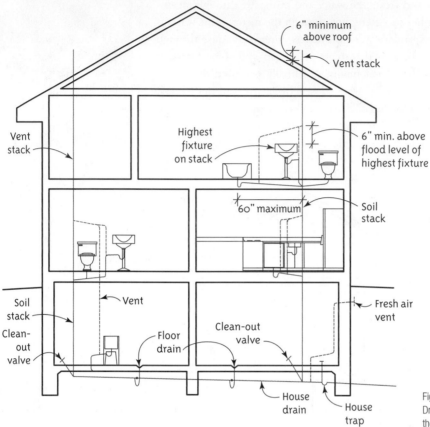

Figure 13-56
Drainage lines. Their contents flow via gravity, so they are angled down at 1/4" per foot.

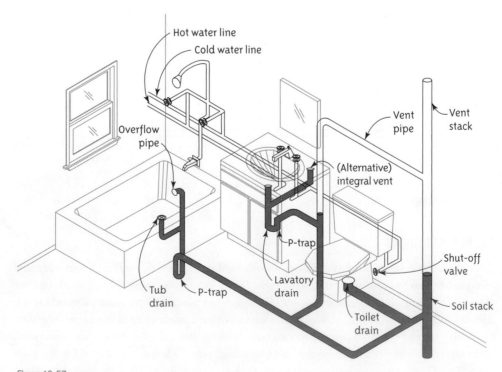

Figure 13-57
Pictorial of bathroom piping.

The house sewer connects to the public sewer system. If a municipal system is not available, the house will use a septic system. A septic system will have separate drawings or have its information documented on the site plan.

Reading the Plumbing Plan

A simple floor plan of walls and architectural features forms the outline. There are no dimensions except for those partition positions that will require the placement of stacks before a slab or foundation is poured. On the floor plan are gas appliances, plumbing and waste removal pipes, built-in vacuum systems, pest control pipes, hydronic heating systems, fixtures, hose bibs, floor drains, roof vents, *joints, valves*, intersections, and control devices. As with mechanical plans, schematic symbols are used to represent the actual items (Figures 13-58, 13-59).

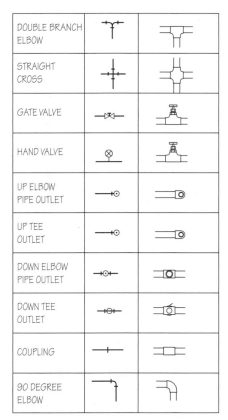

DOUBLE BRANCH ELBOW		
STRAIGHT CROSS		
GATE VALVE		
HAND VALVE		
UP ELBOW PIPE OUTLET		
UP TEE OUTLET		
DOWN ELBOW PIPE OUTLET		
DOWN TEE OUTLET		
COUPLING		
90 DEGREE ELBOW		

Figure 13-58
Plumbing symbols legend.

90 DEGREE TEE		
45 DEGREE LATERAL		
CLEAN OUT LATERAL OR Y		
REDUCER		
FLANGED FITTING		
SCREWED FITTING		
BELL & SPIGOT FITTING		
WELDED FITTING		
SOLDERED FITTING		
EXPANSION JOINT		

METER		
FLOOR DRAIN		
SEPTIC TANK		
SEPTIC TANK DISTRIBUTION BOX		
WATER LINES: COLD HOT		
GAS LINE		
VENT		
SOIL STACK PLAN VIEW		
SOIL LINE: ABOVE GRADE BELOW GRADE		
ICE-WATER LINE		

Figure 13-59
Plumbing symbols legend.

All fixtures, floor drains, and hose bibs are connected with plumbing lines to the water supply, waste discharge, gas line, or HVAC hydronic system that serves them. Hot water lines connect to the house main, to the water heaters, and to each fixture that requires them. They run parallel to cold water lines. Symbols for shut-off valves are drawn on these lines. Thick lines represent wastewater branch lines and connect fixtures with soil stacks. Soil lines are larger than supply lines, so they are drawn thicker. A trap symbol is shown at each fixture. The positions of stack and vertical pipes are shown with a circle inside walls and partitions. The main house drain and sewer lines are connected to the stack symbols, clean-outs, and house trap. The house drain connects to the city sewer line or a septic tank. Septic tanks and fields are usually drawn on the site plan. If gas appliances or hydronic heating systems are used, lines with cutoff valves connect to the main gas line or to gas tanks for each appliance.

Descriptions of pipe types and sizes appear in notes on the plan; dimensions give the locations of underground valves and cleanouts.

The plans only show horizontal positions and dimensions of pipes, valves, and fixtures. *Riser diagrams and elevation drawings* show the pipes' vertical heights and angles above the floor and the flow of fresh water and waste between levels.

Demolition Plans

A **demolition** plan shows existing items designated for removal or relocation (Figure 13-60). It's needed if there are many such items. Otherwise, this information is conveyed with notes on the floor plan. Depending on the specific demolition information, it may be placed on the architectural, electrical, mechanical, or other plan.

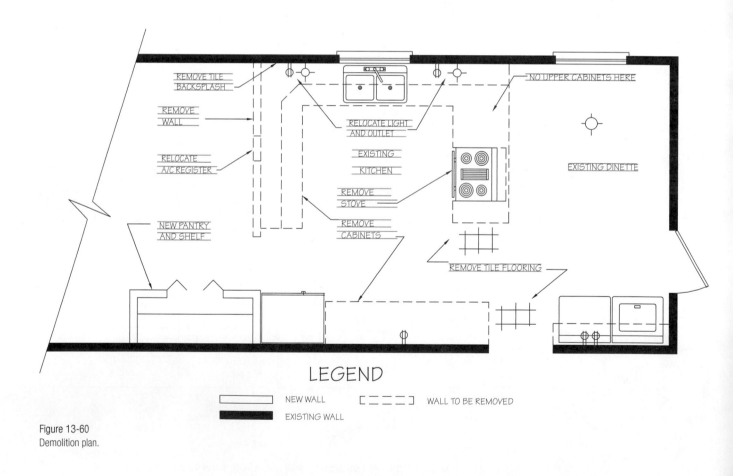

Figure 13-60
Demolition plan.

Site Plans

The **site plan** is a bird's-eye view of the entire property (Figure 13-61), and is prepared by a landscape architect or civil engineer. It shows physical characteristics and most are in symbolic form (Figure 13-62).

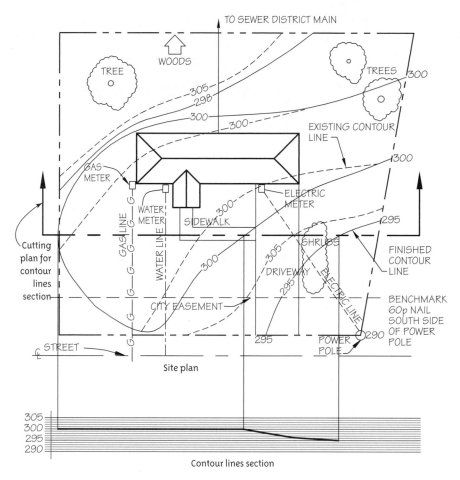

Figure 13-61
Site plan and a section showing its contour lines. A cutting plane is placed through the plan and points projected down to plot the contour lines in section.

The ability to interpret it enables a designer to incorporate natural features such as views, and to better place windows, awnings, trees, and heating/cooling equipment.

These are items typically found on it.

North arrow and engineer scale. The north arrow shows the solar orientation of the house. The engineer scale is typically used to dimension the property because the plan is often too large for the architect's scale. 1' = 10' and 1' = 20' are the most common scales. However, the building itself is still dimensioned with the architect's scale.

Footprint The **footprint** is the building's shape, size and orientation. It may depict an outline, the roof, or interior partitions (to show relationship between the outside and interior spaces). If there are multiple buildings or outbuildings (such as sheds or unattached garages), they are included.

Property lines Also called *boundary* lines, property lines are the property's physical boundaries.

Hard surfaces This includes walks, entries, driveways, access roads, patios, retaining walls, terraces, sport courts, the septic system, and any other features within the property lines. Their surface materials are pochéd. Surrounding property and full or partial outlines of adjacent buildings may be included.

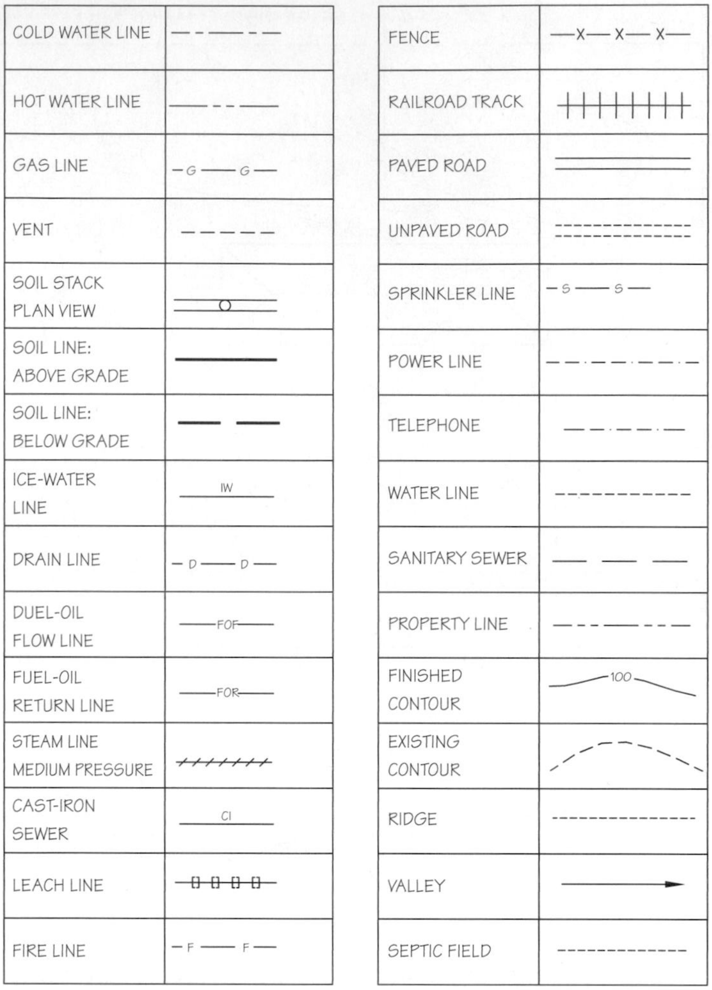

COLD WATER LINE	— · — · —	FENCE	—x— x— x—
HOT WATER LINE	—— · · ——	RAILROAD TRACK	++++++++
GAS LINE	—G——G—	PAVED ROAD	══════
VENT	— — — —	UNPAVED ROAD	╌╌╌╌╌╌
SOIL STACK PLAN VIEW	══o══	SPRINKLER LINE	—s——s—
SOIL LINE: ABOVE GRADE	▬▬▬	POWER LINE	— — · — —
SOIL LINE: BELOW GRADE	▬ ▬ ▬	TELEPHONE	— — · — —
ICE-WATER LINE	—— IW ——	WATER LINE	╌╌╌╌╌╌
DRAIN LINE	—D——D—	SANITARY SEWER	—— · ——
DUEL-OIL FLOW LINE	——FOF——	PROPERTY LINE	— — — —
FUEL-OIL RETURN LINE	——FOR——	FINISHED CONTOUR	⌒100
STEAM LINE MEDIUM PRESSURE	╱╱╱╱╱╱	EXISTING CONTOUR	⌒ ⌒
CAST-IRON SEWER	—— CI ——	RIDGE	╌╌╌╌╌╌
LEACH LINE	▯▯▯▯	VALLEY	———→
FIRE LINE	—F——F—	SEPTIC FIELD	╌╌╌╌╌╌

Figure 13-62
Site plan symbols legend.

Waterways These include rivers, streams, lakes, ponds, or pools that are on, or cross, the property.

Utility lines These include gas, electricity, water, and sewer lines that serve the building. Connection of the house sewer to the public system appears on the site plan.

Vegetation Vegetation comprises trees and shrubs. Trees are drawn at their trunk base locations, at a diameter that represents the size of the trunk.

Setbacks and Easements A setback is the building's distance from the property line. Local zoning laws regulate the exact distance, which may differ between the front, sides, and back property lines. Setback lines are drawn within, and parallel to, the property lines to show the total allowable building area. An easement is a waterway, walkway, or street that lies within the property lines but is designated for public use.

Contour lines Contour lines show ground elevations. They connect points on the land surface that are the same elevation above a **benchmark**, or reference point. A benchmark is an item mounted on a permanent site feature, such as a large nail on a power pole; it may be assigned the U.S. Geological Survey's number for that point's height above mean sea level. The numbers on the contour line are the heights above or below the benchmark and are usually spaced in intervals 5' to 20' apart. Closely spaced contour lines describe a steeply sloping site; lines spaced far apart describe a gradually sloping one. Dashed lines indicate "before" status; the solid ones indicate "after." Some site plans have *spot elevations* instead of contour lines. These are elevations at key points and are used when the property is relatively flat.

Dimensions Overall lot dimensions, size, and location of constructed features and their positions on the property are given.

Legal Description The legal description includes the location of the property lines.

Details Construction drawings of fences, bollards, gutters, sidewalks, and other items relevant to the site are shown.

Related Drawings Related drawings include a *survey,* which is a legal document showing property boundaries; *plats,* which are large-area maps; *landscape plans,* which show vegetation and planter arrangement; and *plot plans,* which incorporate some, but not all, of the features of site plans.

Summary

Many different types of drawings are included in a set of instructions for constructing or renovating a building. These drawings are used by a variety of professionals. Floor plans, elevations, sections, details, electrical, water, and building systems drawings are all necessary to describe the design. While the interior designer may not draw all of them, it is useful to be able to read them and recognize key features.

Suggested Activities

1. Draw a reflected ceiling plan of the ceiling in your classroom.
2. Sketch an electrical plan of the lights, switches, and outlets in your classroom.
3. Identify the pipes in an unfinished basement. Look for clean-outs, shutoff valves, fixture traps, and the sump pump.
4. Call your town's Codes Inspector Office to ask for code requirements for residential electrical, plumbing, and mechanical systems.
5. Identify these parts of the HVAC system in your house or other building: furnace, air conditioner, registers, humidifier, and thermostat.
6. Collect and read manufacturer literature on different plumbing and HVAC products. Compare and contrast them in a report.
7. Visit a plumbing supply store and examine different components used in residential plumbing.
8. Sketch a site plan of a residential or commercial building, noting all its features. Note how the ground is sloped.

Questions

1. Name four items found on an electrical plan.
2. Explain what three-way and four-way switches do.
3. How does a plan show which light or outlet is operated by which switch?
4. What purpose does a reflected ceiling plan serve?
5. Name four items found on a reflected ceiling plan.
6. What is a GFI outlet, and where is it found?
7. Describe the difference between forced air and radiant heat systems.
8. What does an HVAC plan show?
9. Name four items found on a site plan.

Further Reading and Internet Resources

National Electrical Code (2005). Clifton Park, NY: Thomson-Delmar Learning.

http//www.corp.carrier.com
The website of Carrier Corporation, provider of heating, cooling, and refrigeration equipment.

http//ga.water.usgs.gov/
Water usage facts.

http//www.kohler.com
Kohler company website. Contains information on their kitchen and bath products.

http//members.tripod.com/
The "Watertowers" page of this general site shows interesting water towers.

http//www.mercuryplastics.com/
Website of Mercury Plastics. Contains information on PEX tubing.

http//www.stiebel-eltron-usa.com/
Website of Stiebel-Eltron. Contains information on tankless electric water heaters.

http//www.usgs.gov/
Website of the United States Geological Survey office. Contains lots of general science information.

http//www.viega-na.com/
Website of Viega NA. Contains information on plumbing and radiant heat systems.

http//www.warmlyyours.com
Warmly Yours company website. Contains information on their radiant heating floor system.

CHAPTER 14

Drawing Pictorials

OBJECTIVE

- This chapter discusses how to sketch and hard-line pictorial line drawings using geometric construction techniques. Specifically covered are axonometric and perspective drawing.

KEY TERMS

axonometric	height line	rendering
cabinet oblique	horizon line	rotated plan
cavalier oblique	isometric	sight ray
cone of vision	line drawing	station point
diametric	one-point perspective	trimetric
distortion	paraline	two-point perspective
eyeballing	pictorial	vanishing point
general oblique	plan oblique	

What Are Pictorials?

A **pictorial** is a three-dimensional (3-D) drawing (Figure 14-1). All three of an object's physical dimensions—height, width, and depth—are shown. Pictorials are used because they communicate the shape and volume of an object better than two-dimensional drawings. Laypeople can understand pictorials better than orthographic drawings, because a pictorial's skewed viewing angle reflects how we commonly look at space and objects, unlike orthographic views, which are positioned directly on top or in front of their subjects. Designers and field workers also benefit from pictorials, because pictorials can clarify construction details in a way orthographic drawings can't.

There are two major groups of 3-D drawings: **axonometric** and **perspective**. Like orthographic drawings, they are created via a projection method, but a pictorial projection method is used, not a 2-D one. They are created from orthographic drawings such as plans, elevations, and sections.

Pictorials can be **line drawings** or **renderings**. Line drawings are simple drawings that outline an object or space. Renderings contain artistic elements such as shade, shadow, and color. Media varies from ink, pencil, chalk, marker, and paint to software such as Chief Architect, Sketchup, Lightscape, AutoCAD, and Illustrator, all of which create art as electronic files (Figures 14-2, 14-3).

The availability of pictorial-creating software does not eliminate a designer's need to know how to construct pictorials manually. No program can provide the impromptu sketches needed during the conceptual design phase. Knowing the mechanical construction process will make you a better sketcher. Furthermore, the ability to construct a pictorial will also enable you to create drawings that show precise intent, instead of working with software limitations.

Figure 14-1
Two-point perspective in casein and acrylic paints. *Courtesy of artist Reggie Stanton, www.reggiestanton.com, for Smallwood, Reynolds, Stewart, Stewart Architects, Atlanta, GA.*

Figure 14-2
Two-point perspective in casein and acrylic paints. *Courtesy of artist Reggie Stanton, www.reggiestanton.com, for Smallwood, Reynolds, Stewart, Stewart Architects, Atlanta, GA.*

Figure 14-3
Two-point perspective done using Lightscape. *Courtesy of artist Tomasz Biernacki, www.pechara.com.*

Renderings require innate artistic talent, take years of practice, and are time consuming to produce, whether by hand or using software. Because of this, designers usually contract renderings out to professional artists. However, anyone can quickly master the skill of making simple drawings using geometric techniques.

Axonometric Drawing

Axonometric (also called **paraline**) is an umbrella term for different drawing types that share one characteristic: all line sets are parallel (Figure 14-4) Axonometric drawing technique evolved from technical drafters' need to be able to describe an object's shape simply and accurately without artistic talent.

Most axonometrics have a vertical *z-axis* and *x*- and *y*-axes skewed between 0° and 90°. Axonometrics are also characterized by the relationship between the size of the objects and the size at which they are drawn. Unlike perspective drawings, where item size is proportional to distance from viewer, axonometric drawings are scale-drawn, measurable, and have no relation to the distance from the viewer. In fact, the word "axonometric" means "measurable along axis."

However, while axonometric drawings enable the reader to measure for size, there is still **distortion**, which is deviation from the human perception of three-dimensional space. Technical illustrators employ techniques to minimize distortion and achieve the best picture possible. From their efforts, four subcategories of axonometric have evolved: oblique, isometric, diametric, and trimetric.

Frontal Oblique Oblique is the easiest, if crudest, drawing technique. The front of the object is drawn true size and shape (height and width are measured exactly). The depth is drawn at an angle between 0° and 90°. An oblique is really not so much a 3-D drawing as it is a 2-D drawing with forced depth. This forced depth is what gives it an element of realism.

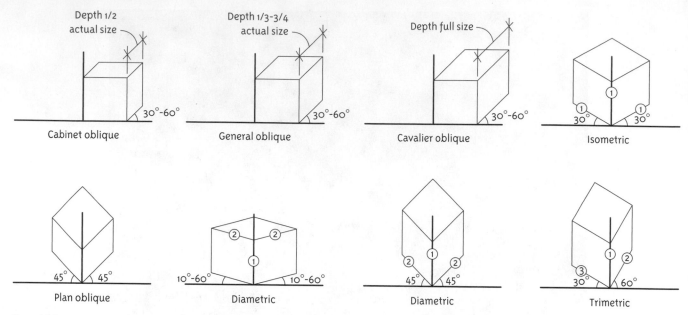

Depth 1/2 actual size — Cabinet oblique — 30°–60°

Depth 1/3–3/4 actual size — General oblique — 30°–60°

Depth full size — Cavalier oblique — 30°–60°

Isometric — ① 30° 30°

Plan oblique — 45° 45°

Diametric — 10°–60° ② ② ① 10°–60°

Diametric — 45° ② ① ② 45°

Trimetric — 30° ③ ① ② 60°

Figure 14-4
Types of axonometric drawings. The circled numbers show differing scales for those axes.

Depending on the type of oblique, the depth is drawn between one-third its size and full size. A **general oblique's** depth is between one-third and three-fourths actual size. A **cavalier oblique's** depth is full size. A **cabinet oblique's** depth is half-size (Figure 14-5). It is probably the most realistic looking of all the obliques, but even so, it is not as realistic or attractive as other pictorials and is mostly used for furniture and cabinet drawings. **Plan obliques** (also called planometrics) show the object's top as true shape and size instead of the front (Figure 14-6). Plan oblique drawings give the clearest, most unobstructed views into interior spaces.

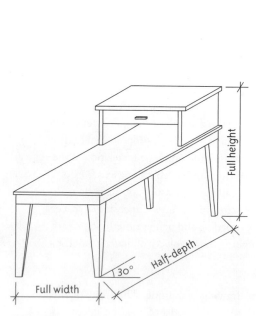

Figure 14-5
Cabinet oblique drawing of a table. This technique is favored for furniture and cabinets.

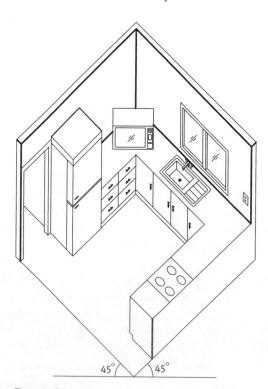

Figure 14-6
Plan oblique (planometric) drawing of a kitchen. The top is seen true shape and size.

Isometric **Isometric** is the most common type of axonometric drawing (Figures 14-7, 14-8). It is drawn at a 30° angle to the horizontal, full-size, and all three axes have the same scale. The word isometric means "equal measure."

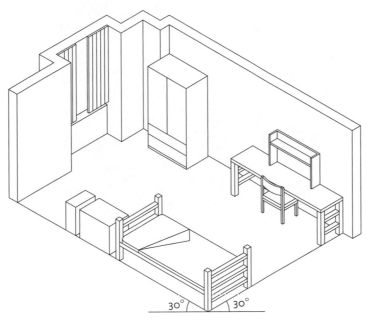

Figure 14-7
Isometric drawing of a dormitory room. Horizontal lines are drawn at 30° angles.

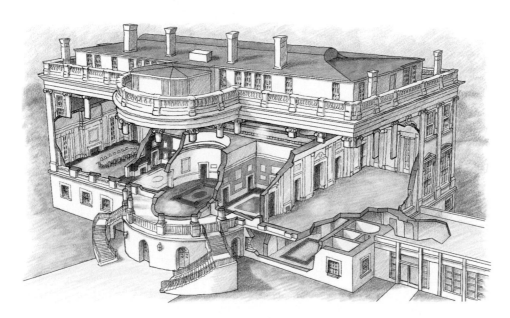

Figure 14-8
Cutaway dimetric done in Fractal Painter. *Courtesy of artist Randal Birkey, www.birkey.com*

Dimetric A **dimetric** drawing is skewed to show one corner of the object facing the viewer (Figure 14-9). No sides are true shape and size. The *x* and *y* axes are drawn at the same angle (between 10° and 60°) and their scale is separate from the vertical axis's scale.

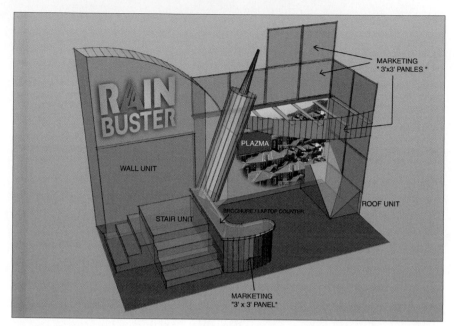

Figure 14-9
Trimetric drawing done using Sketchup. The verticals are slightly tilted to give the illusion of depth. *Courtesy of artist Tomasz Biernacki, www.pechara.com.*

Trimetric Like a dimetric, a **trimetric** drawing has one corner of the object facing the viewer, with no side true shape and size (Figure 14-10). Its difference is that the *x* and *y* axes are set at different angles (between 10° and 60°) and all three axes have different scales.

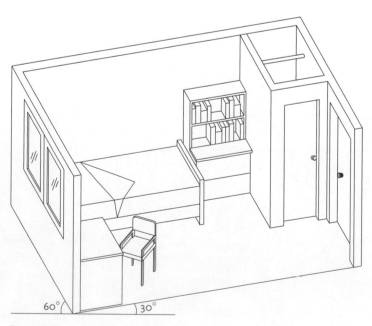

Figure 14-10
Trimetric drawing of a dormitory room. The horizontal axes are drawn at different angles and all three axes have different scales.

For each axonometric type there are further choices, such as exploded vs. non-exploded, intact vs. cutaway, and eye-level vs. bird's-eye view. Decide at the outset which type, angle, and combination of scales will best show the item being drawn. Diametric and trimetric drawing techniques are used more frequently by technical illustrators than by drafters, because they are so time-consuming.

Drawing in Isometric

Block House

Figure 14-11 presents three orthographic views and an isometric pictorial of the block house the orthographic views describe. The following are steps for drawing that pictorial. Refer to the drawings in Figures 14-12 through 14-18 for the lengths and points mentioned in the steps.

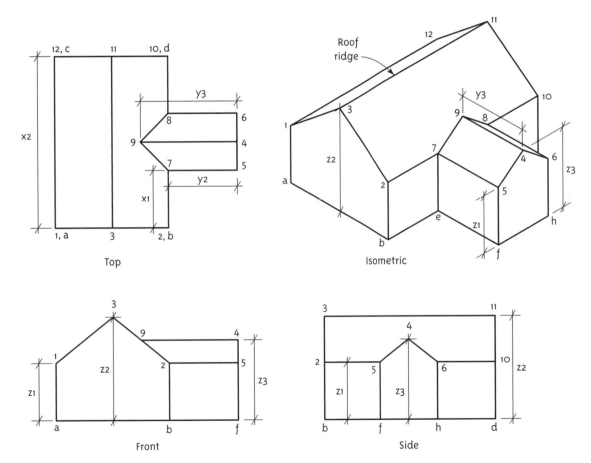

Figure 14-11
Orthographic and pictorial views of a block house with their corresponding points, corners, and distances labeled.

1. *Draw the* x-, y- *and* z-*axes.* The z-axis is vertical; all height measurements are made on it. The x- and y-axes are drawn 30° to the horizontal. All the house's horizontal lines will be drawn on them. All measurements are full size to scale, and the scale is the same on all axes. Use a scale or dividers to measure and mark perimeter *abcd* (the main part of the house) on the x- and y-axes.

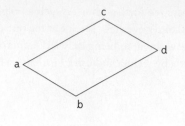

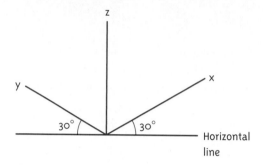

Figure 14-12
Lay out the main part of the house. Exclude the wing.

2. *Draw the wall height.* Project vertical lines up from each corner and mark wall height $z1$.

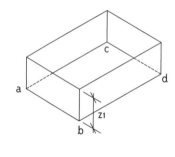

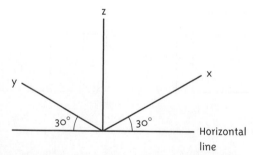

Figure 14-13
Draw the walls.

3. *Find the roof ridge height.* Start by finding the center of the short wall. To do this, connect its corners. Where the diagonal lines intersect is the wall's center. Draw a vertical line through that center up to the height of $z2$. This height is the ridge height; label it point 3, and draw lines connecting it to points 1 and 2.

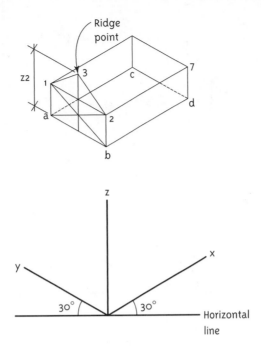

Figure 14-14
Draw a vertical line through the short wall's center to locate
the ridge height.

4. *Draw the roof length.* Starting at point 3, draw a line parallel to line 2-10 (use an adjustable triangle or the sliding triangle method shown in Chapter 2) and mark its length x2. Connect its endpoint (point 11) to point 10.

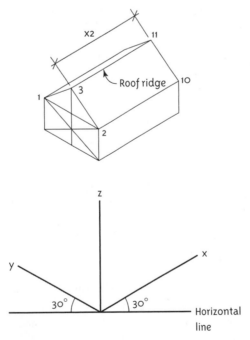

Figure 14-15
Lines 2-3 and 10-11 are parallel to each other. Lines 3-11
and 2-10 are parallel to each other.

5. *Draw the wing.* Draw perimeter *efgh* by marking lengths x1 and y2.

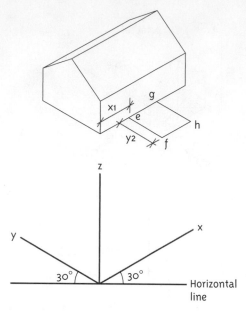

Figure 14-16
Lines e-f and g-h are parallel to each other.

6. *Draw the wall height and ridge height.* Project the wall height from corners *efgh* up height z1. Draw diagonal lines through the corners of the short wall to find its center. Then draw a vertical line through that center up height z3. Label that height point 4.

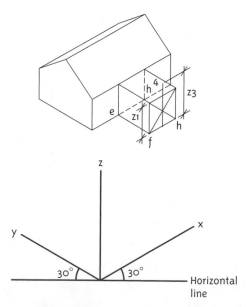

Figure 14-17
Draw a vertical line through the short wall's center to locate the ridge height.

7. *Construct the wing's roof.* Connect point 4 to points 5 and 6. Draw a line from point 4 parallel to the *y*-axis, the length of y3, and call it point 9. This is the length of the wing roof's ridge. Connect point 9 to points 7 and 8. Darken in object lines and erase construction lines.

As seen from the construction of the gable roofs, non-isometric lines and surfaces are drawn simply by locating their endpoints. Note that all vertical lines are parallel to the *z*-axis and parallel to each other. All horizontal lines are parallel to the *x*- and *y*-axes and parallel to each other. This may seem obvious, but a common beginner mistake is to *not* draw these lines parallel to each other.

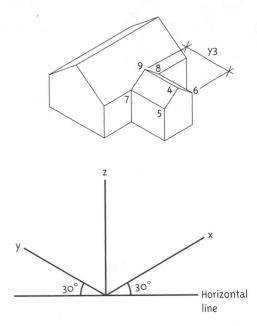

Figure 14-18
Draw the wing's roof to complete the drawing.

Drawing an Isometric Circle

Drawing curved items requires drawing circles, which appear in isometric as ellipses. You can see why by holding a circle template up and slowly rotating it. The circles on the template narrow (foreshorten) and become ellipses. When the template is rotated 90° to your eye, the circles become straight lines. The easiest way to draw is it to use an isometric ellipse template (35°); position one inside a rectangle the size needed and trace (a picture of this is shown in Chapter 4). However, if a template isn't available, you can manually construct one. We'll draw the glass-topped table shown in Figure 14-19 using a method called the "four-center approximate method." You will need a compass. Refer to drawings 14-20 through 14-24 for the steps.

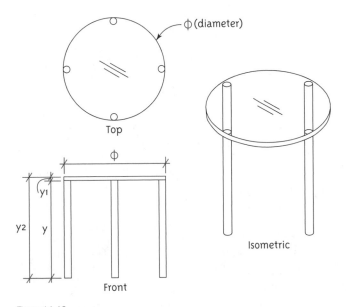

Figure 14-19
Orthographic and pictorial views of a round table.

1. Draw an isometric box with its top the length and width equal to the circle's diameter. The height is whatever the table height is. On the top, draw horizontal, vertical, and diagonal lines that connect the corners.

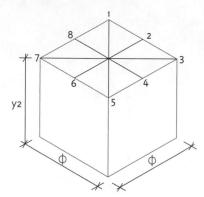

Figure 14-20
Draw a box whose length and width are the circle's diameter.

2. Connect the points as shown. The circled intersections will be the first two swing points used.

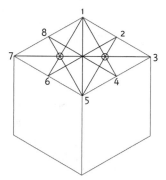

Figure 14-21
Connect points to find intersections that will serve as compass swing points.

3. Swing two arcs.

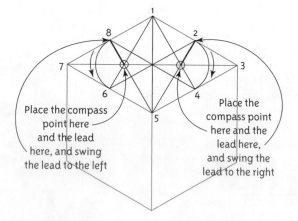

Place the compass point here and the lead here, and swing the lead to the left

Place the compass point here and the lead here, and swing the lead to the right

Figure 14-22
Swing the first pair of arcs.

4. Swing a second two arcs.

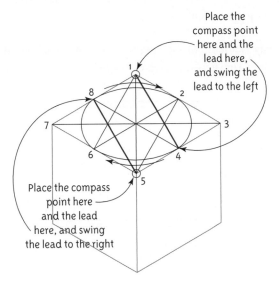

Place the
compass point
here and the
lead here,
and swing the
lead to the left

Place the compass
point here
and the lead
here, and swing
the lead to the right

Figure 14-23
Swing the second pair of arcs.

5. Locate the leg tops by projecting a vertical line the thickness of the tabletop down from one of the ellipse's axes. Then draw the legs at the length described in the orthographic view. A prudent time saver for the small curves at the leg bottoms is to eyeball their shape instead of constructing it.

6. Draw the tabletop thickness by projecting a vertical line down from the intersection shown and then constructing a second ellipse. The distance between the two ellipses will be the tabletop's thickness. An alternative is to mark that thickness with vertical lines along the ellipse and then use an irregular or flexible curve to hard-line a continuous line connecting the points.

   The tabletop in this example is glass, but we have left off its thickness at the table's back for clarity. Note that some of the table legs tops are partially obscured by the tabletop's thickness. Also note that pictorial drawings do not utilize hidden lines.

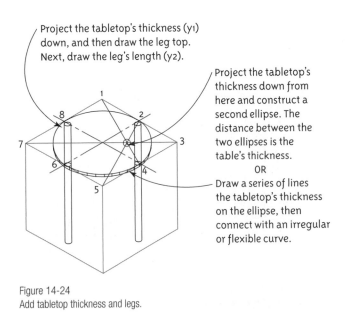

Project the tabletop's thickness (y_1)
down, and then draw the leg top.
Next, draw the leg's length (y_2).

Project the tabletop's
thickness down from
here and construct a
second ellipse. The
distance between the
two ellipses is the
table's thickness.
OR
Draw a series of lines
the tabletop's thickness
on the ellipse, then
connect with an irregular
or flexible curve.

Figure 14-24
Add tabletop thickness and legs.

Creating a Cutaway Isometric Drawing

Cutaway drawings omit or partially remove walls and roof to show what is behind them. It is a popular technique for presentation drawings.

Drawing a cutaway isometric of the single-room floor plan Let's draw the simple one-room floor plan shown in Figure 14-25. Figures 14-26, 14-27, and 14-28 show the steps.

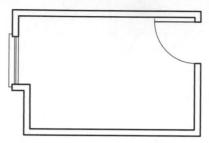

Figure 14-25
Floor plan of a single room.

1. Draw the isometric axis, and then the perimeter of the floor plan. It is good practice to draw the entire rectangular outline first to ensure the overall dimensions are correct, and then add the niche.

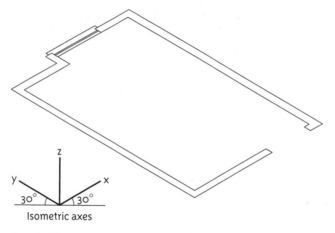

Figure 14-26
Draw the perimeter floor plan parallel to the isometric axes.

2. Measure and mark the height of the walls.

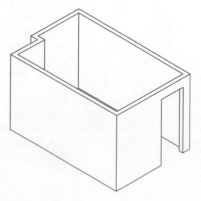

Figure 14-27
Draw the wall height.

3. Cut away the walls as needed to show interior detail. It is best to draw the cutaway lines at angles, as orthogonal cut lines are easily confused with the wall lines. Add line weights.

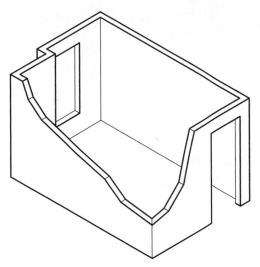

Figure 14-28
Cut away walls as needed to show interior detail and add line weights.

Drawing a cutaway isometric of an entire floor plan Here are suggested steps for drawing a cutaway isometric of the floor plan shown in Figure 14-29. Figures 14-30 through 14-35 show the steps.

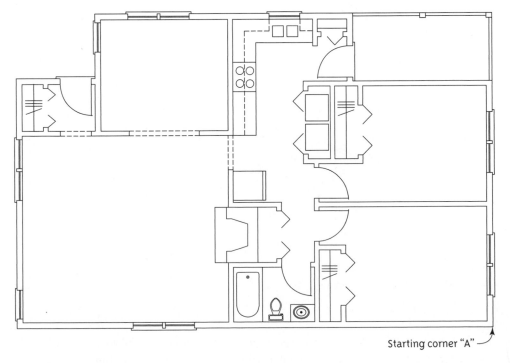

Starting corner "A"

Figure 14-29
Floor plan of house.

1. Draw the three isometric axes. Starting at corner A, measure and mark the perimeter of the floor plan. Then measure and mark the interior walls.

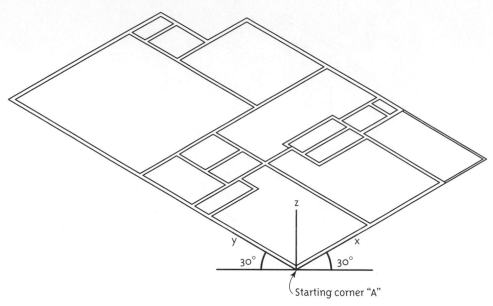

Figure 14-30
Draw the perimeter walls first, then the interior walls.

2. Measure and mark the window and door openings.

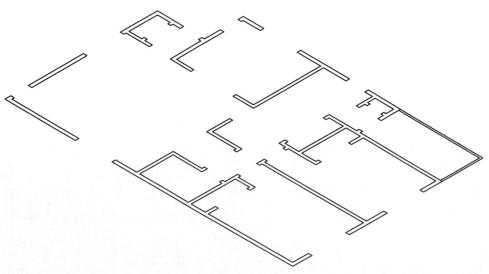

Figure 14-31
Measure and mark window and door openings.

3. Project the walls up, leaving openings where windows and doors will go. Typically they are projected up 1/3 full height to full height. Measure all walls up from the floor line and periodically check different walls to ensure they are the correct height.

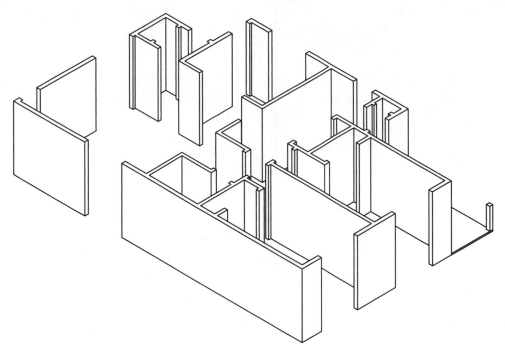

Figure 14-32
Project the walls up.

4. Mark and measure windows and doors.

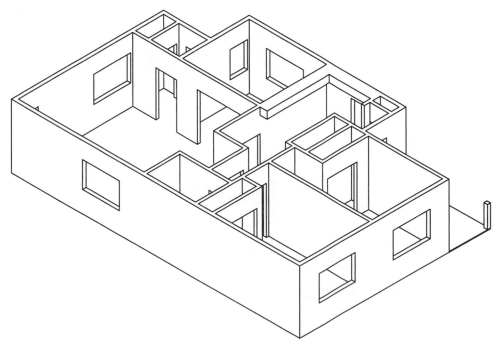

Figure 14-33
Complete the windows and doors.

5. Add furniture, cabinetry, and other relevant details.

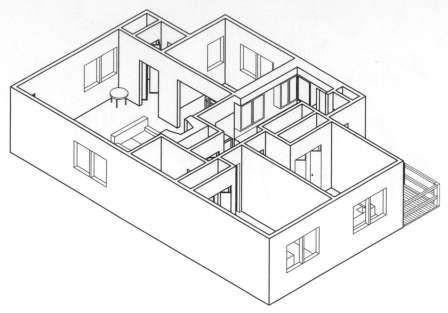

Figure 14-34
Draw interior detail.

6. Cut away walls in front of important features to display them. Again, cut them at angles so they are not confused with the wall lines.

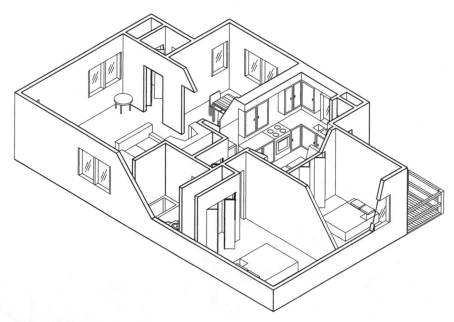

Figure 14-35
Cut away walls as needed to show the interior.

What Is Perspective Drawing?

Perspective drawing represents spaces and objects as we see them, not as they actually are. It is the most realistic of all pictorial drawing types, thus the best for showing ideas. Historically, this drawing type was attempted by early cultures, some of which came up with interesting, if off the mark, techniques. The Chinese used a method where the reader unscrolled a long, horizontal manuscript to see pictures of buildings that resembled para-line projection drawings. The Egyptians drew stacked rows of figures that placed items closest to the viewer at the bottom and items farthest away at the top. It wasn't until the

Renaissance that Italian artists applied Euclidean geometry to their art and perfected perspective drawing as a means of showing item size and shape as a direct function of distance and viewpoint. The Trinity painting by Masaccio circa 1425 is considered the first perspective artwork.

Perspective types are **one-point, two-point,** and **three-point,** as well as **interior, exterior,** and **sectional** (Figures 14-36, 14-37, 14-38).

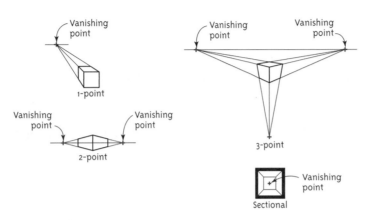

Figure 14-36
Types of perspective drawing.

Figure 14-37
Two-point perspective. Media is pen, ink, watercolor and color pencil washes on watercolor board. *Courtesy of artist Randal Birkey, www.birkey.com.*

Figure 14-38
One-point sectional perspective. Media is Adobe Illustrator and Fractal Painter. *Courtesy of artist Randal Birkey, www.birkey.com.*

With both perspectives and axonometrics, vertical lines are drawn vertically. However, **vanishing points,** which are locations where sets of horizontal lines converge, are a new element. They are located on a **horizon line,** which is an eye-level line formed where the ground meets the sky. Vanishing points and the horizon line are illusions; lines do not really converge and the ground never meets the sky. It just appears this way to us because of the way our brains interpret depth. Due to line convergence, items appear smaller the farther they are from the viewer, and larger the closer they are to the viewer.

There are different perspective drawing construction methods; probably the easiest is the **rotated plan.** This method gets its name from the fact that during the set-up the plan is rotated on the drawing board so the direction of view is straight up.

Drawing a Two-Point Interior Perspective

Sketched floor plans are appropriate to create conceptual design perspectives, but for presentation drawings, you'll need a scaled, hard-line plan. The 1/4" = 1'-0' scale should be the minimum size used, since the smaller the plan, the smaller the final drawing. Also needed are scale elevations for heights and details. Let's draw a **two-point** perspective of the floor plan and elevations in Figure 14-39 using the rotated plan technique. Note that the drawings are to scale and hard-line with all architectural elements and furniture shown.

1. *Select a Station Point and Direction of View* (Figure 14-40). When we look at a room we swivel our heads to build a complete image of it. However, a drawing is made from just one **station point,** which is the viewer location, and it is taken looking in one direction. Because its location determines the drawing outcome, the drafter must understand how different station points create different views. This takes practice, which will become

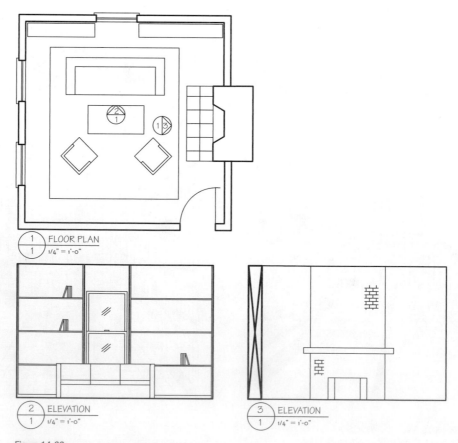

Figure 14-39
Floor plan and two elevations.

evident after you spend time doing drawings that do not offer the view expected. One way to develop an eye for skillful station point selection is to take pictures of existing spaces with a digital camera. Stand in different places in a furnished room and analyze the photos that the different locations produce. Which views cause nearby furniture to loom large, blocking out other things? Photograph a room at different eye levels by standing on stairs or crouching. Keep the camera level with the floor to avoid the vertical line tilting that happens when a camera is pointed up and down.

The station point can be inside or outside the room. The advantage of being outside is that more of the room will fall within the cone of vision. However, if the station point is too far away, the drawing will end up small with a viewed-from-afar appearance. If the station point is too deep inside the room, only a small portion of the room will appear.

Once you choose the station point, mark it with a cross. Do not use a circle, because a cross offers more construction accuracy. Draw the cone of vision lines and an arrow in the cone's center.

We see within a 60° **cone of vision** (COV), which is an angle that defines the limits of our view. To better understand this, hold a 30°/60° triangle parallel to the floor, with the 60° corner touching your nose. You can easily see what lies between the edges of the triangle; the edges define the cone of vision. Items outside it appear in peripheral vision. While the 60° cone of vision is the most natural way to draw a room, you can expand it up to a total of 90°, which will include more of the room. This comes at the expense of distortion proportionate to the amount over 60° that it is expanded.

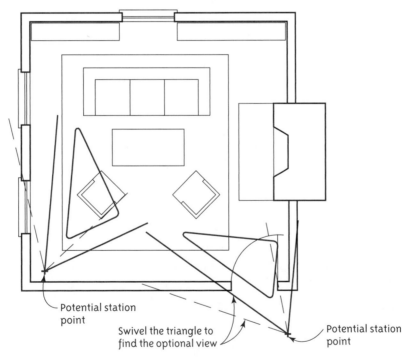

Figure 14-40
Swivel the triangle on the floor plan to find the best view. Everything between the triangle's edges will appear in the drawing.

Choose a station point on the floor plan (Figure 14-41). Then place the 60° corner of a 30°/60° triangle on it and swivel the opposite end. Everything between the triangle's edges is inside the cone of vision, so it will appear in the drawing. By swiveling, you are manipulating the cone of vision to find the most optimal view. Choose one and then draw an arrow in the center of the cone of vision.

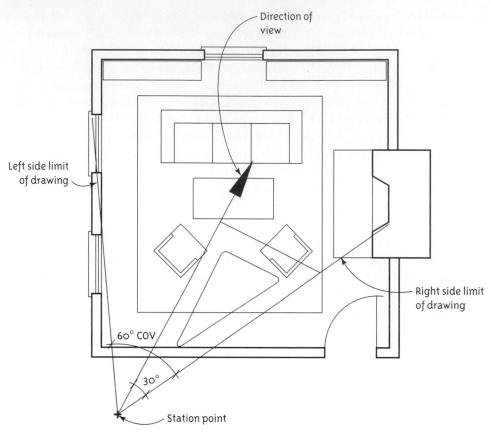

Figure 14-41
Choose a station point and precise view. Then draw an arrow in the cone of vision's center by placing the triangle's 30° corner in the cone's apex.

2. *Tape the paper down* (Figure 14-42). Here is what gives the "revolved" part of this method its name. Physically pick the paper up turn it so the arrowhead points straight up. Then tape the paper to the board, placing it high enough so there is room to construct the perspective drawing below it, but not so high that it's uncomfortable to reach.

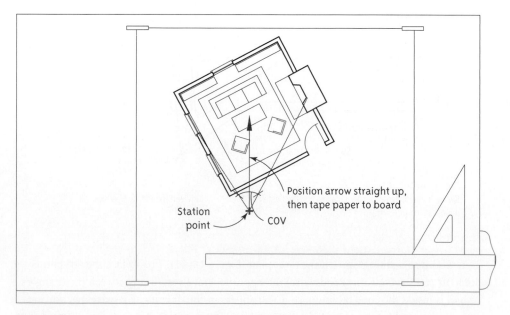

Figure 14-42
Position the paper so the arrow points straight up, then tape the paper to the board.

3. *Draw the picture plane (PP) and horizon line, abbreviated HL* (Figure 14-43). The horizontal picture plane defines the phrase "perspective projection." All **sight rays** (construction lines in the plan view that originate from the station point/eye) are projected up to it and then down to the pictorial drawing. The picture plane can be placed anywhere in the floor plan for different effects, but placing it in the plan's farthest-back corner usually gives the best results. Make sure to place it in the interior corner of the room, not the exterior corner. Next, draw the horizon line. This is a horizontal line placed under the floor plan. This line will be in the center of the pictorial drawing, so ensure there is plenty of blank space above and below it.

4. *Draw the vanishing points (VPs).* Each set of parallel walls needs its own vanishing point. Since the room is rectangular, it only has two sets of walls, so it only has two vanishing points. Draw a line through the station point and parallel to the walls until the line reaches the picture plane. Where the lines intersect the picture plane, project down to the horizon line and mark them LVP (left vanishing point) and RVP (right vanishing point). Again, mark these points with a cross for greatest construction accuracy. All parallel lines will converge to these two points.

5. *Find the cone of vision limits.* Extend both sides of the cone of vision up to the picture plane. Where it intersects, draw vertical lines down. These vertical lines define the pictorial drawing's limits.

Unlike in axonometric drawing, parallel in perspective means convergence.

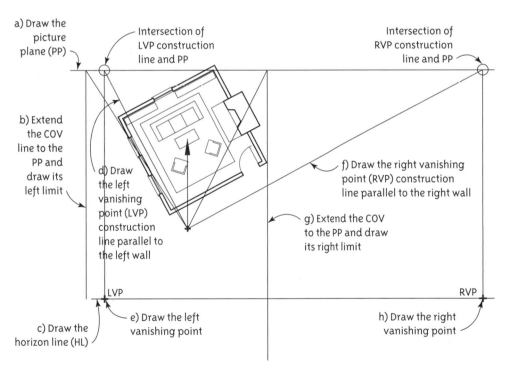

Figure 14-43
Draw lines for the picture plane, horizon, and cone of vision limits. Draw vanishing points.

6. *Draw the height line* (Figure 14-44). This is the "true" line on the perspective drawing, the only one on which measurements can be made. Where the interior corner and picture plane intersect, draw a vertical line down past the horizon line. Starting at the horizon line, measure down the total eye level and make a short horizontal mark. From that mark, measure back up the total ceiling height and make another mark. Darken in the vertical line between those two marks. This is the height line and the room's first corner.

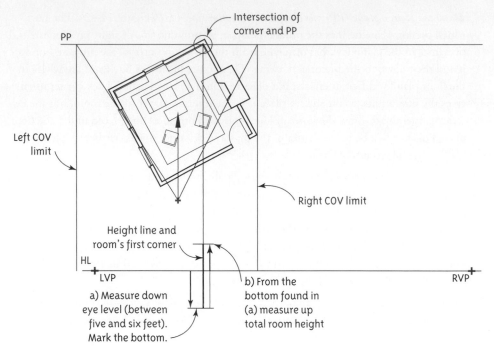

Figure 14-44
Draw the height line. This is the room's first corner.

7. *Draw the ceiling and floor lines* (Figure 14-45). Align a straightedge so that it simultaneously touches the RVP and the height line's top. Going forward from the height line, draw a line until it reaches the cone of vision's left limit. This is the room's ceiling line. Now align a straightedge with the RVP and the bottom of the height line. Going forward from the height line, draw a line until it reaches the cone of vision's left limit again. This is the floor line. Repeat this procedure from the LVP.

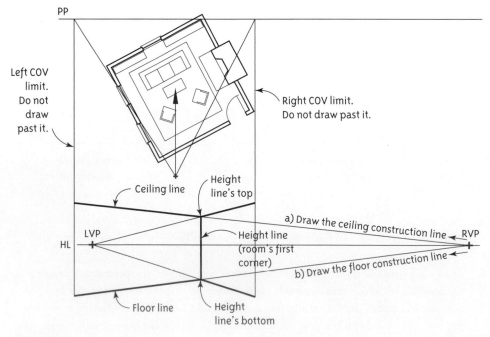

Figure 14-45
Draw the wall and ceiling lines.

8. *Draw the room's second corner* (Figure 14-46). In the plan view, draw a sight ray from the station point through the room's second corner. Extend that line to the picture plane and then drop it down to intersect the ceiling and floor lines drawn in Step 5. The vertical line between the floor and ceiling lines represents the second corner. Note that if this corner were *not* inside the cone of vision, you would not draw it. The ceiling and floor lines would simply continue to the left cone of vision limit. Also note that the vertical cone of vision lines are construction lines and *not* part of the final drawing.

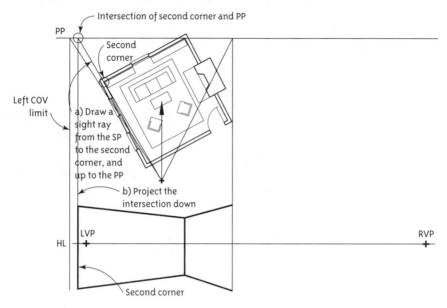

Figure 14-46
Draw the room's second corner.

9. *Draw the third wall* (Figure 14-47). Draw lines from the LVP through the top and bottom of the second corner to define the third wall. Now the whole room is drawn. You can see three walls and two corners, which is exactly what appears inside the floor plan cone of vision selected in Step 1.

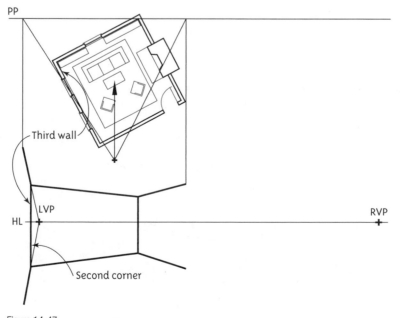

Figure 14-47
Draw the room's third wall.

10. *Draw the window* (Figure 14-48). Draw sight rays from the station point through the left and right interior corners of the window and project them up to the picture plane. Where the rays intersect the picture plane, draw vertical lines down to the wall. Those represent the window's sides. Mark the window bottom and top on the height line, and then draw lines from the RVP through those marks until they intersect the window sides. The rectangle formed is the window.

To divide the window in half vertically, draw diagonal lines connecting its corners and draw a line through their intersection to the RVP (Figure 14-49). Add thickness to

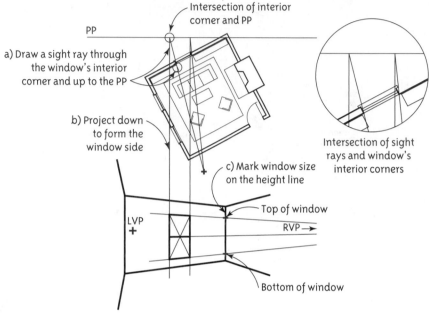

Figure 14-48
Draw the window by locating its interior corners in plan and projecting down.

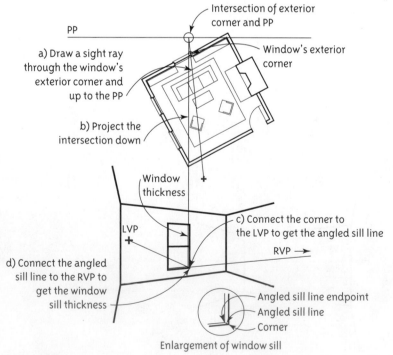

Figure 14-49
Add thickness to the window locating an exterior corner in plan and projecting down.

the window by running a sight ray from the station point through the exterior corner of the window and projecting it down. Draw a line from corner A to the LVP, and connect that intersection to the RVP to draw the sill.

11. Draw the fireplace (Figures 14-50, 14-51). You can see on the floor plan that only half the fireplace is inside the cone of vision. Therefore, only that half will appear in the pictorial. On the plan, draw a sight ray through the left back corner (where the fireplace intersects the wall) and up to the picture plane. Then draw a vertical line at that intersection down to intersect the floor in the pictorial. Draw another sight ray from the station point to the front corner of the fireplace, up to the picture plane, then drop that intersection down to the pictorial.

Now draw a line from the RVP to the back corner/floor intersection and continue it forward until in intersects the vertical line representing the fireplace front corner, which will locate the fireplace front corner. Connect that front corner to the LVP and draw a line that stops at the right side cone of vision.

The elevation shows that the fireplace is the full height of the wall. Repeat this construction process at the ceiling line.

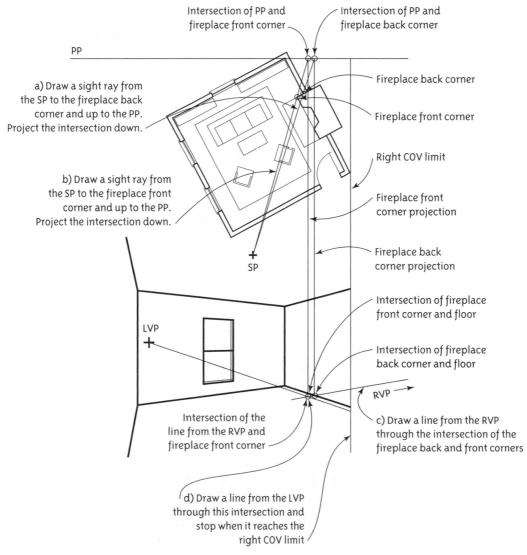

Figure 14-50
Locate the fireplace's position on the wall.

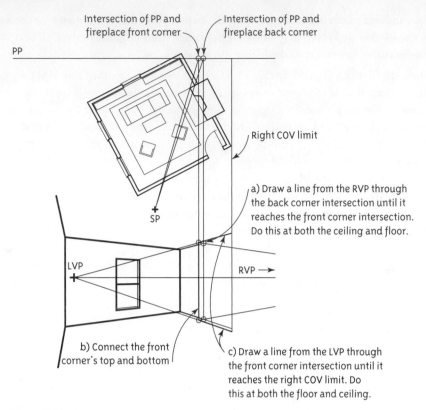

Intersection of PP and fireplace front corner

Intersection of PP and fireplace back corner

PP

Right COV limit

a) Draw a line from the RVP through the back corner intersection until it reaches the front corner intersection. Do this at both the ceiling and floor.

SP

LVP

RVP →

b) Connect the front corner's top and bottom

c) Draw a line from the LVP through the front corner intersection until it reaches the right COV limit. Do this at both the floor and ceiling.

Figure 14-51
Construct the fireplace's front corner.

Find the points of the firebox's angled walls (Figure 14-52). These walls cannot be drawn to the room's vanishing points because they are not parallel to the walls. Instead, draw sight rays from the station point through the angled wall's endpoints to the picture plane, project down to the pictorial, locate their positions as shown and "connect the dots."

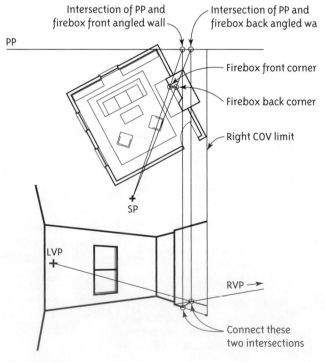

Intersection of PP and firebox front angled wall

Intersection of PP and firebox back angled wa

PP

Firebox front corner

Firebox back corner

Right COV limit

SP

LVP

RVP →

Connect these two intersections

Figure 14-52
Construct the firebox's angled walls.

Find the firebox height (Figure 14-53). Measure and mark its height on the height line and draw a line to it from the LVP. A zigzag line from the back wall to the front of the fireplace must be constructed as shown. The height cannot be taken straight from the height line to the protruding fireplace, because the fireplace is not flush with the wall.

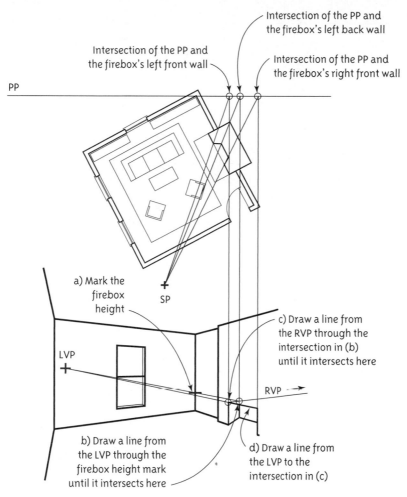

Figure 14-53
Zigzagging a line to construct the firebox's height.

12. *Draw furniture adjacent to the wall* (Figure 14-54). Let's draw the bookcases, because they are adjacent to the wall. Draw sight rays from the station point to their back corners and up to the picture plane. Then draw a third sight ray from the station point through one front corner, up to the picture plane and down to the pictorial as shown to find the bookcases' front corner. Project down to the pictorial and construct the bookcase outline as shown.

You may add shelves (Figure 14-55) by marking their heights on the height line and zigzagging those heights between both vanishing points.

13. *Draw freestanding (not adjacent to a wall) furniture* (Figure 14-56). Let's draw the couch. We'll leave off the bookcases on that wall for clarity. The couch is constructed similarly to adjacent-to-wall furniture, but requires a preliminary step of projecting its footprint (outline) to a wall. To do this, draw construction lines from the couch's sides perpendicular to a wall. Where they intersect that wall, draw sight rays from the station point up to the picture plane, and then drop down vertically to the floor line in the pictorial. Then draw lines from the LVP forward, passing through the intersections of the vertical lines and the floor. The vertical line representing the right couch side will be used later to construct the couch height.

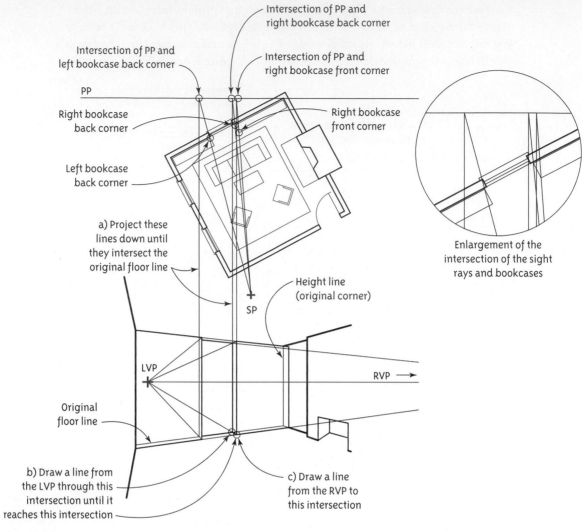

Intersection of PP and
right bookcase back corner

Intersection of PP and
left bookcase back corner

Intersection of PP and
right bookcase front corner

PP

Right bookcase
back corner

Right bookcase
front corner

Left bookcase
back corner

a) Project these
lines down until
they intersect the
original floor line

Height line
(original corner)

SP

Enlargement of the
intersection of the sight
rays and bookcases

LVP

RVP →

Original
floor line

b) Draw a line from
the LVP through this
intersection until it
reaches this intersection

c) Draw a line
from the RVP to
this intersection

Figure 14-54
Construct the bookcase outline.

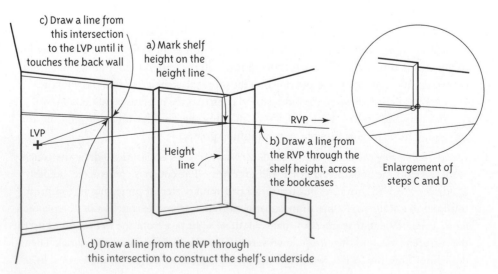

c) Draw a line from
this intersection
to the LVP until it
touches the back wall

a) Mark shelf
height on the
height line

LVP

RVP →

Height
line

b) Draw a line from
the RVP through the
shelf height, across
the bookcases

Enlargement of
steps C and D

d) Draw a line from the RVP through
this intersection to construct the shelf's underside

Figure 14-55
Constructing a bookcase shelf. The viewer will see the top of shelves below eye level and the bottom of shelves above eye level.

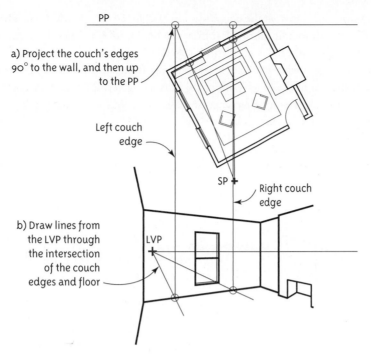

PP

a) Project the couch's edges
90° to the wall, and then up
to the PP

Left couch
edge

SP

Right couch
edge

b) Draw lines from
the LVP through
the intersection
of the couch
edges and floor

LVP

Figure 14-56
To construct a piece of freestanding furniture, first project it at a 90° to a wall.

Now construct the couch's footprint, or outline (Figure 14-57). Project the actual locations of three of the couch's corners up to the picture plane, then down to the pictorial as shown. It is not necessary to project all four corners, because the fourth will be formed by construction line intersections.

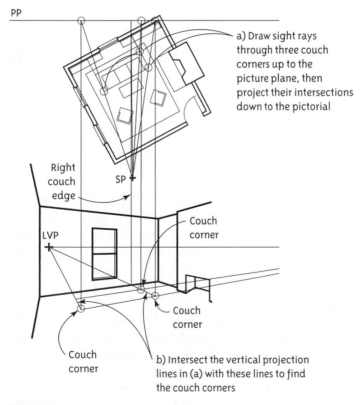

PP

a) Draw sight rays
through three couch
corners up to the
picture plane, then
project their intersections
down to the pictorial

Right
couch
edge

SP

LVP

Couch
corner

Couch
corner

Couch
corner

b) Intersect the vertical projection
lines in (a) with these lines to find
the couch corners

Figure 14-57
Constructing the couch's footprint, or outline.

Now you can construct the couch's height (Figure 14-58). Mark the height on the height line and draw a line through it from the RVP until it reaches the vertical line representing the couch's right side. At this intersection, use the LVP to bring another line forward to intersect with a vertical line emanating from the footprint's right front corner. Construct the rest of the couch's outline by emanating vertical lines from the other three corners. Whittle the resulting rectangular mass into a couch, making sure to draw all lines to their respective vanishing points (Figure 14-59).

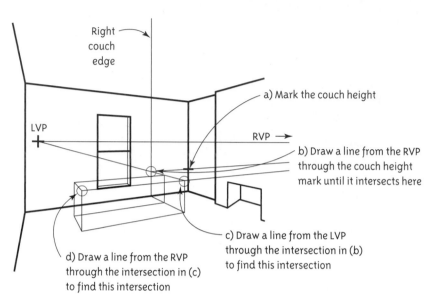

Figure 14-58
Constructing the couch's height. Project vertical lines up from all four corners to intersect with the points shown.

Figure 14-59
All lines on a piece of furniture must go to the vanishing points.

When constructing furniture, use photos in a catalog for guidance. Marking individual heights on the height line for the seat, arm, and other features may be useful at first, but developing the skill of **eyeballing** these heights is a useful timesaver. You can sometimes take artistic license to make objects show up better even if technically this makes them inaccurate. For instance, if a piece of furniture shows up very narrow, widening its footprint a bit to make it read better can enhance the drawing. You can adjust a circle-top table that looks tilted to make it appear more pleasing. As long as all lines go to the correct vanishing points and the fudging is not too obvious, it's acceptable practice.

14. *Draw catty-corner furniture.* Furniture that is not parallel to the walls requires its own vanishing points, making it a mini-problem within the larger problem. Let's draw one of the catty-corner chairs. Find its vanishing points the same way as you did the room vanishing points, by drawing lines parallel to its edges through the station point (Figure 14-60). Mark the vanishing points with the name of the piece to avoid confusion with the room vanishing points. Colored pencils may make it easier to tell each piece of furniture's sight rays apart.

After you find the chair's vanishing points, draw a sight ray from the chair corner closest to the viewer (station point) up to the picture plane, then straight down to the perspective drawing until it intersects the floor line (Figure 14-61). Draw a line from the room's LVP (not the chair's) through that floor intersection. The reason that the room's vanishing point is used is because you are seeking a specific location in the room, a location that happens to coincide with the chair corner. Once that point is located, use the chair's vanishing points as shown.

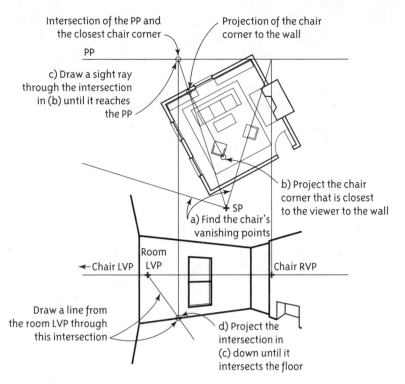

Figure 14-60
Drawing a freestanding catty-corner chair requires projecting a corner at a 90° to the wall and then finding its own set of vanishing points.

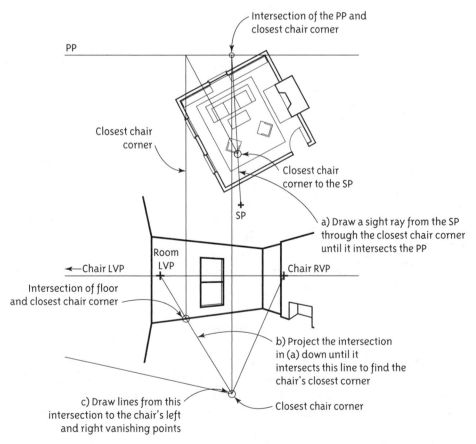

Figure 14-61
Use the room vanishing point to locate the chair's first corner, then use the chair's vanishing points for subsequent corners.

Draw sight rays from the station point through two more chair corners up to the picture plane and then down to the pictorial. Find those corners to finish drawing the chair's footprint (Figure 12-62).

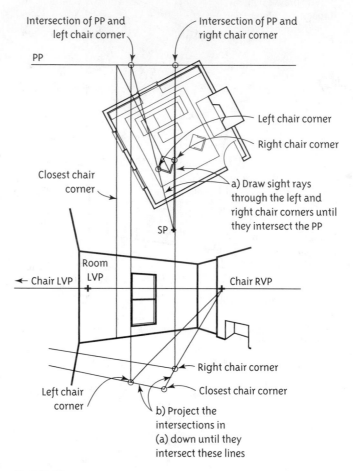

Figure 14-62
Complete the chair footprint.

Once you have constructed the footprint, you can draw the chair's height. Mark its height on the room's height line. Draw a line from the room's RVP through the chair height mark until it intersects the vertical line used to find the first (closest) chair corner. Then use the room's LVP to draw a line through that intersection until it intersects a vertical line emanating from the chair's closest corner outline. That intersection is the chair height (Figure 14-63). Draw lines from the chair vanishing points to the top of that height line. Project points up from the remaining corners. Since we see the chair's back from our station point, whittle the resulting cube into a chair's back side.

15. *Add people* (Figure 14-64). Placing people and accessories in a perspective drawing gives it a recognizable scale. Mark a person's height on the height line and draw a line through it from the RVP. Any vertical line on the wall between that line and the floor will represent that person's height. To bring the person into the room, draw lines from the LVP through the top and bottom of any vertical line drawn on the wall, and bring forward into the room. You can see that the closer the person is to the viewer, the bigger she is. Care must be taken not to obscure important architectural features behind her.

You don't need artistic talent to draw people and accessories, because tracing books of these images are available. When using an image from a tracing book, take care to match the detail of the figure to the detail of the drawing. Taking digital photos of people and groups of people to trace is also useful.

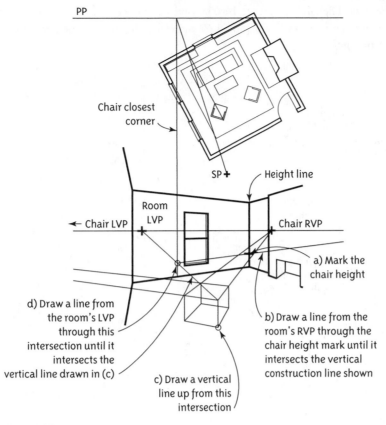

Figure 14-63
Use the room vanishing point first, and then the chair's vanishing points, to draw the chair height.

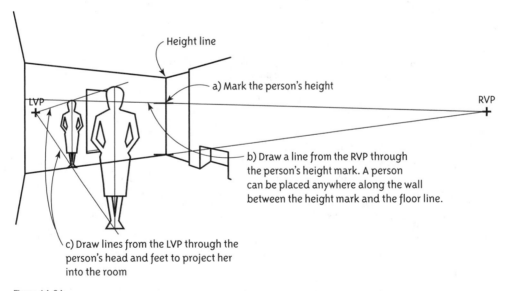

Figure 14-64
Drawing people helps give a room a recognizable scale.

Drawing a Two-Point Perspective of a Circle

As with isometrics, circles in **two-point** perspective will appear as ellipses. There are different ways to construct them; probably the easiest is to enclose the circle in a square, draw the square in perspective, then trace a circle inside. For greater accuracy, draw points on the circle perimeter in plan, then project each point up to the picture plane and down to the pictorial, and connect the dots. The circle in Figure 14-65 represents a circular space; however, the process in the steps shown in Figures 14-66 and 14-67 equally applies to drawing circular tabletops or any other round item.

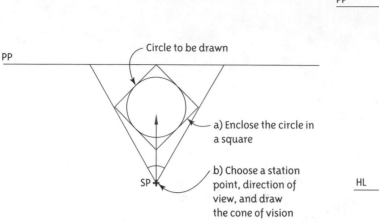

Figure 14-65
To draw a circular room, first enclose it in a square.

a) Enclose the circle in a square

b) Choose a station point, direction of view, and draw the cone of vision

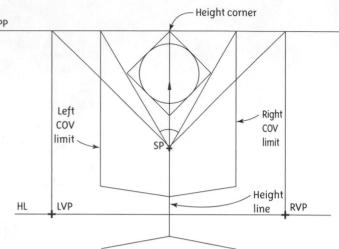

Figure 14-66
Draw a square room. Find its vanishing points, height line, and walls.

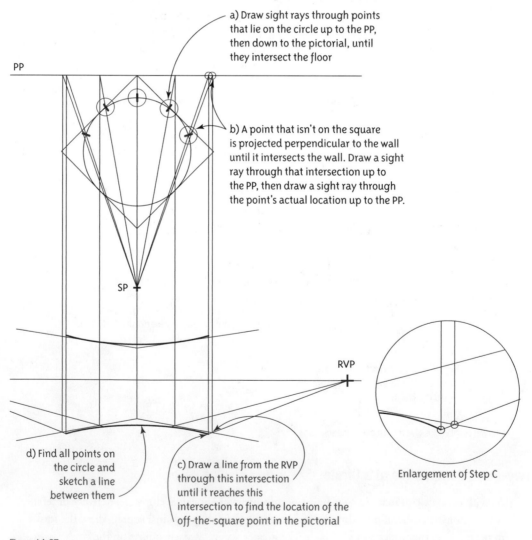

a) Draw sight rays through points that lie on the circle up to the PP, then down to the pictorial, until they intersect the floor

b) A point that isn't on the square is projected perpendicular to the wall until it intersects the wall. Draw a sight ray through that intersection up to the PP, then draw a sight ray through the point's actual location up to the PP.

d) Find all points on the circle and sketch a line between them

c) Draw a line from the RVP through this intersection until it reaches this intersection to find the location of the off-the-square point in the pictorial

Enlargement of Step C

Figure 14-67
Project points on the circle down to the pictorial and sketch a curve.

Drawing a One-Point Perspective

In a **one-point perspective,** the floor plan is oriented so that one entire wall rests on the picture plane. This results in that wall appearing in elevation in the pictorial. Hence, there is no height line; rather there is a height wall (the whole wall serves as the height line) and only one vanishing point. Otherwise, the procedure is similar to that of drawing a two-point perspective. Let's draw a one-point perspective of the room shown in Figure 14-68.

1. *Orient the plan on the board* (Figure 14-69). Tape the plan to the board by aligning a horizontal line with the parallel bar. Draw the picture plane along the entire back wall. (The pictorial will be easier to draw if the plane is aligned with the room wall instead of the offset.) Choose a station point. It may be located inside or outside the room; using one that is centered in the room will result in a symmetrical drawing. Draw the cone of vision and an arrow inside, pointing straight up.

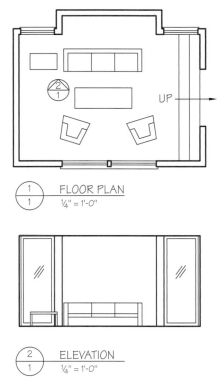

1 FLOOR PLAN
1   ¼" = 1'-0"

2 ELEVATION
1   ¼" = 1'-0"

Figure 14-68
Floor plan and elevation.

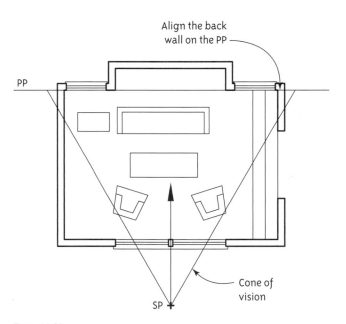

Align the back wall on the PP

PP

Cone of vision

SP

Figure 14-69
Align the back wall with the picture plane to draw a one-point perspective.

2. *Draw the back wall* (Figure 14-70). Draw a horizon line. Find the vanishing point by projecting the station point straight down to it. You will not be able to find a second vanishing point, because a line drawn parallel to the second set of walls will never intersect the picture plane. Project the edges of the wall down. From the horizon line, measure the viewer's eye level down; mark that eye level to define the floor and from that mark measure up the total room height to define the ceiling. Project everything on the plan's back wall down to the pictorial. Here they include two windows and the edges of the offset. All these features go from floor to ceiling, so there is no measuring and marking of their heights.

3. *Draw the adjacent walls* (Figure 14-71). To construct the walls adjacent to the back wall, draw lines that radiate from the vanishing point through the corners of the back wall. Stop at the cone of vision. Remember that the vertical lines denoting the cone of vision are not included in the perspective drawing.

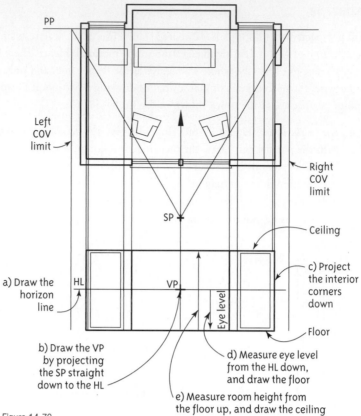

PP

Left
COV
limit

Right
COV
limit

SP

Ceiling

a) Draw the
horizon
line

HL

VP

c) Project
the interior
corners
down

Eye level

Floor

b) Draw the VP
by projecting
the SP straight
down to the HL

d) Measure eye level
from the HL down,
and draw the floor

e) Measure room height from
the floor up, and draw the ceiling

Figure 14-70
Construct the back wall.

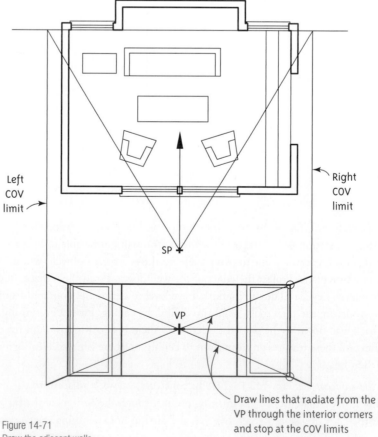

Left
COV
limit

Right
COV
limit

SP

VP

Draw lines that radiate from the
VP through the interior corners
and stop at the COV limits

Figure 14-71
Draw the adjacent walls.

4. *Draw the offset* (Figure 14-72). Because the offset's back wall is not aligned on the picture plane, it will show up in perspective, not in elevation like the rest of the back wall. Draw sight rays from the station point to the offset's back corners. At their intersections with the picture plane, draw vertical lines down to the pictorial. Next, in the pictorial, radiate lines from the vanishing point to the offset's four corners. At the intersection of the radial lines and vertical lines, draw horizontal lines to represent the offset's floor and ceiling lines.

Trim the original ceiling and floor lines to finish the offset, but leave a construction line showing where the original wall was, as you will need it to draw the couch and stairs.

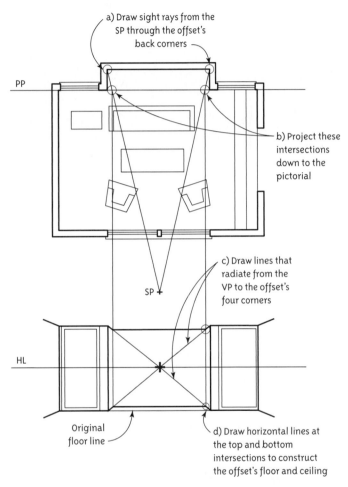

a) Draw sight rays from the SP through the offset's back corners

PP

b) Project these intersections down to the pictorial

c) Draw lines that radiate from the VP to the offset's four corners

SP

HL

Original floor line

d) Draw horizontal lines at the top and bottom intersections to construct the offset's floor and ceiling

Figure 14-72
Draw the offset.

5. *Draw the couch* (Figure 14-73). Project the couch's two front corners from the plan down to the pictorial until they intersect the floor construction line. (This line defined the floor before you drew the offset). At those intersections draw two lines radiating from the vanishing point. Then draw sight rays from the station point through three couch corners, up to the picture plane, and down to intersect the radiating lines. Finish the couch's footprint (Figure 14-74).

Find the couch's height by marking its height on the room's back wall and extending a line from that mark horizontally until it intersects the vertical line shown (Figure 14-75). Draw a line from the vanishing point through that intersection, and bring it forward until it intersects a vertical line emanating from the couch's front corner. Draw the other three corners similarly and whittle a couch inside the resulting rectangular mass.

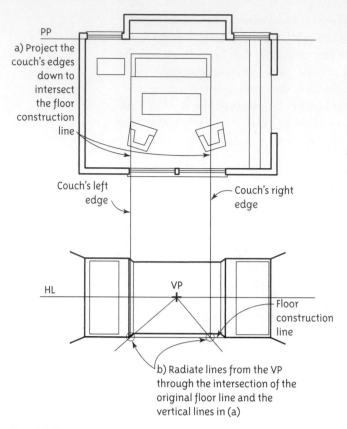

a) Project the couch's edges down to intersect the floor construction line

Couch's left edge

Couch's right edge

HL

VP

Floor construction line

b) Radiate lines from the VP through the intersection of the original floor line and the vertical lines in (a)

Figure 14-73
The couch edges are projected straight down from the plan to the pictorial.

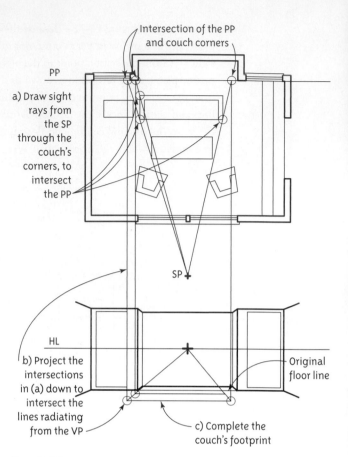

Intersection of the PP and couch corners

PP

a) Draw sight rays from the SP through the couch's corners, to intersect the PP

SP

HL

b) Project the intersections in (a) down to intersect the lines radiating from the VP

Original floor line

c) Complete the couch's footprint

Figure 14-74
Draw the couch's footprint.

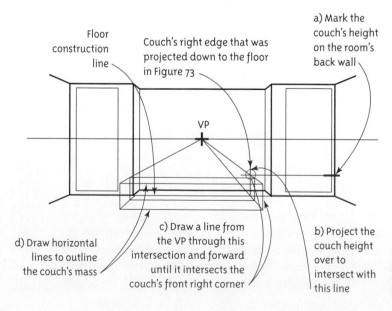

Floor construction line

Couch's right edge that was projected down to the floor in Figure 73

a) Mark the couch's height on the room's back wall

VP

d) Draw horizontal lines to outline the couch's mass

c) Draw a line from the VP through this intersection and forward until it intersects the couch's front right corner

b) Project the couch height over to intersect with this line

Figure 14-75
Mark the height on the back wall. Note that it cannot be marked on the offset's back wall, as the offset's back wall does not lie on the picture plane.

6. *Draw the stairs* (Figure 14-76). Project the tread widths down from the plan until they intersect the pictorial's floor line. Draw a line from the vanishing point and those intersections, then draw lines forward into the room. These lines are the stairs' footprints.

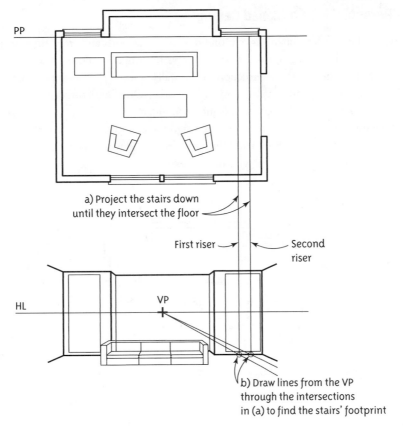

a) Project the stairs down until they intersect the floor

First riser   Second riser

HL   VP

b) Draw lines from the VP through the intersections in (a) to find the stairs' footprint

Figure 14-76
Construct the stairs' footprint.

Mark the riser heights and project them horizontally to intersect with vertical lines emanating up from each tread (Figure 14-77). Then draw lines from the LVP through those intersections forward into the room. Complete the stairs by drawing horizontal tread lines as shown in Figure 14-78.

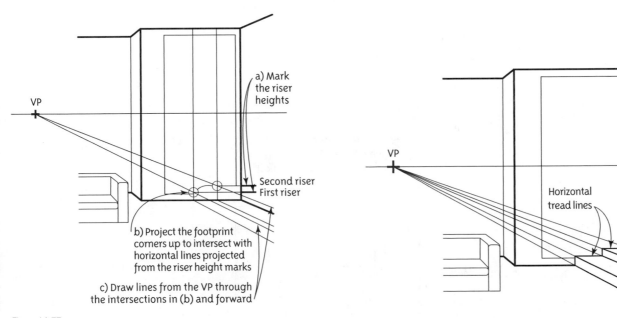

VP

a) Mark the riser heights

Second riser
First riser

b) Project the footprint corners up to intersect with horizontal lines projected from the riser height marks

c) Draw lines from the VP through the intersections in (b) and forward

Figure 14-77
Mark riser heights and project them horizontal over to intersect with vertical lines emanating from each step's footprint.

VP

Horizontal tread lines

Figure 14-78
The completed stairs.

Drawing a One-Point Perspective of a Sloped Ceiling

Let's put a gabled ceiling in this room; one that slopes at both ends to a ridge in the center (Figure 14-79). The dashed line in the plan shows the ridge location. Draw sight rays from the station point through the intersection of the walls and ridge endpoints, project up to the picture plane, and then project down to the pictorial. On the pictorial, draw an outline of the back wall to represent the ceiling's height at the ridge.

Radiate lines from the vanishing point through the endpoints of the ridge height line and extend them until they reach the vertical lines you dropped down from the plan (Figure 14-80). Connect the resulting intersections with the corners of the shorter wall.

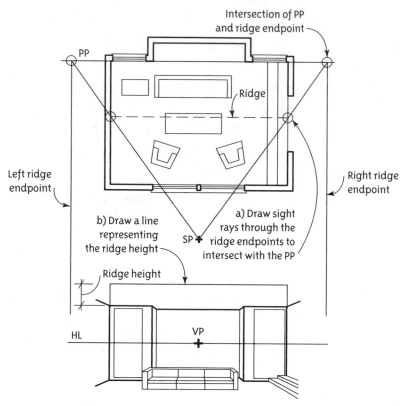

Figure 14-79
To construct a sloped ceiling, draw sight rays through the ridge endpoints in the plan, and draw the ridge height in the pictorial.

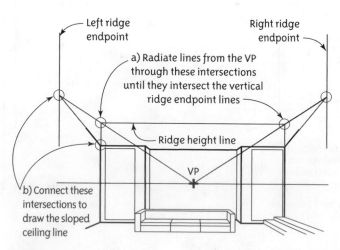

Figure 14-80
Connect the intersections shown to find the sloped ceiling lines.

Mechanically construct a perspective, then place a sheet of tracing paper on it and freehand trace over it to "loosen it up" for an artistically pleasing drawing. You can also do this over photographs of existing spaces.

Troubleshooting

Perspective drawings have many steps and it is easy to make mistakes. Often when you are drafting you may feel something isn't quite right but not know what it is. When this happens, retrace all steps starting at the beginning to analyze what went wrong. The most common mistakes include

- Inadvertently using the vanishing point locations on the picture plane instead of the vanishing point locations on the horizon line.
- Using the cone of vision's intersection with the picture plane as the vanishing point location.
- Not constructing vanishing points through the station point or not constructing them parallel to the walls.
- Using an incorrect height line; that is, using a corner or edge of the plan that does not touch the picture plane.
- Not projecting freestanding furniture perpendicular to the wall first, or projecting it at a non-90° angle.
- Attempting to draw object heights in the pictorial before constructing their footprints.
- Not drawing lines to the two vanishing points, instead drawing them to multiple random points.
- Darkening incorrect wall lines; that is, darkening in construction lines instead of object lines.

Summary

3-D (pictorial) drawing shows spaces and objects close to, or exactly like, the way we see them. Axonometric pictorials are easy to draw because there is no line convergence, but their measurability comes at the price of distortion. Perspective drawing is more complicated and is not measurable, but its photographic quality of representation makes it useful for display and presentation. Knowing how to construct a perspective drawing mechanically facilitates grid sketching, because rudiments like stopping the drawing at the cone of vision and using the height line properly are not readily apparent to someone without prior knowledge of perspective construction techniques.

Questions

1. What is an axonometric drawing?
2. Name three types of axonometric drawings.
3. What is an isometric drawing?
4. What is a perspective drawing?
5. What is the relevance of the cone of vision in a perspective drawing?
6. When should a perspective drawing be used? When should an axonometric drawing be used?

Suggested Activities

1. Take photos of interior spaces and furniture, trace over them, and then make changes and add or subtract things from them.

2. Draw a floor plan and elevation of a room on campus, and use them for a perspective drawing.

3. Draw a floor plan of your home, and do a cutaway isometric of it.

4. Obtain a simple floor plan and do several perspectives of it, experimenting with different cone-of-vision angles and eye-level heights.

5. Obtain furniture catalogs and study the pieces carefully to analyze how they would appear in a perspective drawing.

Further Reading and Internet Resources

Arends, M. (1990). *Interior Presentation Sketching for Architects and Designers.* New York: Van Nostrand Reinhold.

Burden, E. (2003). *Entourage, A Tracing File and Sourcebook* (4th ed.). New York: McGraw-Hill.

Pile, J. (1985). *Perspective for Interior Designers,* New York: Whitney Library of Design.

Stanton, Regie (1979). *Drawing and Painting Buildings: Reggie Stanton's Guide to Architectural Rendering.* New York: Van Nostrand Reinhold

http://www.reggiestanton.com
Website with portfolio of professional artist Reggie Stanton's work.

http://www.birkey.com
Website with portfolio of professional artist Randal Birkey's work.

http:// www.pechara.com
Website with portfolio of professional artist Tomasz Biernacki's work.

GLOSSARY

accessibility The accommodation of people with disabilities in buildings

accordion Refers to a door that operates like a folding door but has multiple leaves

active panel The hinged, or operating, panel in a door

actual size True size of building components; see *nominal size*

air conditioner Appliance that cools, filters, and dehumidifies air

aligned Describes a dimensioning method in which the dimension notes, which are numbers that describe size and location, are drawn parallel to the *dimension lines*

Ames lettering guide Tool that draws multiple rows of guidelines (construction lines that regulate letter height), each separated by a small space

annotation Note

anthropometry The study of human sizes and proportions and their variations and reaches

appliance A general term for any item powered through a plug and a flexible cable

arch A curved structural member

architect's scale Scale that measures in units of feet and inches

area of refuge A place where building occupants can safely wait for emergency assistance

as-builts Printed sets of drawings that have handwritten notes on them showing field changes

awning A window that is hinged at the top and swings out, resembling an awning when open

axonometric General term for different drawing types in which all line sets are parallel; also called *paraline*

balloon framing Framing technique characterized by two-story high studs that run continuously from the foundation to the roof

balustrade Assembly containing balusters, newel posts, and handrail

basement plan Plan that shows walls, footings, grade beams, and pilasters plus interior spaces

bay A window that projects 18" to 24" from the wall

beam Horizontal structural member

benchmark A reference point for contour lines

blind cabinets Cabinets in corners that are bordered by other cabinets and that contain dead (inaccessible) or hard-to-access space

blueprints Copies produced by blueprinting, an old-style, photographic, ammonia-based copying method that produced copies that had white lines on a dark blue background

boiler Device that heats water

bolt Hardware that provides supplemental security to a lock or holds the inactive half of a double door in place

bond An arrangement of brick or blocks in a wall

border lines Thick lines that go around the perimeter of a sheet; also used to underline titles

bow Window that projects from the wall; its panes generally are narrower and more numerous than those of a bay window, and are installed to give the illusion of a curve

box bay Bay window in which the side windows are 90° to the wall

branch circuit Circuit that runs from the service panel or a sub-panel and is routed to outlets and switches

branch pipe Horizontal plumbing pipe that runs from a stack pipe to a fixture

bubble diagram Visual means of organizing thoughts and developing a concept; consists of labeled circles that represent spaces and functions that occur inside; also called a schematic or relationship diagram

buck Metal door frame

building codes Rules that govern the materials and methods of construction for residential and commercial design; also see model code

cabinet oblique Oblique drawing whose depth is half size

callout Symbol that links two drawings (also spelled *call-out*)

call-out See *callout*

case goods Any non-upholstered piece of furniture used to store, or case, household goods

casement Window that is hinged on the side and swings out

casing See *trim*

cast Material (such as concrete) that is poured into forms, whose shape it takes

cavalier oblique Oblique drawing whose depth is full size

cavity wall Wall that consists of two separate, parallel walls built several inches apart

center line Line drawn through the center of a feature

chase Passageway between floors or walls; also, a plumbing wall

chase walls Walls that house plumbing pipes; also called stack or wet walls

chimney Vertical, freestanding structure that carries smoke and gas out of the firebox

circuit Path of an electric current

circular stairway Stairway whose steps are trapezoidal in shape, rise along a circular path, but are not attached to a center pole

circulating heat Heat distributed throughout a room via airflow

cleanout Plugged hole in a pipe or trap that allows access to the inside for unclogging

clear span Interior space unobstructed by columns or load-bearing walls

clear width Distance between jamb protrusions

clerestory Window set high in the wall, usually directly under the roof

closer Hardware that automatically closes a door after it has been manually opened

cabinet code Number and letter combination that describes cabinet size and type

combination Pairing of two or more different window types

combustion Burning

compass Tool used for drawing circles and arcs; has a point at one end and a lead on the other

column Load-bearing, vertical member outside a wall. Runs continuously from foundation to roof

concrete Mixture of cement, water, aggregate, and possibly admixtures (chemicals)

conduction Process whereby an energy source heats the air around it, the warmed air rises, cool air moves in to replace it, and a current is formed

cone of vision Sixty-degree angle emanating from the viewer's eye that defines the limits of his or her view

construction drawings Drafted documents that describe building components and systems

construction lines Light lines used for layout work; they help create object lines but are not part of the object itself

construction plan Plan whose purpose is to show building assembly

contour lines Lines that show ground elevations; they connect points on the land surface that are the same elevation above a benchmark or reference point

convection Process whereby heat moves through a solid material; the denser the material, the better it conducts heat

cooktop Appliance that contains stove burners

course Horizontal row in which bricks are laid

current Flow of electrons

curtain wall Non-load bearing wall hung on a load-bearing frame, whose purpose is to shield and enclose a building

cutting plane line Line drawn on the floor plan that shows where an object is sliced to create a section view

cylindrical break line Also called an s-break; a freehand line placed through a cylindrical object such as a pipe or a column

decking Horizontal cover on a roof or floor

demolition Removal or relocation; a demolition plan shows items that have been designated for demolition

detail A vertical or horizontal cut made through a small portion of the building, such as a window, doors, stairs, level change, or anything else where multiple pieces join

digital copies Black-line copies on white background made by scanning the original; can then be reproduced onto whatever media is wanted

dimension lines Also called stringers; linear dimensions strung together to form a line, or string; run perpendicular to extension lines and contain the dimension notes

dimension notes Numbers on a drawing that describe size and location

dimensional mill lumber Lumber cut to specific sizes such as 2" × 4"

dimensioning Process of indicating size and location in space

dimensions Numbers that describe the size of features and their distance from other features

dimetric Drawing that is skewed to show one corner of the object facing the viewer; no sides are drawn at their true shape and size

distortion Deviation from the human perception of three-dimensional space

dividers Tool that resembles a compass, but has points on both legs; used for transferring distances where numerical measurement is not needed

double door Refers to two swing doors hung from opposite jambs in one opening

double hung Also called vertical slide windows, these windows have two sashes that slide vertically in grooves inside a frame

double action Refers to a door that swings both ways and is common in high-traffic areas such as between a restaurant kitchen and dining room

drafting Also known as graphic communication; the art of putting ideas to paper in picture form to create instructions

drafting board A large, smooth surface atop a base

drafting brush Tool used to remove eraser crumbs from the paper or vellum

drafting pencil Wood pencil made in a 2.0 mm size

drafting tape Affixes paper to the drawing board; available in rolls and dots

ducts Distribution and return-air paths in a forced-air system; large quantities of air for heating and cooling are moved through them

Dutch Refers to a swing door that is cut into top and bottom halves that operate independently

egress Means of exit

electrical plan Plan that shows lights, switches, circuits, and outlets; also called a lighting plan

elevation symbol Sometimes referred to as a callout; symbol that indicates where, exactly, on the floor plan an elevation drawing is made, and of which wall

elevation Vertical, or height, drawing of walls; an orthographic view projected onto a vertical plane

engineered wood product Wood veneers and fibers that have been laminated (adhered and pressed together) to produce longer-spanning, greater-load-bearing pieces than dimensional lumber

enlargement box Area where a close-up is provided; encircled by heavy dashed lines

ensuite Bedroom connected to a bathroom

environmental Relating to the interior environment's physical conditions that affect occupant health and safety, such as air quality and circulation

equipment plan Plan that shows equipment used for heating and cooling

eraser shield Metal template that has openings of different sizes and shapes; allows the drafter to do fine erasing without removing surrounding lines

ergonomics Applied science of using design and anthropometric data to make the work environment more comfortable

exhaust fan Appliance that removes polluted and humid air from a space, allowing room for fresh air to enter

exit device Latch released by pressing a panic bar (also called a crossbar) to allow a crowd quick, easy egress in an emergency

exit stairs Consists of a stairs, a protected enclosure of fire-rated walls, and (if applicable) doors, which swing in the direction of the exit discharge

extension lines Lines that emanate from the end points or center of an object being dimensioned

exterior elevation Vertical drawing of an outside wall of a building

eyeballing Visually estimating

fabricated member Engineered building component; a trussed rafter is one example

fan Half-circle window

fenestration Arrangement, proportioning, and design of windows and doors in a building

film Polyester or plastic sheets used for ink or plastic lead work; Mylar® is the brand name for a popular film

firebox Combustion chamber where a fire is contained

firebrick Heat-tempered brick

fireplace Framed opening in a chimney that holds an open fire

fittings Faucets and controls

fixed Type of window that does not open

fixture Plumbing device that is permanently attached to a building, such as a sink, toilet, or shower, and that draws freshwater and discharges wastewater

flexible curve A piece of rubber that can be shaped to any curve; used for any curved line that cannot be drawn with a compass or circle template

floor plan Orthographic, two-dimensional drawing made by inserting a horizontal cutting plane through a building or room 4'-0" above the ground

flue Metal or clay-lined vertical path to the roof inside a chimney

flush A door with a smooth, even face. Also refers to components, such as walls, that are aligned (as opposed to offset)

folding Refers to a door that is part hinged and part sliding

footing Widened bottom of a foundation wall, pier, or column; supports and distributes the weight of the loads resting on it

footprint A building's shape, size, and orientation on its site

forced-air Type of heating and cooling system in which a furnace or air conditioner draws room air through ductwork and returns the warmed or cooled air via other ductwork; also called central heat or air

foreshortened Shorter than true, or actual, length

foundation Base upon which a building is placed; it provides a level surface to build on and carries the building's loads; see spread footing and slab foundation

foundation plan Plan that shows elements of the foundation, such as footings and grade beams

French door A double door with inset windows

French curve See irregular curve

frost line Level at which the ground no longer freezes; varies with geographic location

full A full section is a vertical cut through the entire length or width of the building and from foundation to roof

full bath Bathroom that has a lavatory, water closet, and tub or a tub/shower combination

full overlay Cabinet door type in which 1/8" or less of cabinet frame shows around each door and drawer head and between doors; the one option available for frameless cabinets

furnace Appliance that produces heat

galley kitchen Layout where cabinets, appliances and counters run parallel to each other along two walls. May also refer to a kitchen where all cabinets and counters run along just one wall

garden Bay window that projects 12" to 18" from the wall

general note Note that applies to everything on a sheet

general oblique Oblique drawing whose depth is between 1/3 and 3/4 actual size

GFI Ground fault interrupter; an electrical receptacle with a safety feature installed in the circuit that monitors the amount of current going to the appliance and compares it to the current coming back

girder Large beam that supports smaller beams

glazing Glass

glider Also called a slider; a two-sash window that slides horizontally

grab bars Bars anchored into the wall over tubs, toilets and in showers to help provide accessibility

grade Ground

grade beam The portion of a slab that is thicker than the rest. Its purpose is to support load-bearing walls above it

grid paper Paper with horizontal and vertical lines placed at regular increments, facilitating to-scale sketching; also called graph paper

ground Electrical connection to the earth; its purpose is to terminate electrical and lightning protection systems

G-shape Also called *peninsula,* this is a modified U-shape kitchen. It offers the benefits of the u-shape plus additional storage, food preparation, and appliance areas. This design usually only has one opening, which means less pass-through traffic

guard Item that prevents people from falling over the edge of a staircase or balcony; is required on any elevation over 30" where there is no adjacent wall. A handrail and guard are often combined as one assembly; the guard portion comprises the vertical balusters below the handrail

gypsum board Common interior wall finish; consists of gypsum powder pressed between two sheets of building paper

half-bath bathroom that has a water closet and lavatory; also called a *powder room*

half-diagonal rule Rule that requires the minimum distance between two exits to be at least one-half of the longest diagonal distance within the building or room that the exits serve

handle Hardware that operates a latch that opens a door; lever handles are used in commercial buildings

handrail The balustrade's horizontal or "rake" member; sits on top of the balusters, is supported by newel posts, and runs level and continuously along the stairs and landing

handrail Railing that helps people steady themselves on stairs; usually consists of a single rail installed at a specified height

hard-lining Using straight-edged tools to draw a presentation or construction-ready product

hardware Everything that operates and holds a door in place; includes hinges, locks, handles, knobs, exit devices, bolts, closers, holders, and stops

head Top of a door or window

header Lintel made of wood

headroom Clear vertical distance between the tread and the ceiling, measured vertically along a sloped plane

hearth Floor of a fireplace

heat pump Appliance that heats and cools a building

height line The "true" line on the perspective drawing, the only one on which measurements can be made

hidden object line Line that defines an item that is not visible in the reader's current view but must be acknowledged

hinge Hardware that holds a door to a wall

holder Hardware attached to a door and frame; keeps a door in an open or quasi-open position

hollow core Door that has interior made of honeycombed wood strips; may have a veneered exterior

hopper Window that is hinged at the bottom and swings into the house so rain won't enter

horizon line Eye-level line formed where the ground meets the sky

hose bib Outdoor faucet

humidifier Appliance that adds moisture to the house

ID label Label that goes under each drawing; usually consists of a 3/4" diameter circle with an attached horizontal line

inlet Cold air return; a component of forced-air systems. Also called return register

insert Heating unit that fits inside an existing fireplace to convert it into an efficient zone heater

interior elevation Height drawing of a wall inside the building or of the interior side of an exterior wall

irregular curve Tool that allows the drafter to hard-line arcs; also called French curve

island Stand-alone work center, usually located in a kitchen's center

isometric An axonometric drawing in which horizontal lines are drawn at a 30 degree angle to the horizontal, parallel to each other

jalousie Window that has rows of narrow, horizontal glass slats fastened with clips to a frame; also called a louvered window

jamb Side of a door or window

joinery Connection technique

key Type of legend

kiosk Independent stand from which merchandise is sold

knob Hardware that operates a latch that opens a door. Used on residential construction

L kitchen Kitchen layout where cabinets, appliances and counters form an L-shape

L stairs Stairway that has a landing and a turn, in an L-shaped configuration

lamp Light bulb

landing A level rest area on a staircase

lavatory Bathroom sink

layout tape Thin, opaque tape used to represent walls on presentation drawings or as a border

lazy Susan Rotating platform that maximizes hard-to-access space inside a cabinet

lead Stick of powdered graphite (a form of carbon) mixed with a binder

leader line Has an arrow or slash mark at one end and a local note at the other end

left-hand Door on which the knob is on your left and the door swings away from you; also called an in-swing door

left-hand reverse Door on which the knob is on your left and the door swings toward you, also called an out-swing door

legend Chart that explains symbols on a plan view drawing

lettering Drafted note-making on a drawing

light Window in a door; also called a *lite*

line Fundamental symbol of graphic communication; a thin (relative to its length) geometric object

line drawing Simple drawing that describes an object or space strictly with lines, not shade, shadows or other rendering

line quality Appearance of lines; good-quality ones are a consistent width from end to end, are the proper width for their relative importance, and are crisp and dark

line type A pattern that a line takes to represent a specific concept

line weight Thickness of a line

lintel Beam over a door, window or fireplace opening

lipped Door that has a rabbet (groove) cut all the way around the door on the back edge

load-bearing A wall that carries weight besides its own, such as an upper story or roof

load Weight

local note Describes the feature a leader line points to

lock Hardware that provides security to a door; its three basic components are a latch bolt, a lock strike, and a cylinder

long break line Line used to end a feature when drawing it in its entirety is not necessary; used after the feature's defining characteristics have been shown

longitudinal section Drawing made by slicing parallel to the roof ridge

main Underground pipe used to transport potable (drinkable) water from a water utility to individual consumers

maintenance The work required to sustain the condition of products and materials over their lifespan

mantel The frame surrounding a fireplace; also describes just the shelf over the fireplace

mark Symbol by which an elevation or schedule is referenced in a floor plan; see *callout*

masonry Units of brick, concrete block, stone, glass block, structural clay tile and terra-cotta

masonry opening Hole in the wall for an interior door in a concrete or brick building

match lines Lines that show where to align a large drawing that spans two or more sheets of paper

means of egress Occupants' route to exits, especially in an emergency

mechanical eraser Hollow, refillable stick in which long, thin erasers are placed; also called a click eraser

mechanical pencil Multi-piece, refillable tool available in 0.3 mm, 0.5 mm, 0.7 mm, 0.9 mm, and 2 mm sizes, which refers to the tip's opening diameter

mechanical plan Plan that shows heating, ventilation, and air conditioning systems, also called the HVAC plan

medicine cabinet Small bathroom wall cabinet

meter Device that measures the amount of energy the customer uses, which in turn tells the power company how much to charge; an electric meter is also called a watt-hour meter

metric scale Scale used for the same purpose as an architect's scale; the difference is that the metric scale measures system international (SI) or metric units

millwork General term for finished wood items manufactured in a lumber mill (although polyurethane, paper veneers over substrate, foams, and MDF products used for the same purpose are also referred to as millwork)

model code Code written by a standards organization separate from the jurisdiction that adopts and enforces the code

modular Construction technique that uses as many standardized and mass-produced components as possible, which helps eliminate cutting and waste

moldings Decorative lengths of wood that hide joints at intersections of the floor, wall, and ceiling

mudroom Residential back entrance that offers a place to clean up

multi-view drawings Another term for orthographic drawings; so called because multiple views are typically needed to describe the object

nominal sizes Sizes in which building materials such as concrete block, brick, and dimensional lumber are named, which differ from their actual sizes

non–load-bearing Wall that does not bear any of a building's weight; hence not factored into structural computations; also called *non-structural* or *non-seismic*

notes Notes are an integral part of a drawing; see *general note* and *local note*

O.C. Abbreviation for "on center"

object line A line that defines the idea being communicated

oblique A drawing technique where the front of the object is drawn true size and shape and the depth is drawn at an angle between 0 degrees and 90 degrees

occupancy A category that describes a building's use

occupant load Number of people a space is expected to hold

on center Distance from the center of each face of a structural member; abbreviated O.C.

one-point interior perspective Perspective drawing technique for interior spaces that utilizes one convergence point. The room's entire back wall rests on the picture plane

orientation Compass location

orthographic drawing Drawing that shows two of an object's three dimensions

orthographic projection Drawing technique that deconstructs a 3-D object into multiple 2-D views

outlet Connection device in a circuit that allows electricity to be siphoned off for appliances or lights via plug-in cords; also called a *receptacle*

overlay The amount of a cabinet's front frame covered by the door and drawer

Palladian Three-part window composed of a large, arched central section flanked by two narrower, shorter sections with square tops; also called a *venetian* window

pane Framed sheet of glass within a window

panel door Door that has rectangular raised or recessed areas framed by rails and stiles (vertical and horizontal framing elements)

paraline See *axonometric*

parallel bar Long, straight tool installed on a drafting board used for drawing horizontal lines

partial wal Not full height

partial section A partial section is a vertical cut through a small portion, such as one wall

partition Interior, non-structural wall that only carries its own weight

passive panel Door panel that may only be opened once the active panel is opened

peninsula See G-shape

perspective theory A drawing technique that represents spaces and objects as we see them; objects are drawn smaller commensurate with their distance from the viewer, and parallel lines converge to a common point

perspective sketching Freehand drawing using perspective theory

pictorial Three-dimensional (3-D) presentation of an item

picture plane Flat (2-D) surface onto which a suspended object is projected

pier Short post found under buildings, as in crawlspaces or porches

pilaster Post or column attached to a wall; its purpose is to strengthen the wall where heavy beams will rest

pipes Tubes, round in cross-section, that serve as the distribution method for plumbing and hydronic systems

plan north Convention of drawing all plans parallel to the edges of the paper, with north at the top of the sheet; enables the drafter to give simple names to interior and exterior elevations

plan oblique Oblique drawing that shows an object's top in its true size and shape; also called planometric

plan Orthographic view projected onto the horizontal plane

plate Horizontal board; bottom plates evenly distribute loads placed on them, and top plates tie studs together

platform framing Framing standard for most modern homes; studs run from level to level and floors are built on top of them

plumbing plan Drawing that describes a plumbing system, which delivers fresh water to a building and removes dirty water and waste

poché French word for a repetitive, textural pattern that describes the material of which an object is made

pocket A pocket door is a single sliding door that retracts into a wall pocket

post Vertical, load-bearing member outside a wall that supports beams. Posts run from level to level

post-and-beam framing The oldest building technique, whereby trees were cut, hewn into square cross-sections, and connected with joinery techniques

power Electric power or electricity is the process of freeing electrons (negatively charged particles) in a conductor and moving them to a device where they are put to work

power/telephone/data/plan Plan that shows where electrical outlets, phone jacks, data ports, computers, video equipment, and communications systems are located

precast Items that are molded off-site and transported to the building site

presentation drawing Also called a general purpose drawing; a show drawing, typically used to sell ideas to clients or as an exhibit

presentation plan Plan whose purpose is to be an exhibition or selling tools, so it may display color, furniture and decor

production drawings Also called contract documents or construction or working drawings; aimed at the tradespeople who will implement the ideas; consist of floor plans plus other descriptive drawings such as details, sections, and elevations

profile gauge Tool that measures complex moldings in place; consists of a magnetic handle and moving pins that take on the contour of whatever object they're pressed against

program Written statement that describes environmental, maintenance, and sustainability issues and lays out the problem, scope, goals, requirements, and constraints

programming Process whereby information is gathered, analyzed, and synthesized

projection lines Construction lines along which points and features are transferred from one plane to another

property line A property's physical boundary

proportional scale Scale that enables a picture to be enlarged or reduced by a specific amount. Consists of a small wheel overlaid on a large wheel

protractor Tool used for drawing angles

proxemics Describes set, measurable distances among people as they interact; encompasses spatial relationships, body language, boundaries, colors, physical territory, and personal territory

radiant Type of heat that emanates from a stove

radiant hydronic Describes a heat system that uses hot water or steam from a boiler to distribute heat through the house

radiation Process whereby heat travels as waves through space

rafter Inclined beam that supports a roof

rail Horizontal structural component on a window or door

rake Inclined edge of a sloped roof over a wall

ramp A sloped surface whose purpose is to make a building accessible

range Oven

rebar Steel reinforcing bar that is placed inside concrete to strengthen it

receptacle Plug-in outlet

recessed panel A recessed panel door looks like a picture frame; a wood frame around the edge surrounds a panel in the middle; also called a framed door

reflected ceiling plan view of the ceiling as if it were reflected onto a mirror that is flat on the floor

register outlet in a forced-air system through which treated air is returned to a room

rendering Decorative detail on an object; also, drawing that contains artistic elements such as shade, shadow, and color

reveal The exposed part of the frame on a cabinet face

ribbon board Board or length of boards nailed to, or inset inside, walls studs to support floor and ceiling joists

right hand Door on which the knob is on your right and the door swings away from you, also called an out-swing door

right-hand reverse Door on which the knob is on your right and the door swings toward you, also called an in-swing door

rigid frame Two columns and a beam or truss combined with steel skeleton framing and flat steel panels for the roof and walls; also called a *bent*

riser Vertical board at the front of the step

riser diagram Document that supplements and clarifies plumbing plan

roof ridge The top horizontal member of a sloped roof

rotated plan Perspective drawing method in which during set-up the plan is rotated on the drawing board so the direction of view is straight up

rough opening Hole in the wall for a door or window

R-value Energy conservation property

sash Window glass plus the frame that holds it

schedule Charts that give detailed information about components shown in the plan

schematic A drawing that utilizes simple line types and symbols that do not resemble the physical appearance of the items represented

section drawing A drawing showing a cut through a building or a portion of a building

section lining Angled line patterns that indicate an object has been cut; also called hatching

section symbols Sometimes referred to as callouts; symbols that indicate where, exactly, on the floor plan a section drawing is made

septic tank Underground tank used for processing solid waste

service panel Receives electricity into the house from the service entrance or meter box and distributes it throughout via branch circuits; consists of a large metal box inside of which are circuit breakers or fuses

sheathing Vertical covering of boards on exterior walls that goes under the final finish

shop drawing Also referred to as a fabrication or millwork drawing; its purpose is to show the designer precisely how the manufacturing of the ideas communicated in the production drawings will be done

short break line A freehand break line that serves the same purpose as a long break line

sight rays Construction lines in the plan view that originate from the station point/eye

sill Bottom of a door or window

single hung Window that has one sliding sash and one fixed sash

site plan Bird's-eye view of the entire property; prepared by a landscape architect or civil engineer

site visit Information-gathering trip to a construction area

sketching Freehand drawing

slab door Hung on a cabinet, it appears made of one solid piece and has no raised or recessed profile. Also called flat panel or frameless

slab-on-grade A foundation type that consists of a slab poured over gravel directly on the ground; also called a monolithic foundation

sliding Refers to doors that are hung from a metal track in the frame head

solid core Doors made of solid wood or solid wood blocks and covered with a veneer; normally used for exterior entrances

space frame Three-dimensional truss

space planning Process of designing a space to make it functional for the occupants

spandrel Steel girder that spans from one perimeter column to another

specifications Written instructions for projects that describe the quality and type of the materials shown in the drawings

spiral Staircase that rises around, and is connected to, a pole

spread footing A foundation system that consists of a wall built on top of a footing (a wide base); also called a t or perimeter foundation

stack Vertical plumbing pipe used for ventilation or to carry away wastes

stairs Form of vertical access

standard overlay Door option for framed cabinets in which the face frame has a full reveal around the door and drawer perimeters; also called semi-full

station point The viewer's location in the room

stationary panel Door panel that is fixed

steel section Cross-sectional shape

stile Vertical structural component of a window or door's top and bottom edge

stop Hardware that protects walls and equipment from the door as it swings open; typically metal with rubber bumpers on the top or front

stove Freestanding external fireplace in a metal container

straight run Stairs that have no turns

stringers See *dimension lines*

structural backing Supporting layer to which a veneer is attached

stud Vertical load-bearing member inside a wall

substructure Everything below grade in a building

sump pump Pump that drains basement water accumulated from ground water

superstructure Everything above grade in a building; comprises the building's structural system and everything that is hung on it

surround Refers to the immediate border of the face around the firebox opening; also, a non-combustible/masonry decorative frame around the whole fireplace

sustainability Using resources in a manner that does not deplete them and will have the least long-term effect on the environment

switch An electrical device that opens and closes the circuit in an electrical circuit and is operated by the user

synthesis The process of putting together parts to form a whole

T square Tool used for drawing horizontal lines; consists of a head and blade and is not installed on the board

tambour Cabinet door that has its own frame and is made of many separate pieces attached to a flexible backing sheet

thermostat Device that automatically regulates temperature in a furnace or air conditioner via sensors and activating switches

three-quarter bath Bathroom that has a water closet, lavatory, and shower

title block Square or rectangular box typically placed in a sheet's lower right-hand corner or vertically along the right side (in which case it is called a title strip) that contains information about the sheet it is on

toekick Indented space at the bottom of a cabinet

transom Fixed window above a door

transverse section Section drawing cut perpendicular to the roof ridge

trap A curved or S-shaped section of pipe under a drain; also, a device that prevent sewer gas backflow into branch lines

tread Horizontal board at the top of the step

triangle Tool used for drawing vertical or angled lines

trim General term for different types of finish millwork; also, a separate frame, sometimes called a casing, installed around a window or door

trimetric Drawing that has one corner of the object facing the viewer, with no side drawn at its true shape and size

true length Actual length

trussed rafter Engineered, load-bearing member placed at the roof; also called truss

two-point interior perspective Perspective drawing technique for interior spaces that utilizes two convergence points

typical detail Abbreviated TYP; detail that shows a repetitive feature

U kitchen Kitchen layout where the arrangements of cabinets, appliances and counters forms a U shape

U stairs Stairway that consists of two parallel flights; also called a scissors stairs

universal design Philosophy of making buildings and products usable to as many people as possible, regardless of age, height, gender, ability or disability

valve A mechanical device that controls the flow of water or gas

vanishing point Place where sets of parallel lines appear to converge

vanity Cabinet that contains a sink

vaulted ceiling Ceiling that angles up on one or both sides

vellum Semi-opaque, high-quality cotton paper used for ink or pencil work

veneer Thin, parallel layers of wood; also, non-load bearing, aesthetic masonry facing attached to, and supported by, a structural backing

Venetian See *Palladian*

vent An opening that allows combustion byproducts, gasses and pressure to escape, and fresh air to enter

visible object line Line that defines a physical item's outline

visual inventory Written, sketched, measured, and photographic documentation of existing conditions

volt Unit of electric potential

wall section A vertical slice through a wall from footing to roof; its purpose is to show how the wall is put together

warming drawer Slide-out appliance that fits in the cabinetry and keeps plates and cooked food warm

water closet Toilet

water heater Appliance that heats and stores water

water pipe Tube that emanates from a water heater and run throughout a building; also called a water line

watt Unit of power, calculated as energy per unit of time, that measures the work done by electricity

well Water source obtained by digging into an aquifer, which is an underground layer of porous, water-bearing rock

work triangle Area between the stove, sink, and refrigerator

wythe Continuous vertical section of a masonry wall, one unit in thickness

zone Specific area heated or cooled by one unit

APPENDIX 1

Industry Resources

There are many resources for looking up the product information needed when designing. One is Sweet's Catalogs, a compilation of product literature in a multi-volume set of hardbound books and a CD. Information and downloadable, editable AutoCAD drawings are also accessible via the http://www.sweets.com website.

Another resource is http://www.4specs.com, a directory of construction products. Users are redirected to company websites for product information.

Both of these resources organize their information in the Master Format. This is a categorization system devised by the Construction Specifications Institute (CSI), an association of commercial design and construction professionals. It organizes the vast array of building products and activities in an easy-to-find manner. The Master Format is an industry standard used by government and private entities for specifications writing, bid documents, and product literature organization. Therefore, it is useful to understand what its numbers mean.

Subjects are arranged into 49 divisions, with each division containing three levels. Some divisions are labeled "Reserved," meaning nothing has been assigned to them yet, to provide room for expansion. Divisions 8 (Openings), 9 (Finishes), 10 (Specialties), 11 (Equipment), and 12 (Furnishings) are probably most relevant to interior designers.

Each title has a six-digit number. For example:

13 59 13 Panel-Hung Component System Furniture

The first two digits are the division number, or Level 1. The next pair of digits is Level 2 and the third pair is Level 3. Sometimes a dot and another pair of digits is added at the end when the amount of detail needs another level of classification.

On page 406 there is a sample page from the Master Format, as published by the Construction Specifications Institute (CSI) and Constructions Specifications Canada (CSC). Page 407 shows a list of all 49 Master Format divisions. For a more in-depth explanation of the Master Format 2004, as well as a downloadable file, visit www.csinet.org/.

12 56 53	Laboratory Furniture
12 56 70	Healthcare Furniture
12 56 83	Custom Institutional Furniture
12 57 00	**Industrial Furniture**
12 57 13	Welding Benches
12 57 16	Welding Screens
12 57 83	Custom Industrial Furniture
12 58 00	**Residential Furniture**
12 58 13	Couches and Loveseats
12 58 13.13	Futons
12 58 16	Chairs
12 58 16.13	Reclining Chairs
12 58 19	Dining Tables and Chairs
12 58 23	Coffee Tables
12 58 26	Entertainment Centers
12 58 29	Beds
12 58 29.13	Daybeds
12 58 33	Dressers
12 58 33.13	Armoires
12 58 36	Nightstands
12 58 83	Custom Residential Furniture
12 59 00	**Systems Furniture**
12 59 13	Panel-Hung Component System Furniture
12 59 16	Free-Standing Component System Furniture
12 59 19	Beam System Furniture
12 59 23	Desk System Furniture
12 59 83	Custom Systems Furniture
12 60 00	**MULTIPLE SEATING**
12 61 00	**Fixed Audience Seating**
12 61 13	Upholstered Audience Seating
12 61 16	Molded-Plastic Audience Seating
12 62 00	**Portable Audience Seating**
12 62 13	Folding Chairs
12 62 16	Interlocking Chairs
12 62 19	Stacking Chairs
12 63 00	**Stadium and Arena Seating**
12 63 13	Stadium and Arena Bench Seating
12 63 23	Stadium and Arena Seats
12 64 00	**Booths and Tables**
12 65 00	**Multiple-Use Fixed Seating**
12 66 00	**Telescoping Stands**
12 66 13	Telescoping Bleachers
12 66 23	Telescoping Chair Platforms
12 67 00	**Pews and Benches**
12 67 13	Pews
12 67 23	Benches
12 68 00	**Seat and Table Assemblies**
12 68 13	Pedestal Tablet Arm Chairs
12 70 00	*Reserved*

12 - 5

CONSTRUCTION SPECIFICATION INSTITUTE MASTER FORMAT

GENERAL REQUIREMENTS

Div 01   General Requirements

FACILITY CONSTRUCTION

Div 02   Existing Conditions

Div 03   Concrete

Div 04   Masonry

Div 05   Metals

Div 06   Wood, Plastics, and Composites

Div 07   Thermal and Moisture Protection

Div 08   Openings

Div 09   Finishes

Div 10   Specialties

Div 11   Equipment

Div 12   Furnishings

Div 13   Special Construction

Div 14   Conveying Equipments

Div 15–19   [Reserved for future expansion]

FACILITY SERVICES

Div 20   [Reserved for future expansion]

Div 21   Fire Suppression

Div 22   Plumbing

Div 23   Heating, Ventilating and Air Conditioning

Div 24   [Reserved for future expansion]

Div 25   Integrated Automation

Div 26   Electrical

Div 27   Communications

Div 28   Electronic Safety and Security

Div 29   [Reserved for future expansion]

SITE AND INFRASTRUCTURE

Div 30   [Reserved for future expansion]

Div 31   Earthwork

Div 32   Exterior Improvements

Div 33   Utilities

Div 34   Transportation

Div 35   Waterway and Marine

Div 36–39   [Reserved for future expansion]

PROCESS EQUIPMENT

Div 40   Process Integration

Div 41   Material Processing and Handling Equipment

Div 42   Process Heating, Cooling, and Drying Equipment

Div 43   Process Gas and Liquid Handling, Purification and Storage Equipment

Div 44   Pollution Control Equipment

Div 45   Industry-Specific Manufacturing Equipment

Div 46–49   [Reserved for future expansion]

Courtesy of the Construction Specifications Institute.

APPENDIX 2

Websites

ACI
American Concrete Institute
http://www.concrete.org

ADDA
American Design Drafting Association
http://www.adda.org

AFMA
American Furniture Manufacturers Association
http://www.ahfa.us/

AHMA
American Hardware Manufacturers Association
http://www.ahma.org

AIA
The American Institute of Architects
http://www.aia.org

AIBD
American Institute of Building Design
http://www.aibd.org

AAHID
American Academy of Healthcare Interior Designers
http://aahid.org/

AISC
American Institute of Steel Construction, Inc.
http://www.aisc.org

ASFD
American Society of Furniture Designers
http://www.asfd.com

ASID
American Society of Interior Designers
http://www.interiors.org

ADA
Americans with Disabilities Act Information Office
http://www.ada.ufl.edu

APA
Architectural Precast Association
http://www.archprecast.org

AWI
Architectural Woodwork Institute
http://www.awinet.org

BIA
Brick Industry Association
http://www.bia.org

BIDA
British Interior Design Association
http://www.bida.org/

BSI

C2ED

CWC
Chartered Society of Designers
http://www.csd.org.uk/

CAH
Center for Accessible Housing School of Design
http://www.design.ncsu.edu/cud/

CIDEA
The Center for Inclusive Design and Environmental Access
http://www.ap.buffalo.edu

Construction Specifications Institute
http://www.csinet.org

Council for Interior Design Accreditation (formerly FIDER)
http://www.accredit-id.org

Dept. of Defense Ergonomics Group
http://www.ergoworkinggroup.org/

DIA
Design Institute of Australia
http://www.dia.org.au/

Engineered Wood Association
http://www.apawood.org

Ergoweb, Inc.
http://www.ergoweb.com

HFES
Human Factors & Ergonomics Society
http://www.hfes.org

IDAHK
Interior Design Association of Hong Kong
http://www.hkida.com/

ICIAD
International Council of Interior Architects and Designers
http://www.iciad.org/

IDEC
Interior Design Educators Council
http://www.idec.org

IDC
Interior Designers of Canada
http://www.interiordesigncanada.org

IFDA
International Furnishings and Design Association
http://www.ifda.com

IIDA
International Interior Design Association
http://www.iida.org

ISO
International Organization for Standardization
http://www.iso.org

ISP
Institute of Store Planners
http://www.ispo.org/

KCMA
Kitchen Cabinet Manufacturers Association
http://www.kcma.org

NAAMM
National Association of Architectural Metal Manufacturers
http://www.naamm.org

NAHB
National Association of Home Builders
http://www.nahb.org

NAHBRC
National Association of Home Builders Research Center
http://www.nahbrc.org

NARI
National Association of the Remodeling Industry
http://www.nari.org

NCIDQ
National Council for Interior Design Qualification
http://www.ncidq.org

NFPA
National Fire Protection Association International
http://www.nfpa.org

NHFA
National Home Furnishings Association
http://www.nhfa.org/

NIBS
National Institute of Building Sciences
http://www.nibs.org

NIST
National Institute of Standards and Technology
http://www.nist.gov

NKBA
National Kitchen and Bath Association
http://www.nkba.org

OBD
Organization of Black Designers
http://http://www.obd.org/

RAIC
Royal Architectural Institute of Canada
http://www.raic.org

Metric Conversion Chart

Follow this chart to convert from one measurement system to the other.

WHEN YOU KNOW	MULTIPLY BY	TO FIND
inches	25	millimeters
feet	30	centimeters
yards	0.9	meters
miles	1.6	kilometers
millimeters	0.04	inches
centimeters	0.393	inches
meters	1.1	yards
meters	3.3	feet
meters	1.1	yards
kilometers	0.6	miles
ounces	28	grams
pounds	0.45	kilograms
short tons	0.9	metric tons
grams	0.035	ounces
kilograms	2.2	pounds
metric tons	1.1	short tons
fluid ounces	30	milliliters
pints, US	0.47	liters
pints, Imperial	.568	liters
quarts, US	0.95	liters
quarts, Imperial	1.137	liters
gallons, US	3.8	liters
gallons, Imperial	4.546	liters
milliliters	0.034	fluid ounce
liters	2.1	pints, US
liters	1.76	pints, Imperial
liters	1.06	quarts, US
liters	0.88	quarts, Imperial
liters	0.26	gallons, US
liters	0.22	gallons, Imperial

1 meter (m) = 10 decimeters (dm) = 100 centimeters (cm) = 1,000 millimeters (mm). 1,000 meters = 1 kilometer (km).

Related Websites

http://www.worldwidemetric.com/
This website has a metric conversion calculator for length, weight, volume, pressure, and temperature.

http://www.hgtv.com/
Tile, sheetrock, wallpaper, and other calculators can be found on the HGTV website.

Conversion of Fractions to Decimals and Percentages

$1/100 = .01 = 1\%$

$1/10 = .1 = 10\%$

$1/5 = 2/10 = .2 = .20 = 20\%$

$3/10 = .3 = .30 = 30\%$

$2/5 = 4/10 = .4 = .40 = 40\%$

$1/2 = 5/10 = .5 = .50 = 50\%$

$3/5 = 6/10 = .6 = .60 = 60\%$

$7/10 = .7 = .70 = 70\%$

$4/5 = 8/10 = .8 = .80 = 80\%$

$9/10 = .9 = .90 = 90\%$

$1/4 = 25/100 = .25 = 25\%$

$4/3 = 75/100 = .75 = 75\%$

$1/3 = .33\frac{1}{3} = 33\frac{1}{3}\%$

$2/3 = .66\frac{2}{3} = 66\frac{2}{3}\%$

$1/8 = .125 = .12\frac{1}{2} = 12\frac{1}{2}\%$

$3/8 = .375 = .37\frac{1}{2} = 37\frac{1}{2}\%$

$5/8 = .525 = .62\frac{1}{2} = 62\frac{1}{2}\%$

$7/8 = .875 = .87\frac{1}{2} - 87\frac{1}{2}\%$

$5/6 = .83\frac{1}{3} = 83\frac{1}{3}\%$

$1 = 1.00 = 100\%$

$2 = 2.00 = 200\%$

$3\frac{1}{2} = 3.5 = 3.50 = 350\%$

APPENDIX 4

Appendix 4 contains to-scale floor plans for general classroom use.

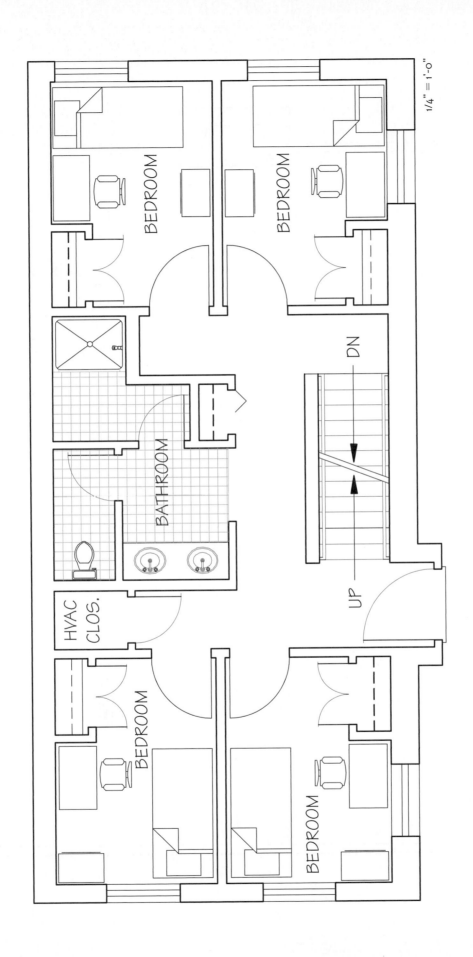

1/4" = 1'-0"

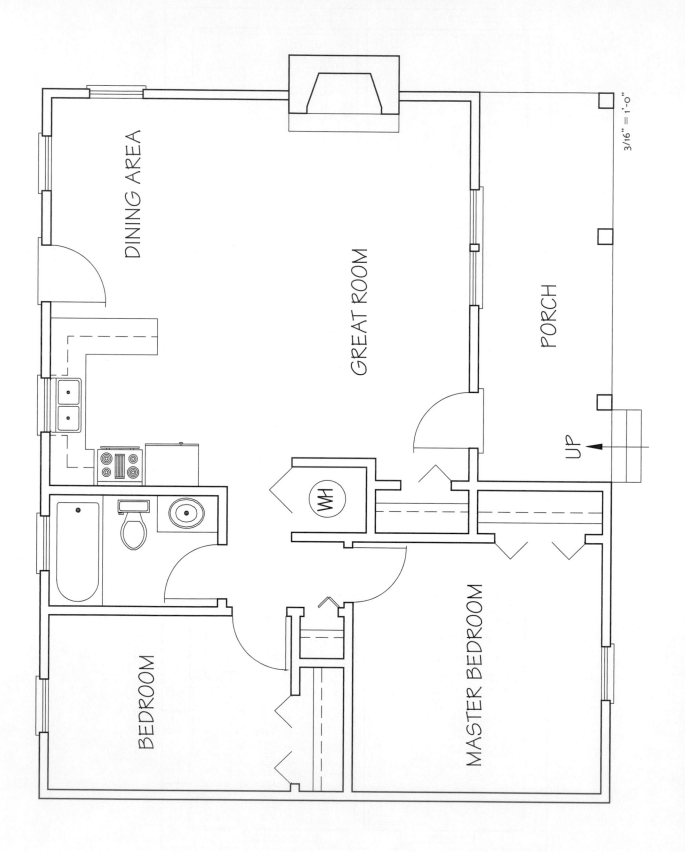

3/16" = 1'-0"

DINING AREA

GREAT ROOM

PORCH

WH

UP

BEDROOM

MASTER BEDROOM

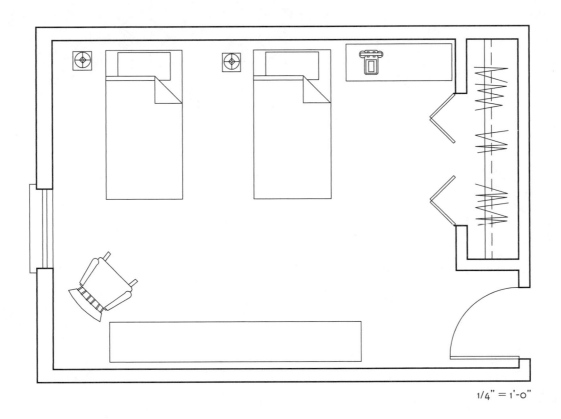

1/4" = 1'-0"

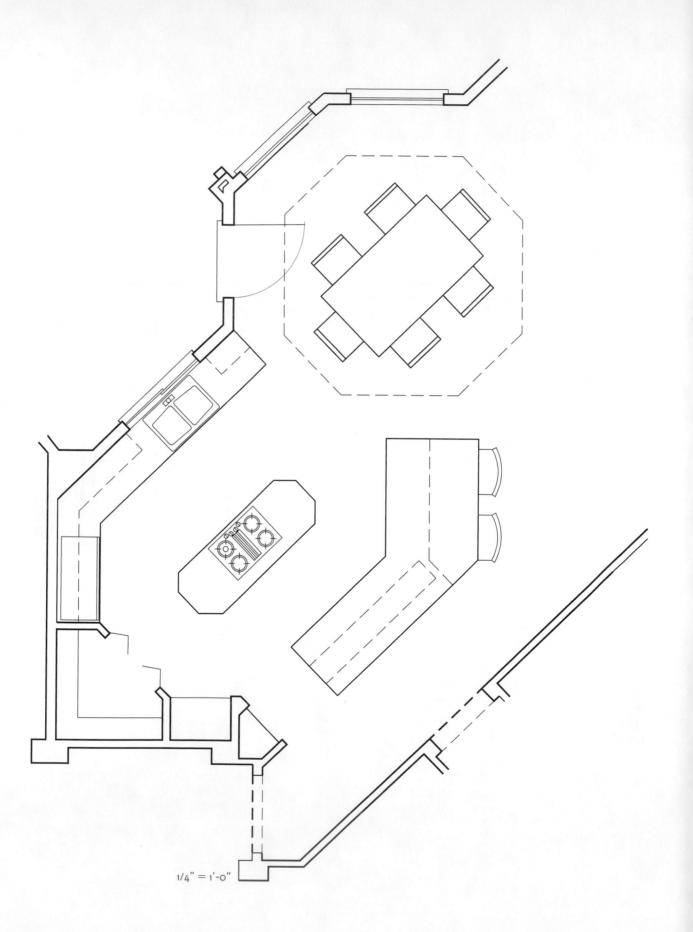

1/4" = 1'-0"

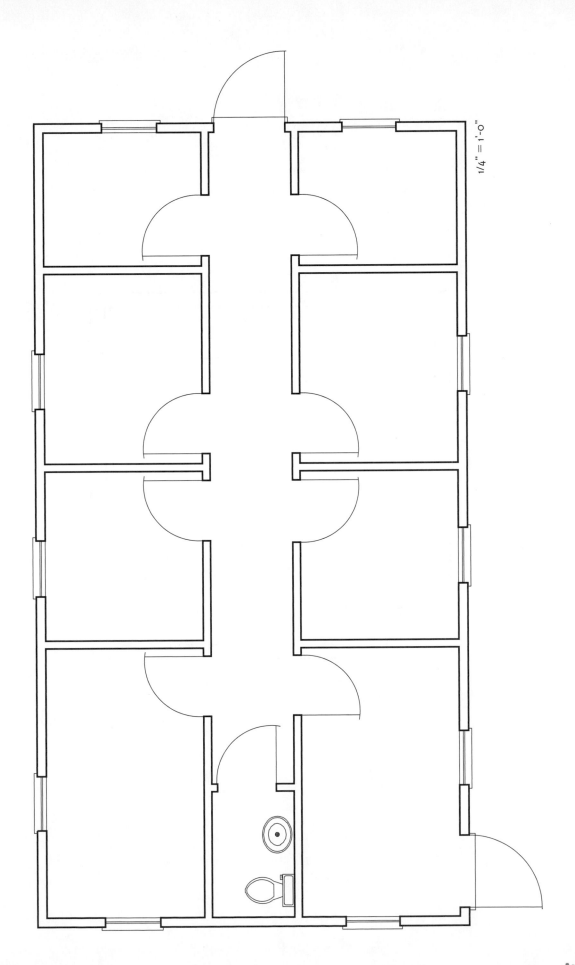

1/4" = 1'-0"

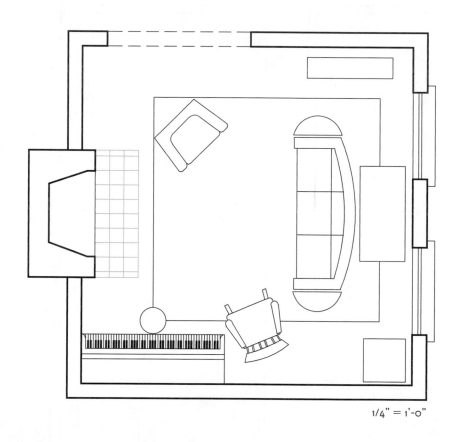

1/4" = 1'-o"

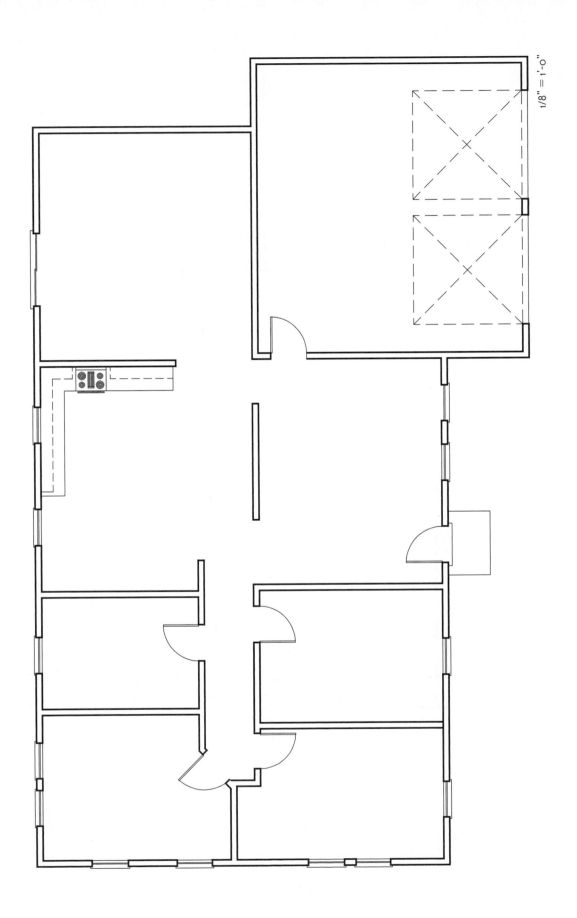

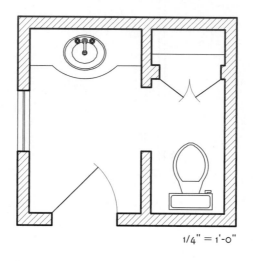

1/4" = 1'-0"

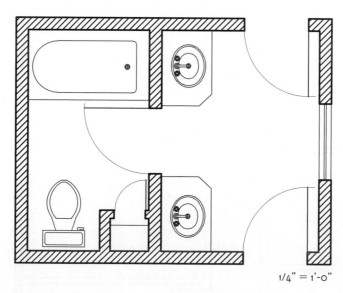

1/4" = 1'-0"

INDEX

WORKSHEETS

Measure the lines in meters with the metric scales given.
Center and letter each answer 1/8" above the line.

1.

├─────────────────────────────────────┤

1:1 m

2.

├───┤

1:20 m

3.

├───────────────────────────────────────┤

1:50 m

4.

├──────────────────────────────┤

1:200 m

5.

├────────────────────────────────────┤

1:100 m

6.

├─────────────────────┤

1:30 m

7.

├─────────────────────┤

1:400 m

| NAME: | SECTION: | PROJECT: | DATE: |

Redraw this 60" × 80" queen bed twice its size. Mark its dimensions with dividers, "walk" the dividers once to double the dimensions, and then use straightedges to draw.

1/4" = 1'-0"

Measure each angle.

1.

2.

3.

4.

5.

6.

| NAME: | SECTION: | PROJECT: | DATE: |

Draw four symbols for this plan: One two-sided section symbol, two elevation symbols, and an ID label. Place them appropriately.

Trace this plan in ink, using line weights. Walls are thickest, doors,
fixtures, and furniture are thinner (and the same weight),
door arcs, tiles, and windows are the thinnest.

1/4" = 1'-0"

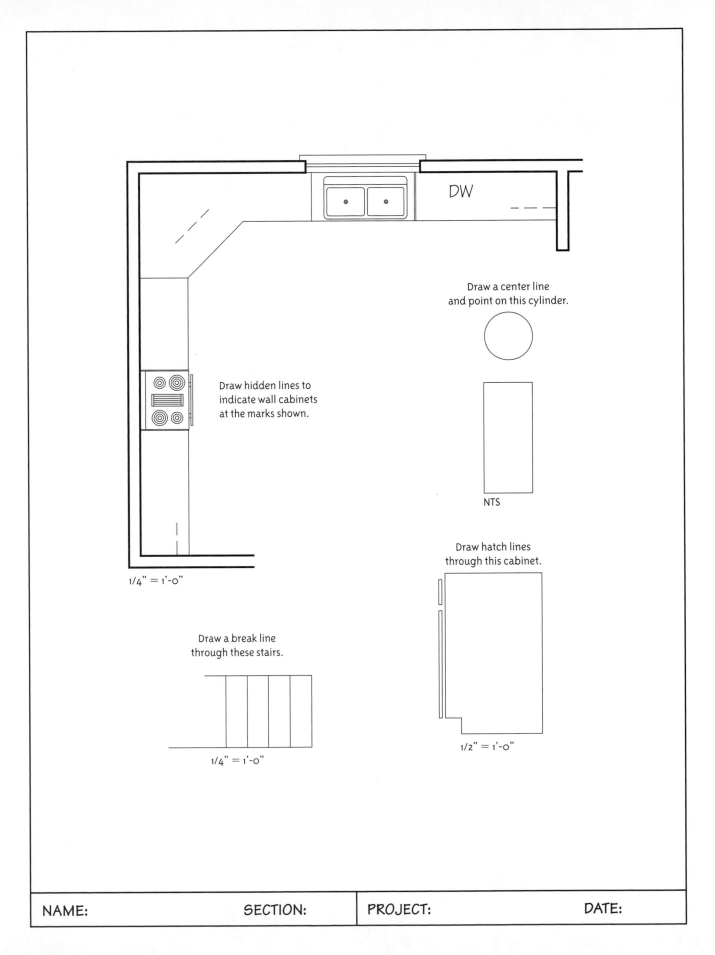

DW

Draw a center line
and point on this cylinder.

NTS

Draw hidden lines to
indicate wall cabinets
at the marks shown.

1/4" = 1'-0"

Draw hatch lines
through this cabinet.

1/2" = 1'-0"

Draw a break line
through these stairs.

1/4" = 1'-0"

NAME: SECTION: PROJECT: DATE:

WS-11

Complete the front view and draw the right-side view of this island cooktop. Include three shelves in the center section of the right side view. 3/4" = 1'-0".

NAME: SECTION: PROJECT: DATE:

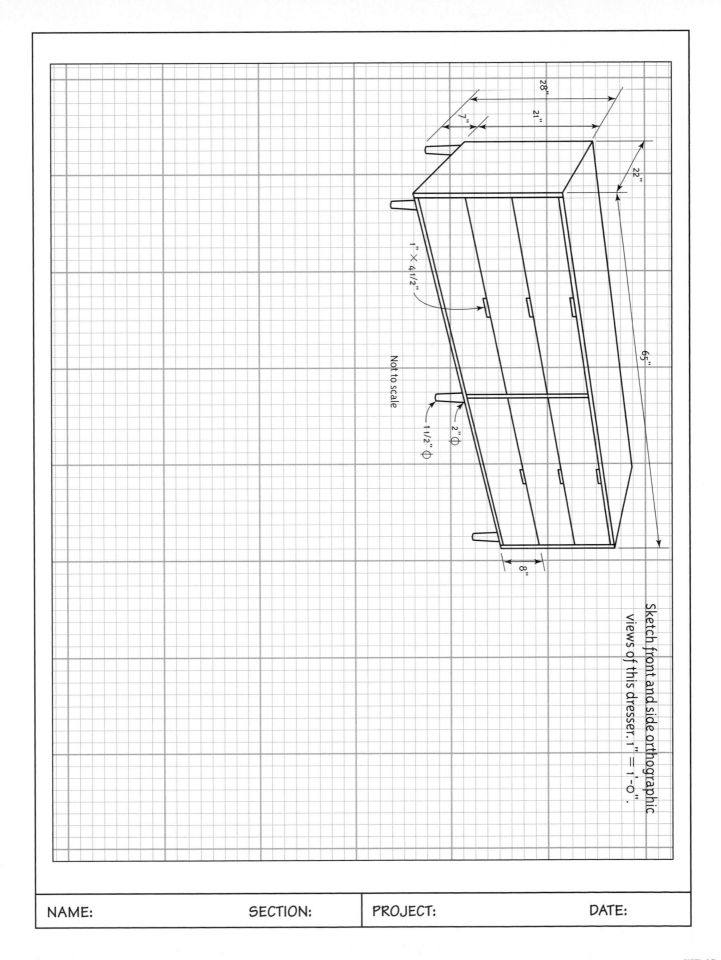

28"

7"

21"

22"

1" × 4 1/2"

65"

Not to scale

1 1/2" ⌀

2" ⌀

8"

Sketch front and side orthographic views of this dresser. 1" = 1'-0".

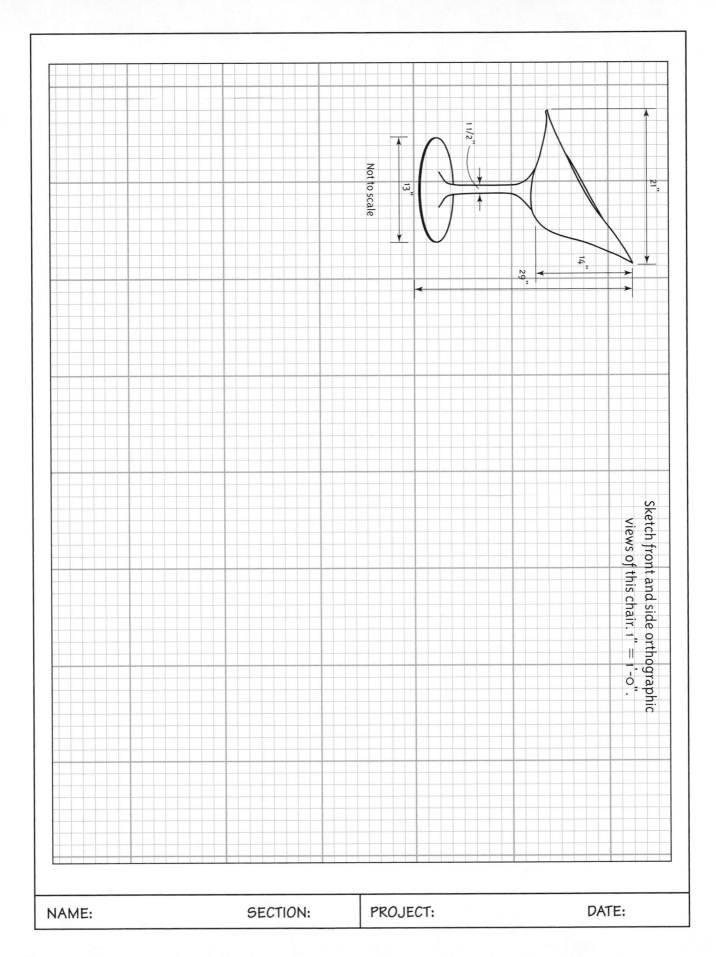

Not to scale

13"

1 1/2"

21"

14"

29"

Sketch front and side orthographic
views of this chair: 1" = 1'-0".

NAME: SECTION: PROJECT: DATE:

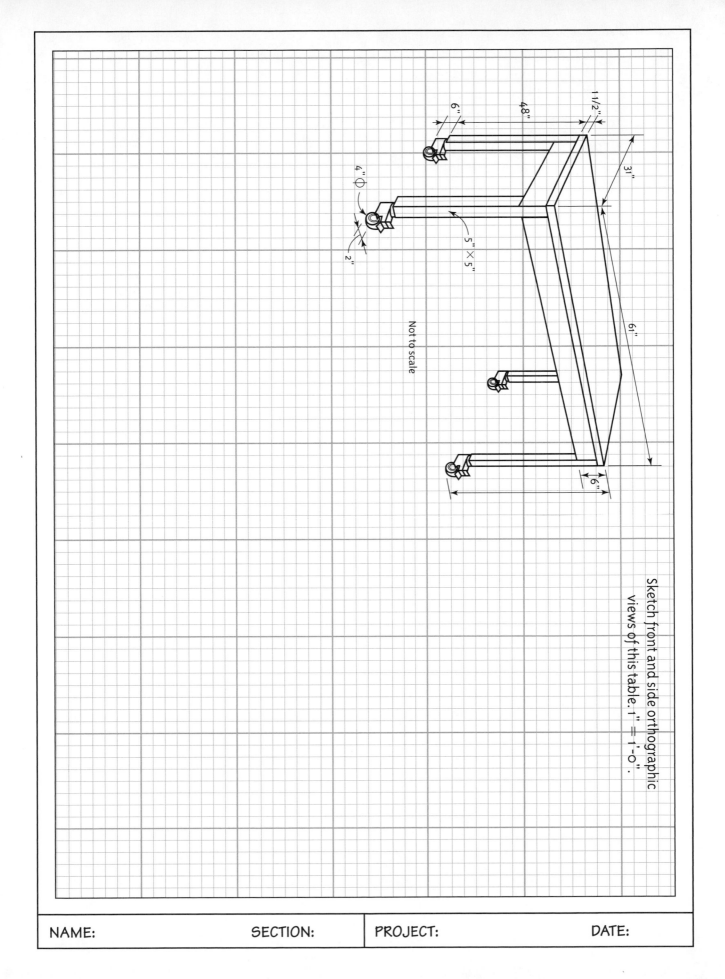

6"

48"

11/2"

31"

4" ⌀

5" X 5"

2"

61"

Not to scale

6"

Sketch front and side orthographic views of this table. 1" = 1'-0".

WS-15

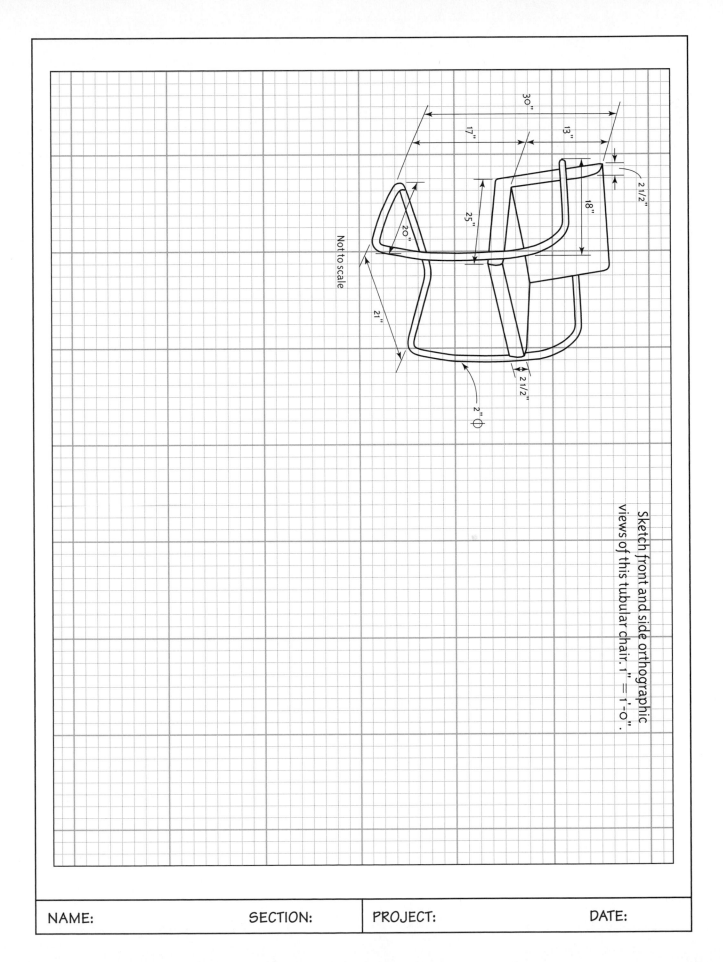

30"

17"

13"

2 1/2"

18"

25"

20"

Not to scale

21"

2 1/2"

2" ⌀

Sketch front and side orthographic views of this tubular chair. 1" = 1'-0".

NAME: SECTION: PROJECT: DATE:

Below is the top view of a 6'-0" tall, 5-shelf wood
bookshelf. Draw a front view. 1/2" = 1'-0".

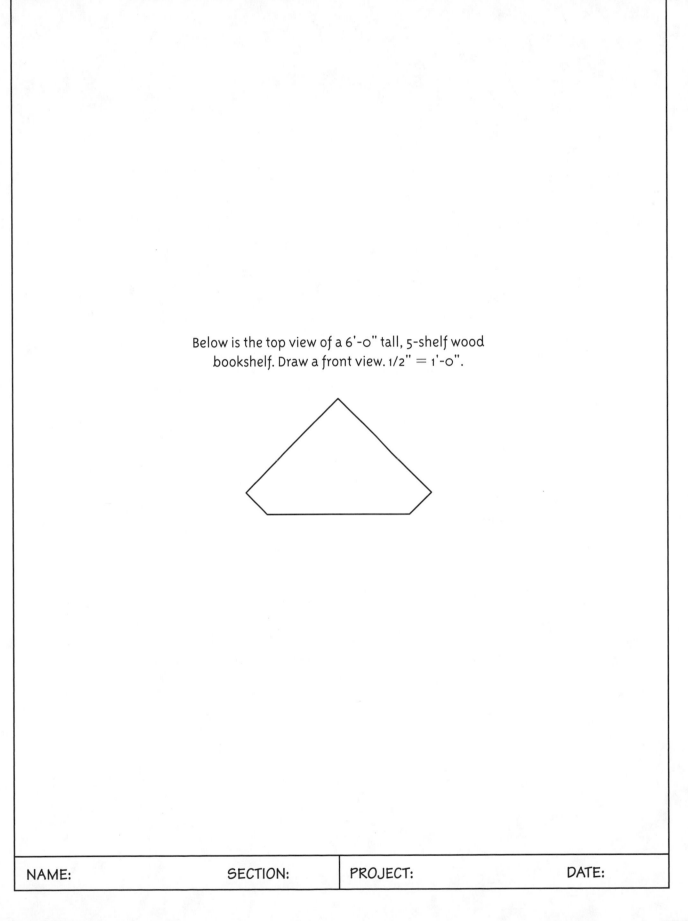

NAME: SECTION: PROJECT: DATE:

Draw top and front views of this cabinet. Obtain
measurements from the isometric drawing. 1/2" = 1'-0".

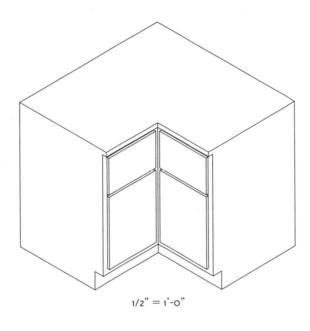

1/2" = 1'-0"

NAME:	SECTION:	PROJECT:	DATE:

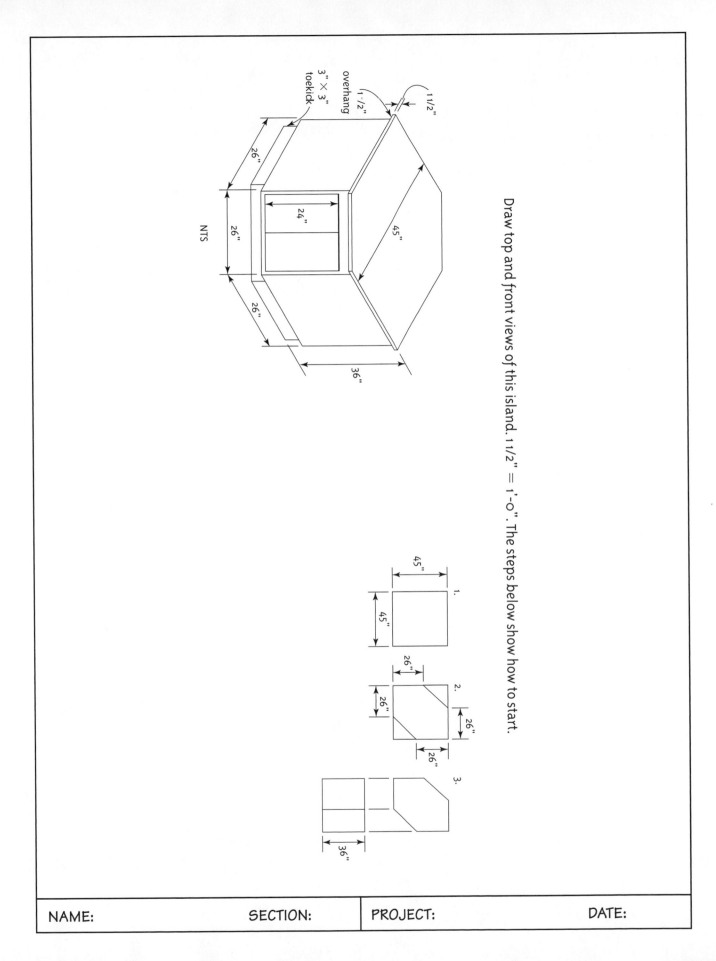

Draw top and front views of this island. 1 1/2" = 1'-0". The steps below show how to start.

NTS

overhang 1 -1/2"

1 1/2"

3" × 3" toekick

26"

26"

24"

45"

26"

26"

36"

1.

45"

45"

2.

26"

26"

26"

26"

3.

36"

NAME: SECTION: PROJECT: DATE:

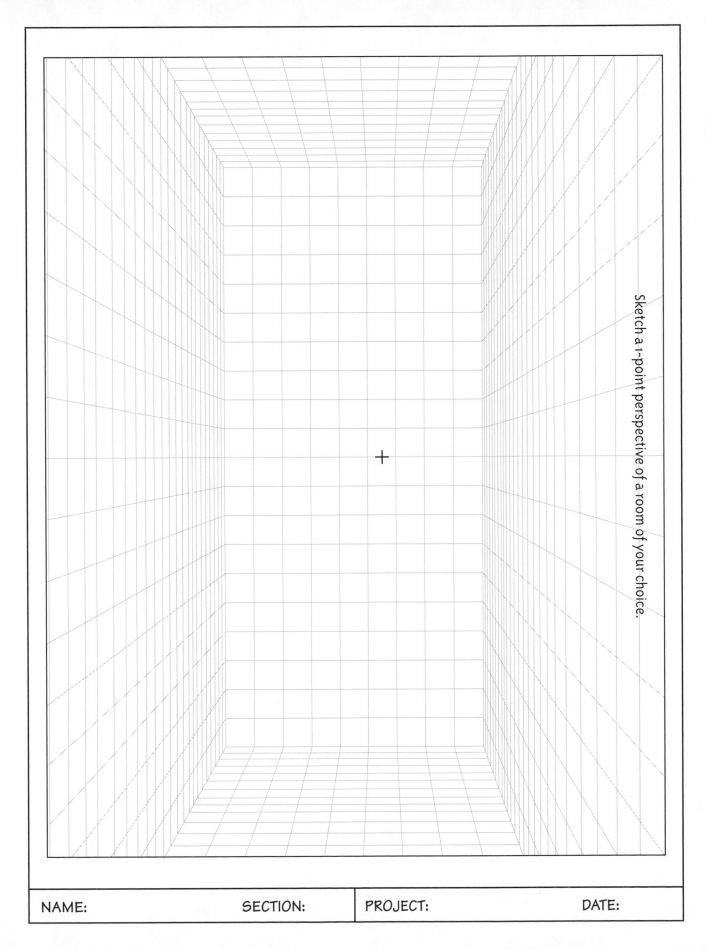

Sketch a 1-point perspective of a room of your choice.

NAME: SECTION: PROJECT: DATE:

Sketch an isometric drawing of an object of your choice.

NAME: SECTION: PROJECT: DATE:

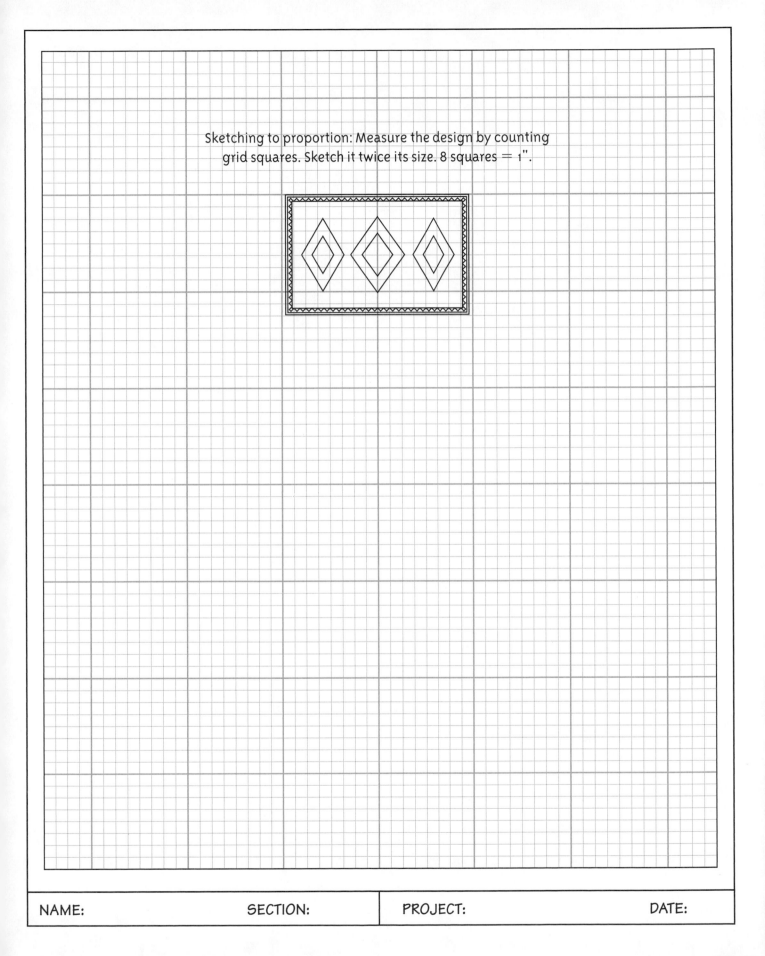

Sketching to proportion: Measure the design by counting grid squares. Sketch it twice its size. 8 squares = 1".

NAME: SECTION: PROJECT: DATE:

WS-22

Draw a section of this foundation wall at 3/4" = 1'-0".

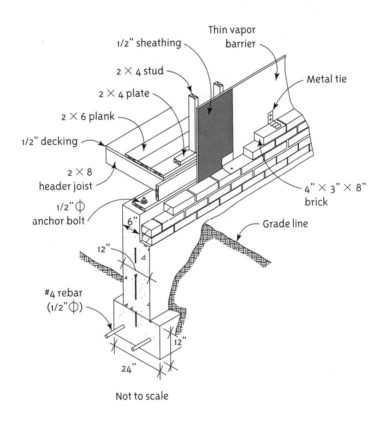

Thin vapor barrier

1/2" sheathing

2 × 4 stud

2 × 4 plate

2 × 6 plank

1/2" decking

2 × 8 header joist

1/2" ⌀ anchor bolt

#4 rebar (1/2"⌀)

Metal tie

4" × 3" × 8" brick

Grade line

6"

12"

12"

24"

Not to scale

NAME: SECTION: PROJECT: DATE:

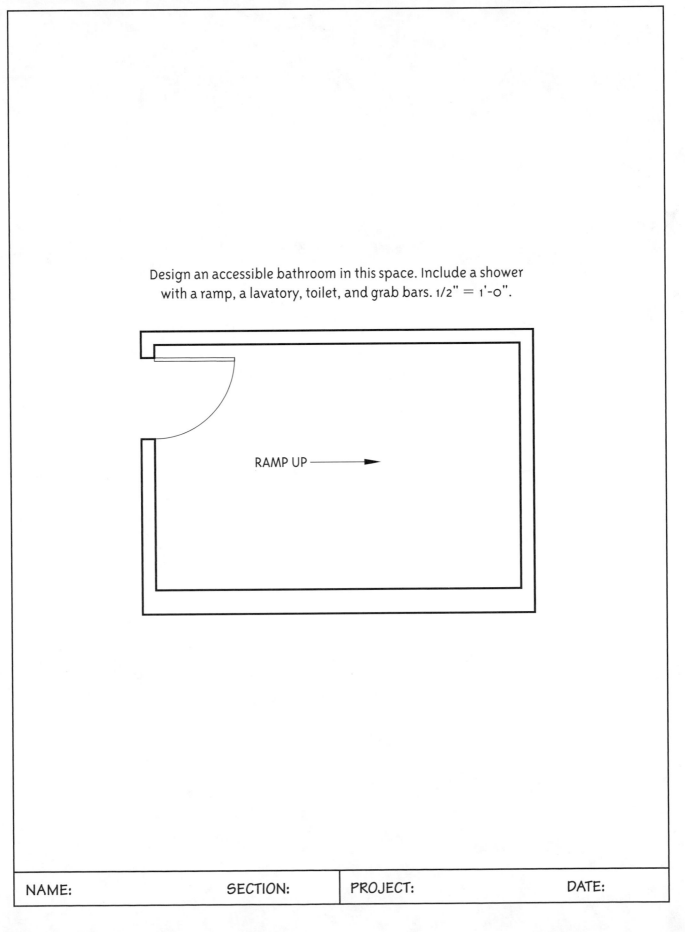

Design an accessible bathroom in this space. Include a shower with a ramp, a lavatory, toilet, and grab bars. 1/2" = 1'-0".

RAMP UP ———→

| NAME: | SECTION: | PROJECT: | DATE: |

WS-24

Place furniture and fixtures in this ensuite. 1/4" = 1'-0".

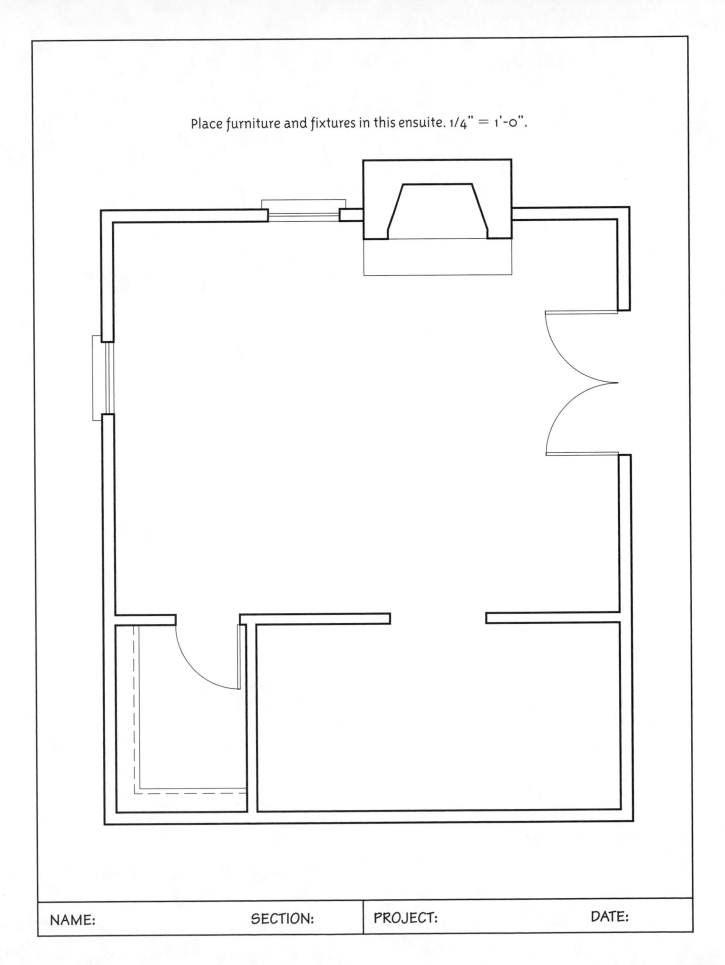

Design a layout for this second floor. Include three bedrooms and one bathroom. 1/4" = 1'-0".

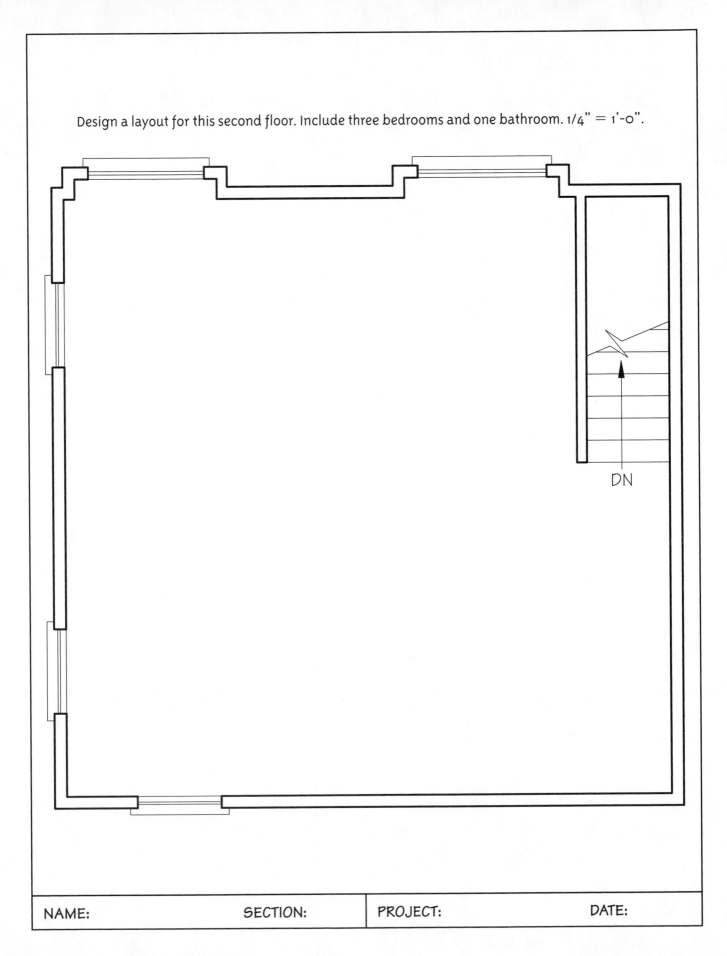

DN

NAME: SECTION: PROJECT: DATE:

Place furniture in this bedroom. Include a queen-size bed, two nightstands, a dresser, seating, and a screen to hide storage items. 1/4" = 1'-0".

Draw the wall thickness and poché indicated. 1/2" = 1'-0".

6" wood frame plus 1/2" gypsum
board both sides

4" brick veneer on 6"
wood frame with 2" as space

4" brick on 8" concrete
block

2" plaster on 6"
wood frame

6" wood frame

6" wood frame
(alternative)

| NAME: | SECTION: | PROJECT: | DATE: |

Draw fixtures in each bathroom type below. Provide proper clearances. 1/4" = 1'-0".

Full bath

3/4 bath

Powder room

Design a layout for this studio apartment. Include a U-shape kitchen, one full bathroom, eating, sleeping, living, and closet areas. Sleeping area contains a single bed, desk, and nightstand, living area contains seating and tables. Include a closet for a water heater and furnace. 1/4" = 1'-0".

Dimension this wood frame plan to architectural construction standards. 1/4" = 1'-0".

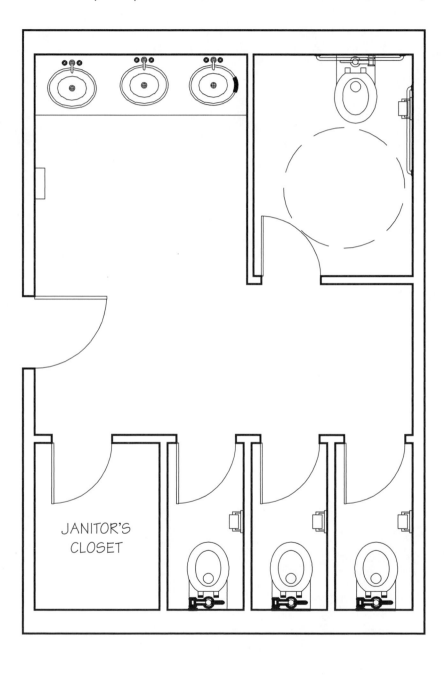

JANITOR'S
CLOSET

NAME:　　　　　　SECTION:　　　PROJECT:　　　　　　　DATE:

Dimension this wood frame plan to architectural construction standards. 1/4" = 1'-0".

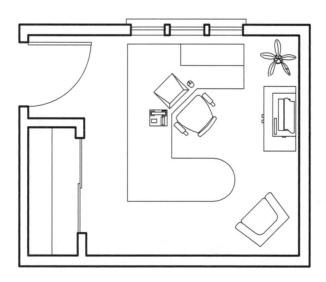

NAME: SECTION: PROJECT: DATE:

Dimension this wood frame plan to architectural
construction standards. 1/4" = 1'-0".

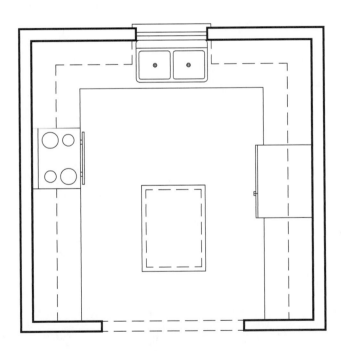

Dimension these partial plans to architectural construction standards. 1/4" = 1'-0".

Wood frame

Masonry

Dimension this floor plan to cabinet industry standards. 1/2" = 1'-0".

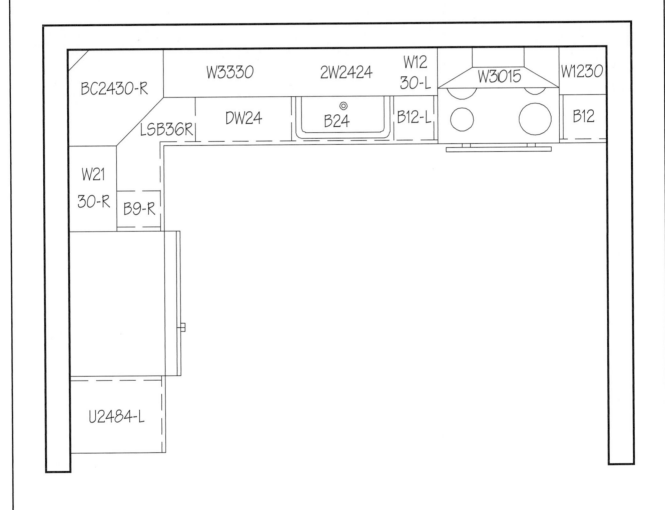

Dimension this plan to cabinet industry standards. 1/2" = 1'-0".

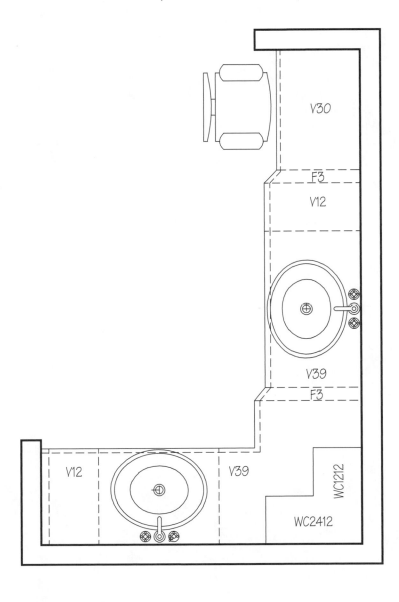

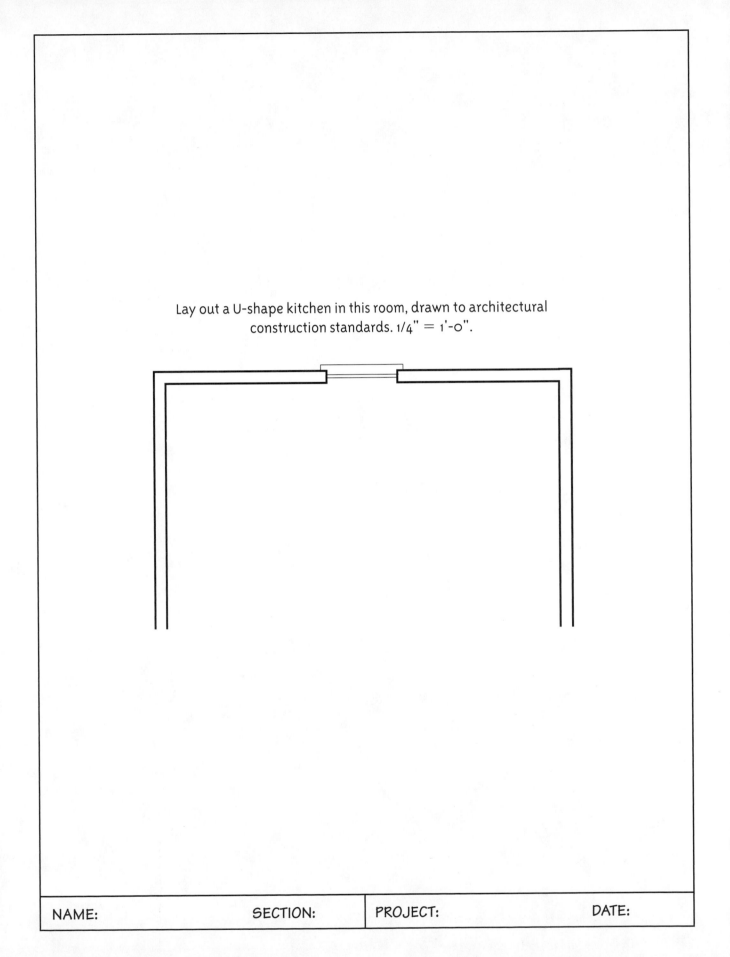

Lay out a U-shape kitchen in this room, drawn to architectural construction standards. 1/4" = 1'-0".

Draw the floor plan symbol for each door type. The corridor side is denoted by "C". 1/4" = 1'-0".

1.

LH exterior swing

2.

C

RH Dutch

3.

Pocket

4.

Double-action

5.

C

6.

Double bi-fold

7.

Accordion

8.

Sliding

9.

C

LHR and RHR French

10.

Exterior glass sliding

NAME:	SECTION:	PROJECT:	DATE:

Draw the floor plan symbol for each door type. 1/4" = 1'-0".

1.

Exterior

LHR swing   Revolving   RHR swing

2.

7'-0" tall garage door

3.

3-part glass sliding door

| NAME: | SECTION: | PROJECT: | DATE: |

Identify the parts.

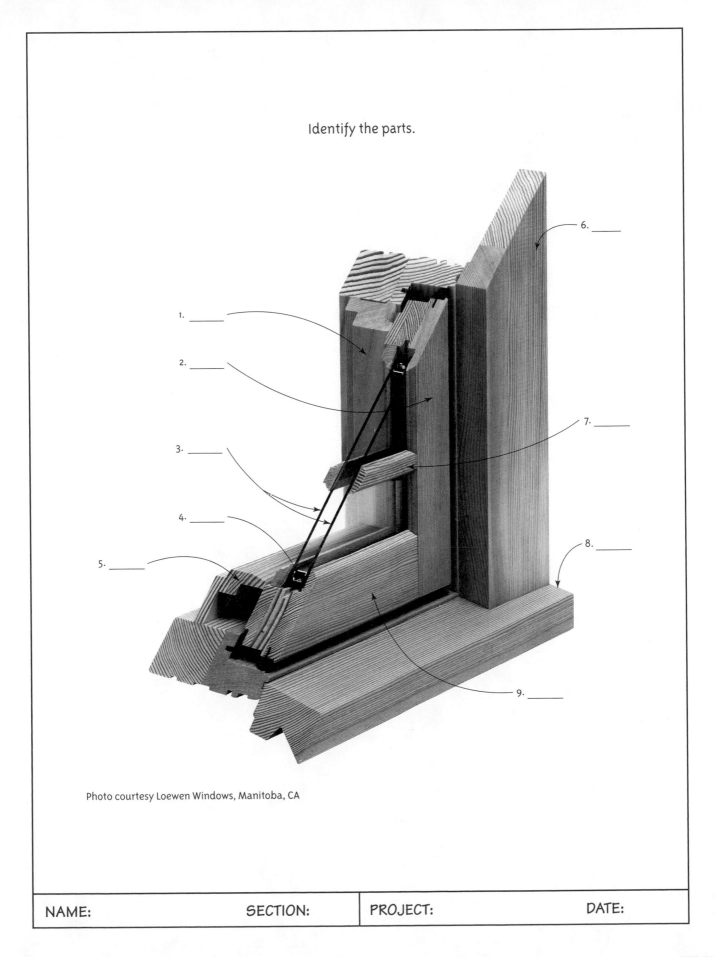

1. _____
2. _____
3. _____
4. _____
5. _____
6. _____
7. _____
8. _____
9. _____

Photo courtesy Loewen Windows, Manitoba, CA

| NAME: | SECTION: | PROJECT: | DATE: |

Draw the floor plan symbol for each window below. 1/4" = 1'-0".

1. Double-hung

2. (2) LH and RH casements

3. Glider

4. Awning

5. Fixed

6. Hopper

7. 45°, 2'-0" deep bay

8. 4-sash bow

| NAME: | SECTION: | PROJECT: | DATE: |

Draw the door elevations. All are 2'-6" × 6'-8". 1/2" = 1'-0".

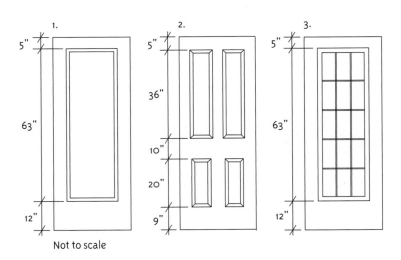

1.

5"

63"

12"

Not to scale

2.

5"

36"

10"

20"

9"

3.

5"

63"

12"

1.

2.

3.

L   L   L

Draw an elevation for each window type. Height and width includes a 2" casing. 1/4" = 1'-0".

1.
5'-0" tall double-hung

4.
4'-6" tall jalousie

2.
4'-0" tall casement, 2-sash

5.
3'-6" tall fixed

3.
4'-6" tall slider

6.
3'-6" tall hopper

NAME:	SECTION:	PROJECT:	DATE:

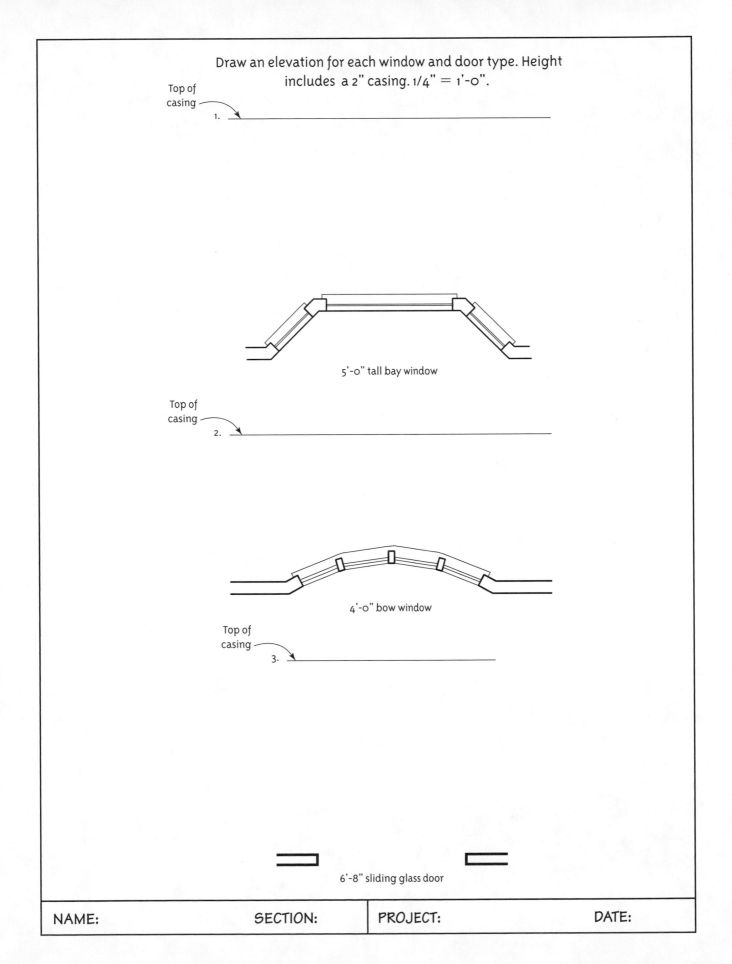

Draw an elevation for each window and door type. Height includes a 2" casing. 1/4" = 1'-0".

Top of casing

1.

5'-0" tall bay window

Top of casing

2.

4'-0" bow window

Top of casing

3.

6'-8" sliding glass door

| NAME: | SECTION: | PROJECT: | DATE: |

Complete the finish schedule for this floor plan.

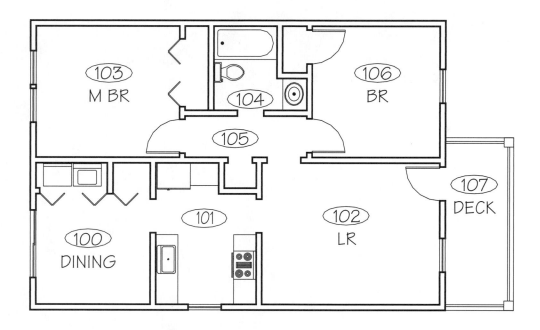

FINISH SCHEDULE

NO.	ROOM	FLOOR	TRIM	COUNTERS	WALLS	CEILING

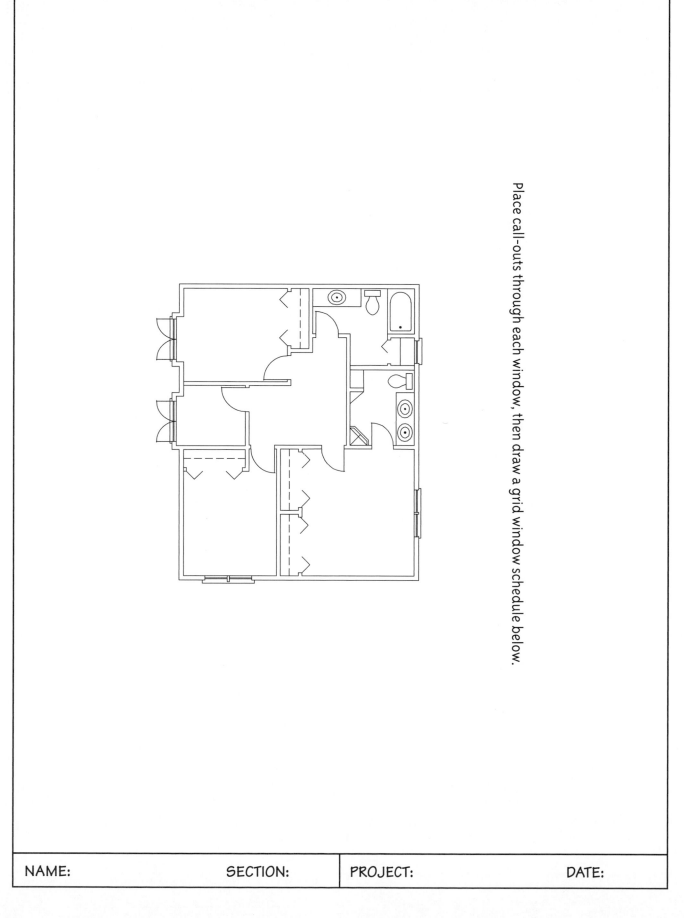

Place call-outs through each window, then draw a grid window schedule below.

NAME:　　　　　　　SECTION:　　　　PROJECT:　　　　　　　　DATE:

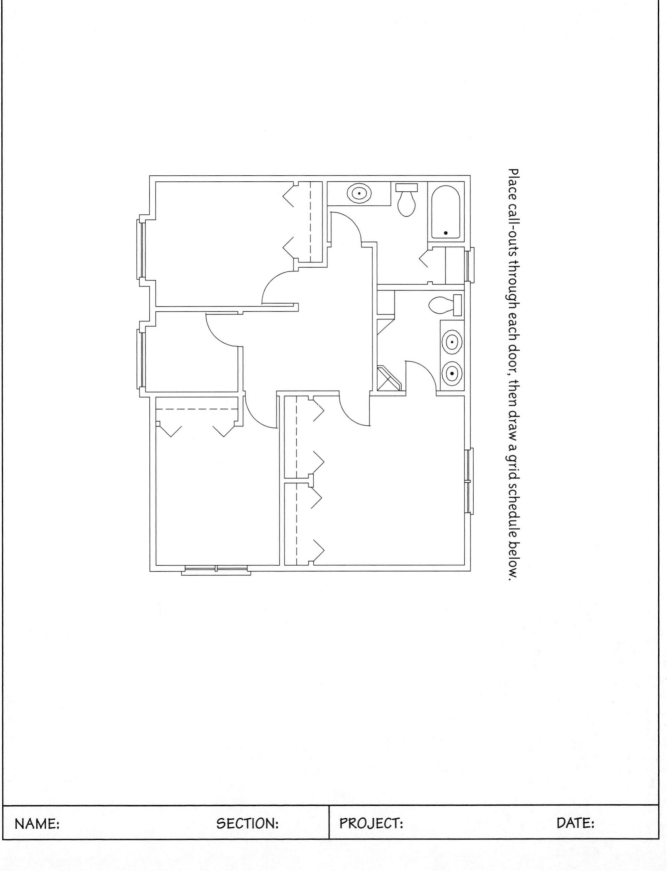

Place call-outs through each door, then draw a grid schedule below.

| NAME: | SECTION: | PROJECT: | DATE: |

Fill in the blanks with the component names.

1. _____

2. _____

3. _____

4. _____

5. _____

6. _____

7. _____

8. _____

9. _____

10. _____

| NAME: | SECTION: | PROJECT: | DATE: |

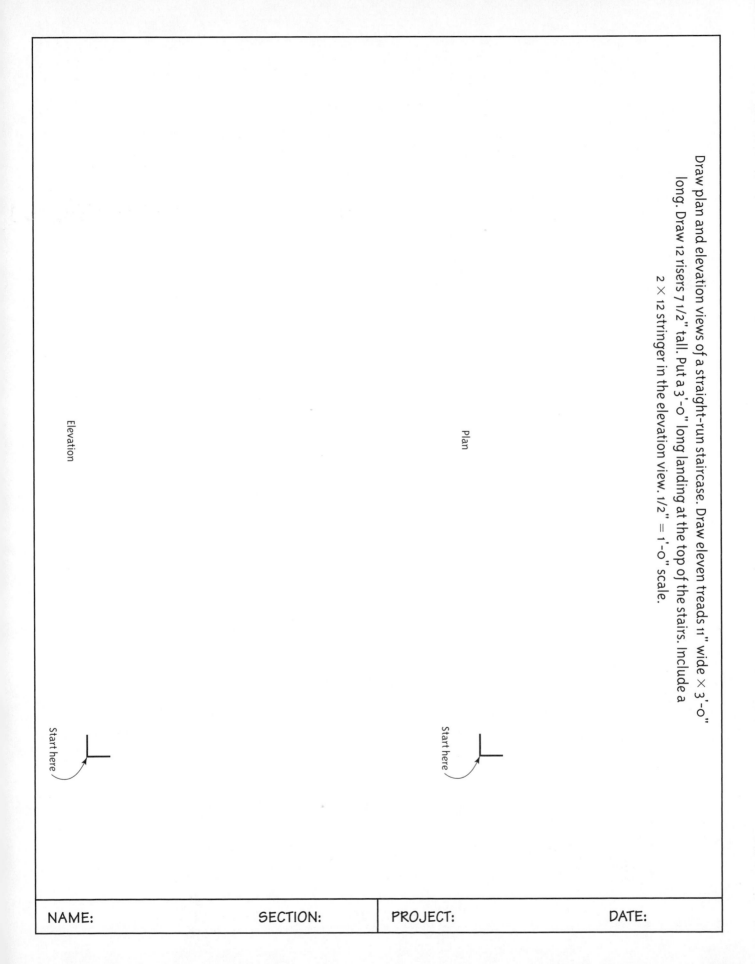

Draw plan and elevation views of a straight-run staircase. Draw eleven treads 11" wide × 3'-0" long. Draw 12 risers 7 1/2" tall. Put a 3'-0" long landing at the top of the stairs. Include a 2 × 12 stringer in the elevation view. 1/2" = 1'-0" scale.

Plan

Start here

Elevation

Start here

NAME: SECTION: PROJECT: DATE:

Center a 7'-6" × 7'-6" sunken area in this living room. Draw two 10'-6" wide steps on all sides leading into it. Draw a 10 1/2" wide step at the room entrance. 1/4" = 1'-0".

Draw plan and elevation views of a single-face fireplace. Draw its face flush with the wall. Opening dimensions are 36" wide, 24" high, 18" deep. Total height is 3'-0". Include a 12" wide surround and a mantel shelf.

Fireplace opening

Plan

Elevation

NAME: SECTION: PROJECT: DATE:

Draw an elevation of the wall indicated. Ceiling height is 9'-0". 1/4" = 1'-0".

Draw an interior elevation of the wall indicated. Ceiling height is 9'-0". Window bottom is 18" above the floor and is 5'-6" tall. 1/4" = 1'-0".

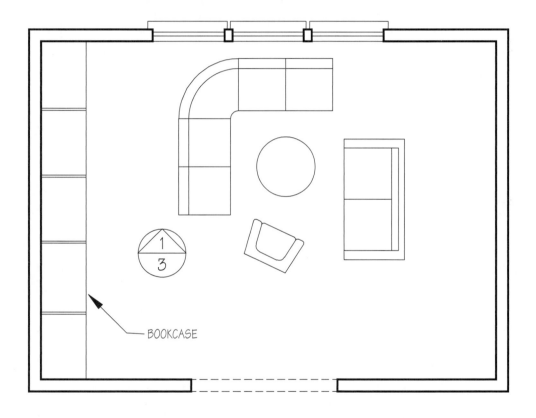

BOOKCASE

Draw the poché indicated.

CONC. BLOCK SECTION	FROSTED GLASS SECTION	FINISH BOARD SECTION	DIM. LUMBER SECTION
BRICK SECTION	GLASS BLOCK SECTION	BATT INSULATION SECTION	EARTH SECTION
CAST CONC. SECTION	CARPET ON PAD SECTION	BLOCK SECTION	PLASTER SECTION
RUBBLE STONE ELEVATION	FINISH BOARD ELEVATION	WOOD SIDING SECTION	BRICK ELEVATION

2 X 4 STUD WALL WITH 1/2" GYP. BOARD BOTH SIDES. 3/4" = 1'-0" SECTION

WALL TO BE REMOVED SECTION

NAME:　　　　　　　SECTION:　　　　　PROJECT:　　　　　　　　DATE:

Draw an interior elevation where indicated.
Ceiling height is 9'-0". 1/4" = 1'-0".

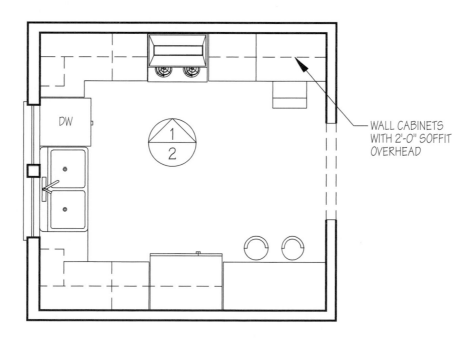

WALL CABINETS
WITH 2'-0" SOFFIT
OVERHEAD

DW

NAME:	SECTION:	PROJECT:	DATE:

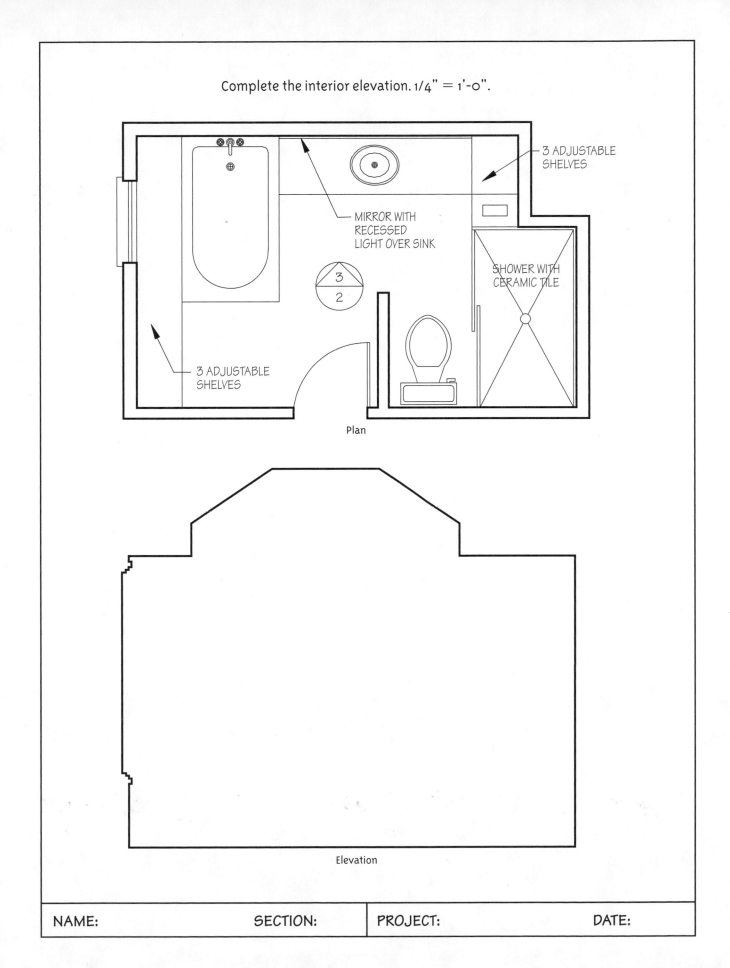

Complete the interior elevation. 1/4" = 1'-0".

3 ADJUSTABLE
SHELVES

MIRROR WITH
RECESSED
LIGHT OVER SINK

3
2

SHOWER WITH
CERAMIC TILE

3 ADJUSTABLE
SHELVES

Plan

Elevation

NAME:

SECTION:

PROJECT:

DATE:

Draw the plan symbols.

CLG MOUNTED LIGHT	RECESSED CLG MOUNTED LIGHT	CLG FAN	VENTILATION FAN
DUPLEX OUTLET	SERVICE PANEL	SPLIT WIRE OUTLET	WEATHERPROOF OUTLET
220V OUTLET	SINGLE-POLE SWITCH	THREE-WAY SWITCH	PUSH BUTTON SWITCH
OUTLET WITH SWITCH	3 SINGLE-POLE SWITCHES	PHONE JACK	FLUORESCENT LIGHT

| NAME: | SECTION: | PROJECT: | DATE: |

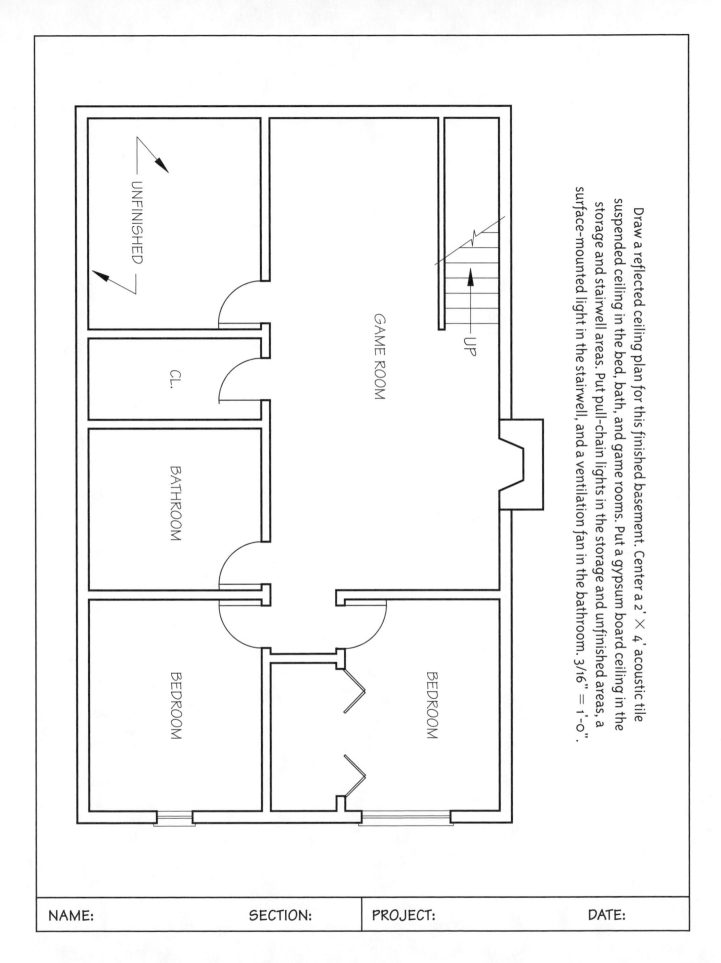

Draw a reflected ceiling plan for this finished basement. Center a 2' × 4' acoustic tile suspended ceiling in the bed, bath, and game rooms. Put a gypsum board ceiling in the storage and stairwell areas. Put pull-chain lights in the storage and unfinished areas, a surface-mounted light in the stairwell, and a ventilation fan in the bathroom. 3/16" = 1'-0".

UNFINISHED

CL.

BATHROOM

GAME ROOM

UP

BEDROOM

BEDROOM

NAME: SECTION: PROJECT: DATE:

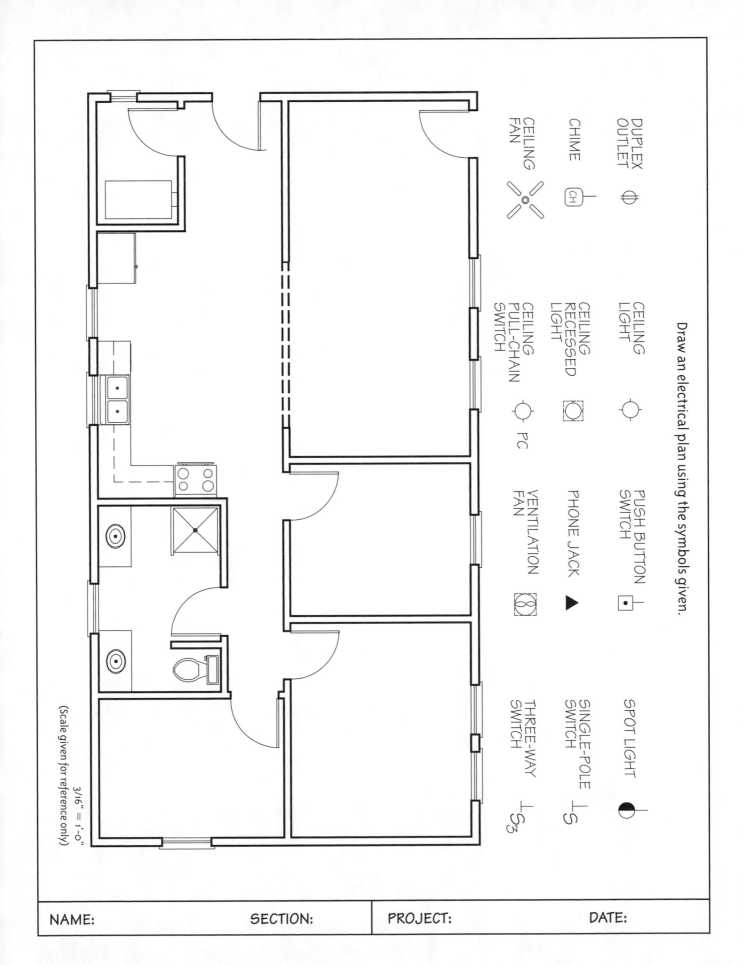

Draw an electrical plan using the symbols given.

DUPLEX OUTLET ⊕

CHIME [CH]

CEILING FAN

CEILING LIGHT

CEILING RECESSED LIGHT

CEILING PULL-CHAIN SWITCH

PC

PUSH BUTTON SWITCH

PHONE JACK ▶

VENTILATION FAN

SPOT LIGHT

SINGLE-POLE SWITCH ⊥S

THREE-WAY SWITCH ⊥S₃

3/16" = 1'-0"
(Scale given for reference only)

NAME: SECTION: PROJECT: DATE:

WS-61

Overlay vellum on this plan and freehand sketch a reflected ceiling plan. Include lights, ventilation/ceiling fans, and air registers. Optional: Sketch extension and dimension lines emanating from the center of each fixture, in both directions, to the interior walls.

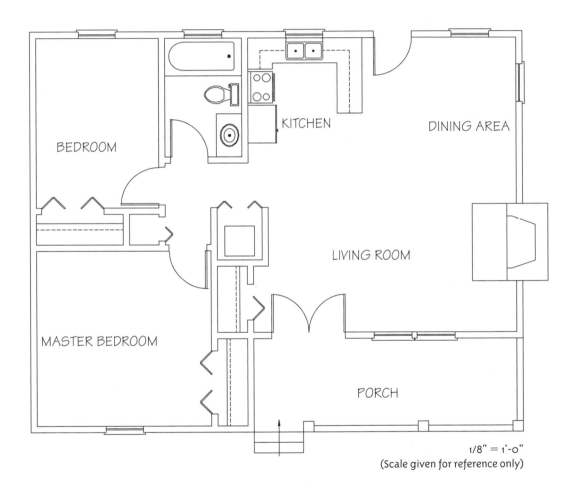

BEDROOM

KITCHEN

DINING AREA

LIVING ROOM

MASTER BEDROOM

PORCH

1/8" = 1'-0"
(Scale given for reference only)

Draw an electrical plan for this garage using the symbols given. 1/4" = 1'-0".

CEILING LIGHT		PUSH BUTTON SWITCH		WALL LIGHT	
FLUORESCENT LIGHT		GFI OUTLET			
THREE-WAY SWITCH	S_3	GARAGE DOOR OPENER	GO		

UP

GFI

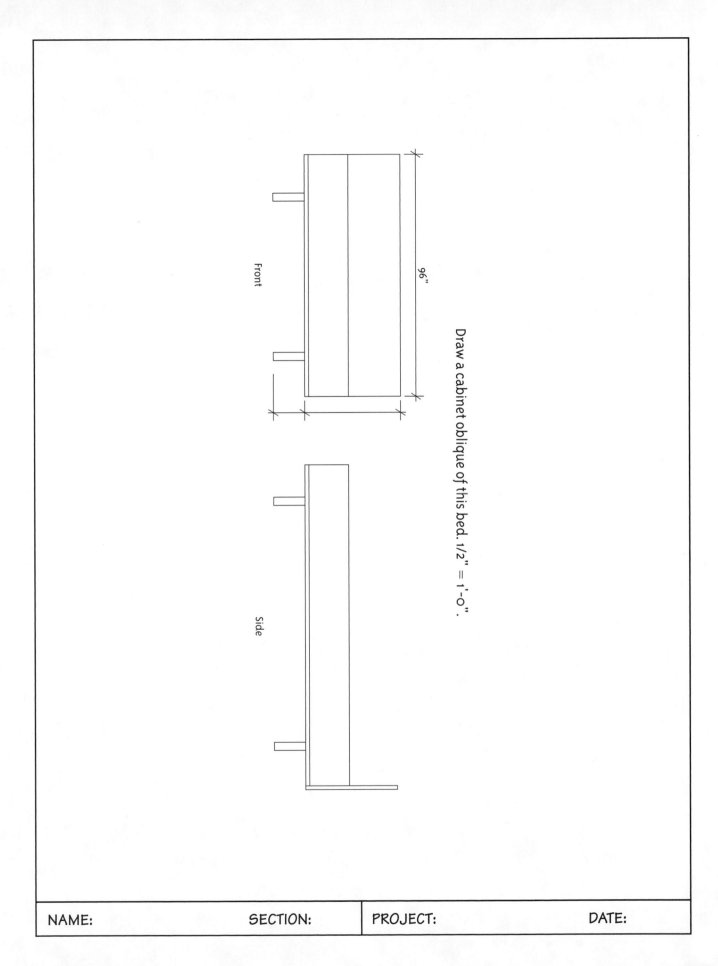

Front

96"

Draw a cabinet oblique of this bed. 1/2" = 1'-0".

Side

NAME: SECTION: PROJECT: DATE:

Draw a cutaway isometric of this studio apartment. 1/4" = 1'-0".

NAME: SECTION: PROJECT: DATE:

Draw a 1-point perspective of this conference room. 1/4" = 1'-0".

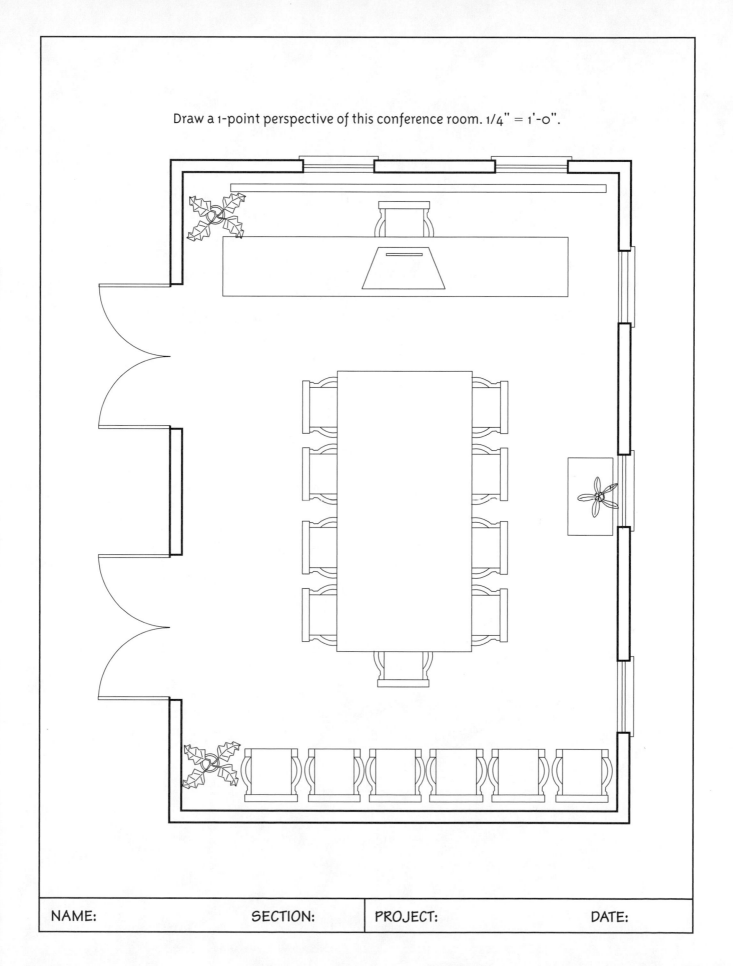

NAME: SECTION: PROJECT: DATE:

Draw a 2-point perspective of this kitchen. 1/4" = 1'-0".

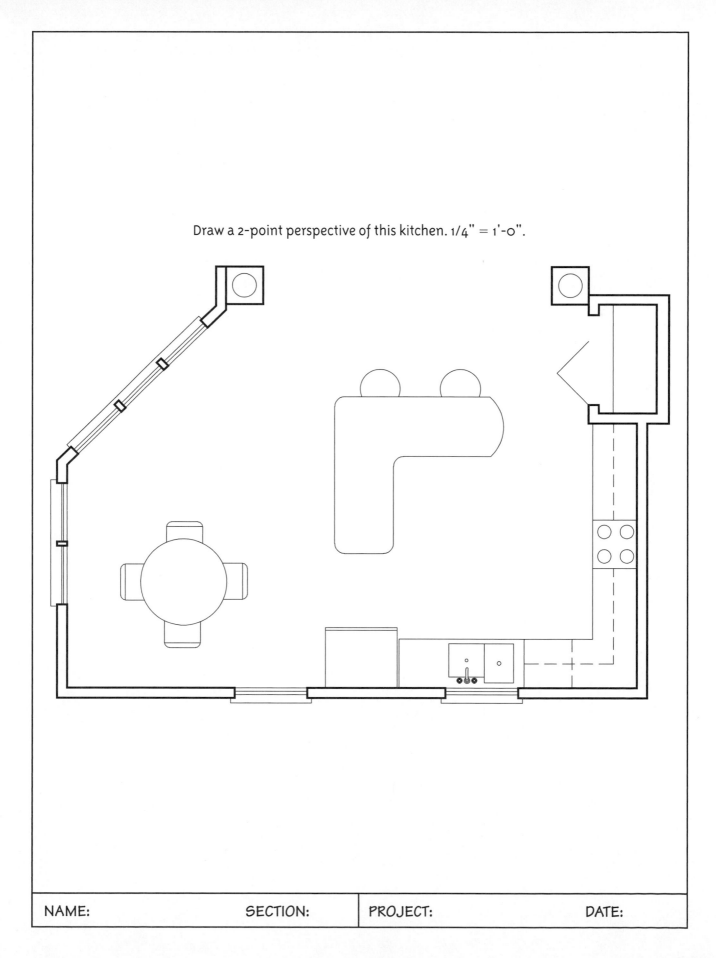